THE OTHER SIDE OF EDEN

Hugh Brody was born in 1943 and educated at Trinity College, Oxford. He taught social anthropology at the Queen's University, Belfast. He is Honorary Associate of the Scott Polar Research Institute at the University of Cambridge and an Associate of the School for Comparative Literature at the University of Toronto.

In the 1970s he worked with the Canadian Department of Indian and Northern Affairs, and then with Inuit and Indian organisations, mapping hunter-gatherer territories and researching Land Claims and the indigenous rights in many parts of Canada. He was an advisor to the Mackenzie Pipeline Inquiry, a member of the World Bank's famous Morse Comission and chairman of the Snake River Independent Review, all of which took him to the encounter between large-scale development and indigenous communities. Since 1997 he has worked with the South African San Institute on Bushman history and land rights in the Southern Kalahari.

BY THE SAME AUTHOR

BOOKS

Gola: The Life and Last Days of an Island Community (with F. H. A. Aalen)
Indians on Skid Row
Inishkillane
The People's Land
Maps and Dreams
Living Arctic
Means of Escape

FILMS

The People's Land (with Michael Grigsby)
Treaty 8 Country (with Anne Cubitt)
People of the Islands
Nineteen Nineteen (with Michael Ignatieff)
On Indian Land
England's Henry Moore (with Anthony Barnett)
Hunters and Bombers (with Nigel Markham)
Time Immemorial
The Washing of Tears

The Other Side of Eden

HUNTER-GATHERERS, FARMERS AND THE SHAPING OF THE WORLD

HUGH BRODY

faber and faber

First published in Canada in 2000
by Douglas & McIntyre Publishing Group
2323 Quebec Street, Suite 201, Vancouver BC V5T 4S7

First published in the United Kingdom in 2001
by Faber and Faber Limited
Bloomsbury House, 74–77 Great Russell Street, London WC1B 3DA
This paperback edition first published 2002

Printed and bound by CPI Group (UK) Ltd, Croydon, CR0 4YY

A CIP record for this book is available from the British Library

ISBN 978–0–571–20502–8

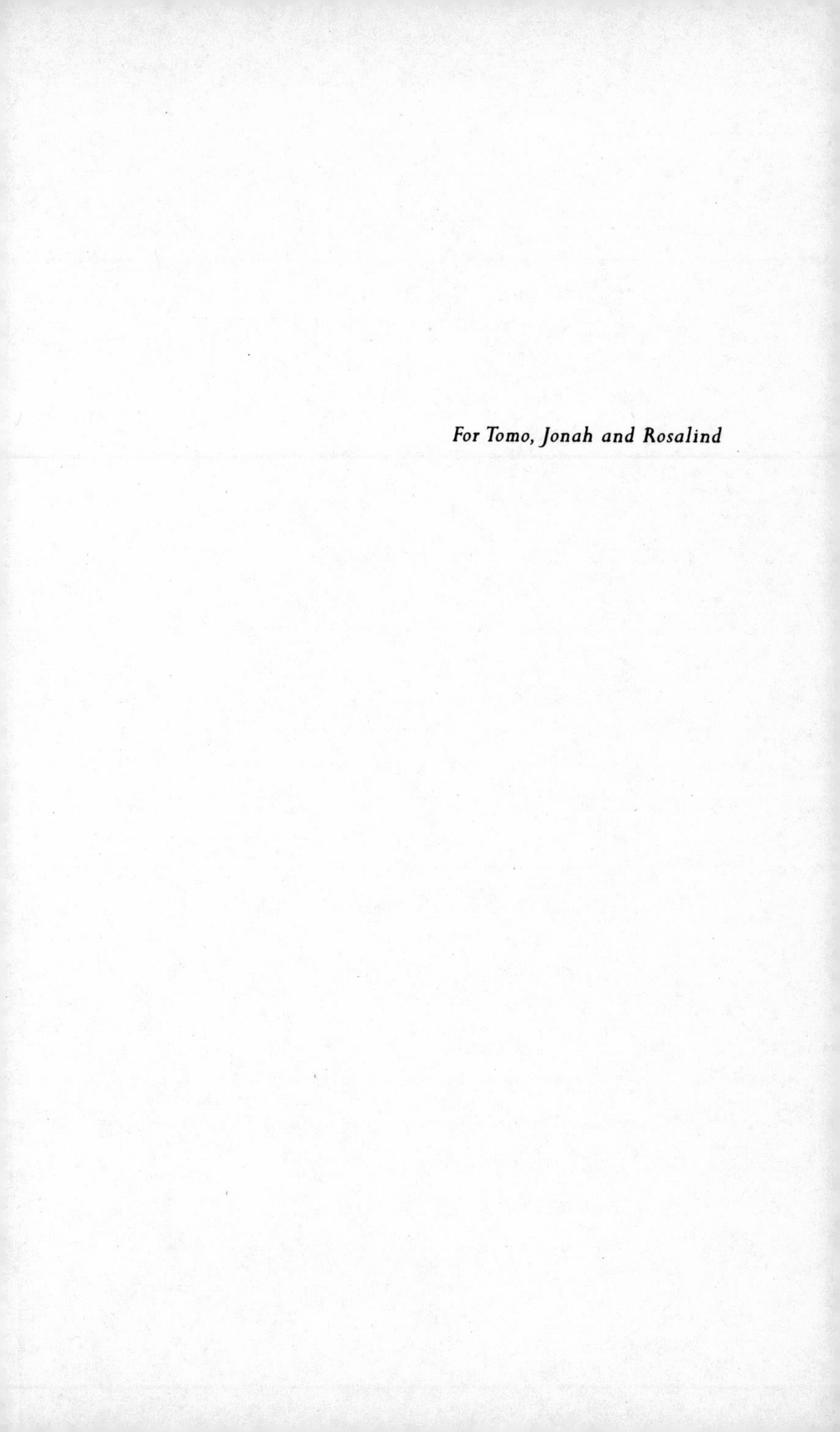

For Tomo, Jonah and Rosalind

CONTENTS

Acknowledgements / ix
Maps / *xi*
OPENING / 3

ONE: **INUKTITUT** / 9

TWO: **CREATION** / 65

THREE: **TIME** / 103

FOUR: **WORDS** / 165

FIVE: **GODS** / 221

SIX: **MIND** / 271

Notes / 315

Bibliography / 355

Index / 362

ACKNOWLEDGEMENTS

The Other Side of Eden began in letters to my friend and colleague Ted Chamberlin. Ted's urging, encouragement and support got this book going and kept it going. I owe him an immense debt of gratitude. Our exchange of letters was itself the result of a Connaught Transformational Grant from the University of Toronto, thanks to which it was possible to make journeys in pursuit of the ideas in this book.

As the exchange of letters with Ted grew into the first draft of this book, I found myself drawing upon my diaries and notebooks, especially for events that took place in the early days of my field work. I have also drawn from transcripts of filmed interviews. Most of these have not been used before—are taken, as it were, from the cutting room floor. But there may be readers who recognise anecdotes and ideas that appear in other books of mine, especially *The People's Land* and *Maps and Dreams.*

My debt to the elders and others in the communities where I lived and worked cannot be overstated. Without the hospitality, support and wisdom of Simon Anaviapik, Ulajuk Anaviapik, the Aragutainak family, Peter and Alick Kattuk, Thomas Hunter, Abalie Field, Jimmy Field, Mary Johnson, Neil Sterritt, Don Ryan, Rod Robinson, Alvin McKay, Harold Wright, George Gosnell, Pien Penashue, Mary Adele Andrew, Alex Andrew, Elisabeth Penashue and Arlene Laboucane, my work would have been impossible. To them, and to many others, I owe both great happiness and a large measure of whatever understanding I have achieved.

In the early stages of this book, Frances Coady gave immense editorial help. Barbara Pulling has been tireless in her efforts as editor at every subsequent stage. I also must thank Walter Donohue and Rebecca Saletan, my editors at Faber and Faber and Farrar, Straus and Giroux; my agent, Georgia Garrett; and Bill Kemp for all his work on the maps. Along the way, I have received great intellectual and moral support from Anthony Barnet, Glenn Bowman, Mike Brearley, Arnold Cragg, Heather Jarman, Olivia Harris, Patrick de Maré, Felix Padel and Leslie Pinder. And I thank Juliet Stevenson for the immense contribution she has made to every aspect of writing this book.

MAPS

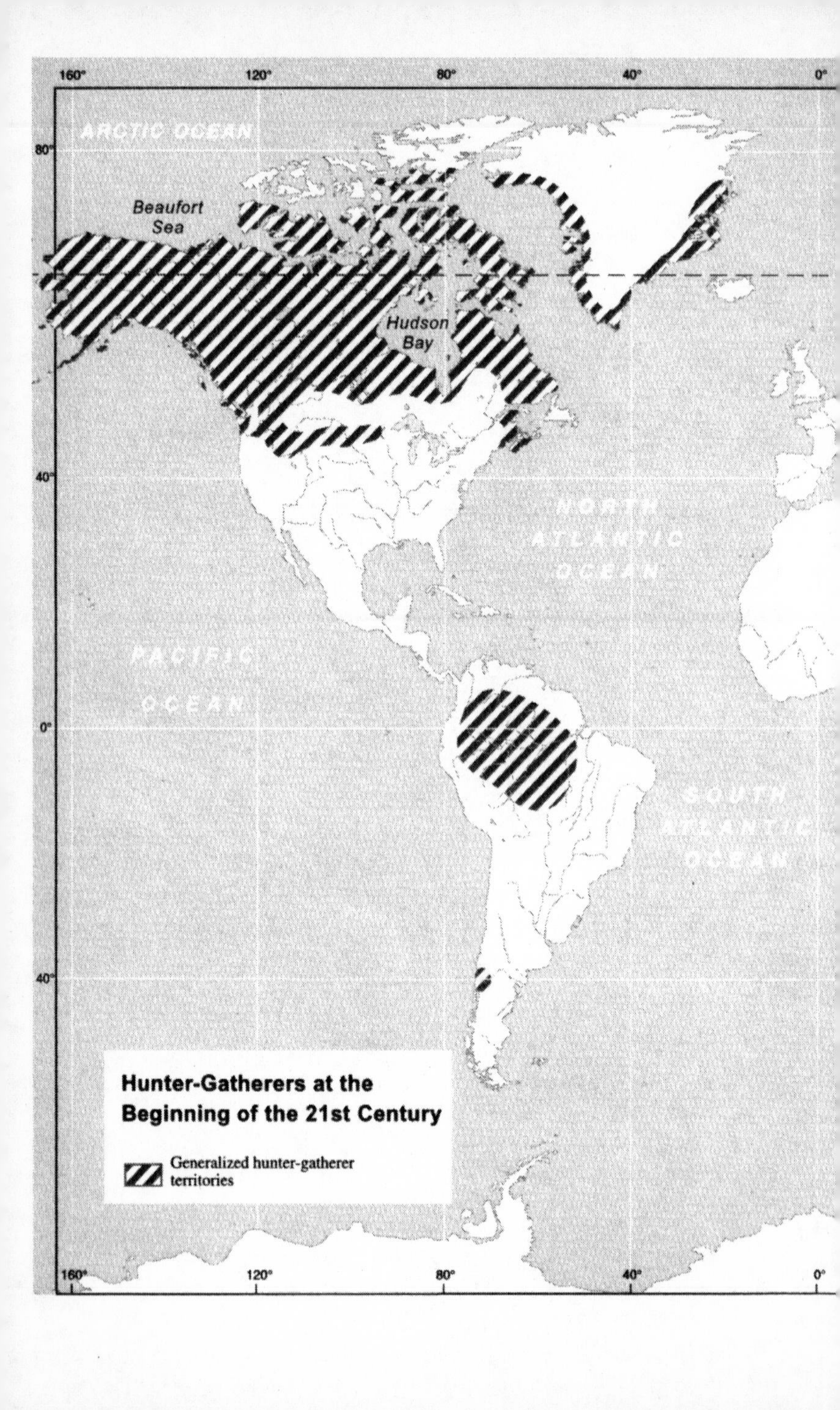
160°
120°
80°
40°
0°
ARCTIC OCEAN
80°
Beaufort Sea
Hudson Bay
40°
NORTH ATLANTIC OCEAN
PACIFIC OCEAN
0°
SOUTH ATLANTIC OCEAN
40°
Hunter-Gatherers at the Beginning of the 21st Century
Generalized hunter-gatherer territories
160°
120°
80°
40°
0°

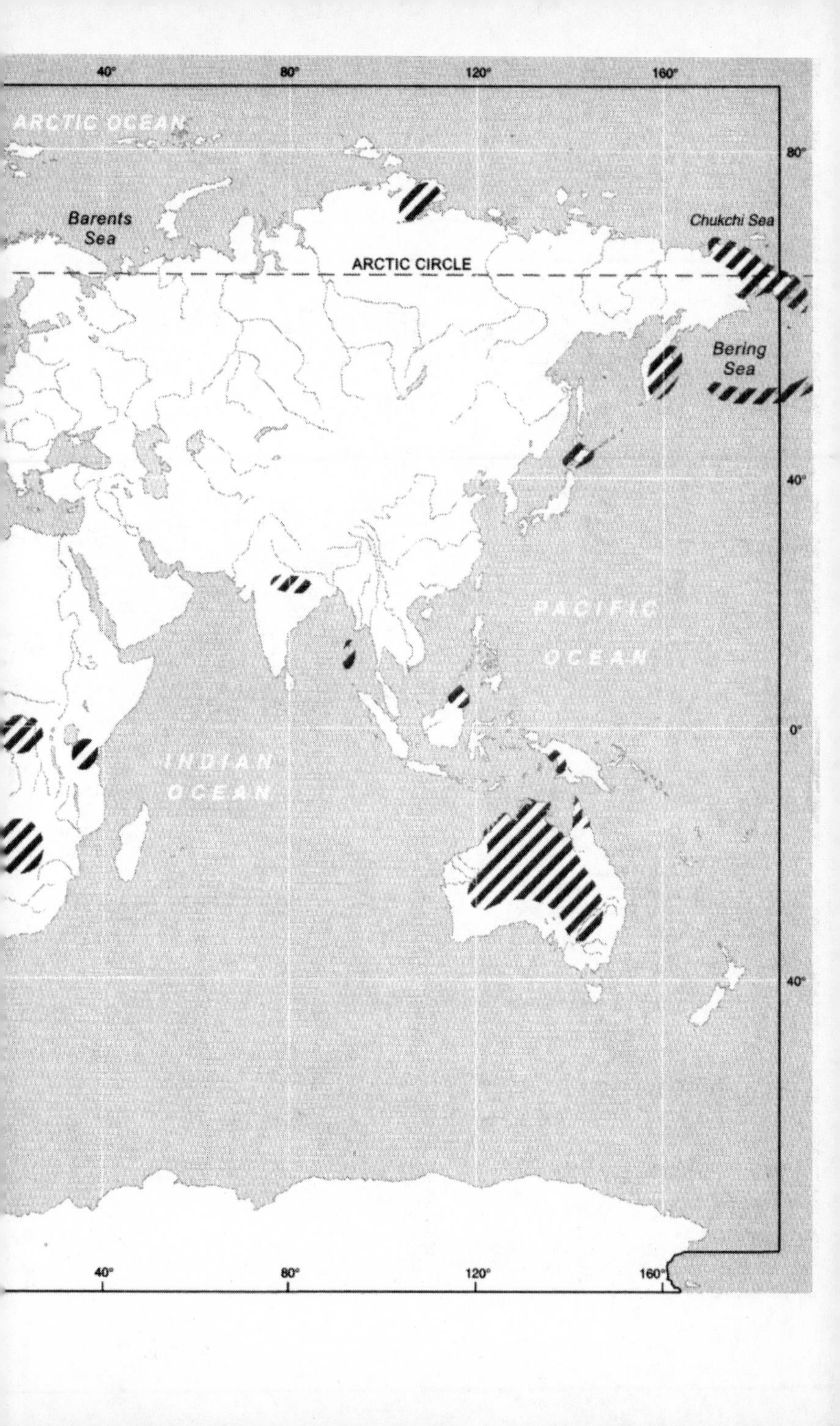

40°
80°
120°
160°
ARCTIC OCEAN
80°
Barents Sea
Chukchi Sea
ARCTIC CIRCLE
Bering Sea
40°
PACIFIC OCEAN
0°
INDIAN OCEAN
40°
40°
80°
120°
160°

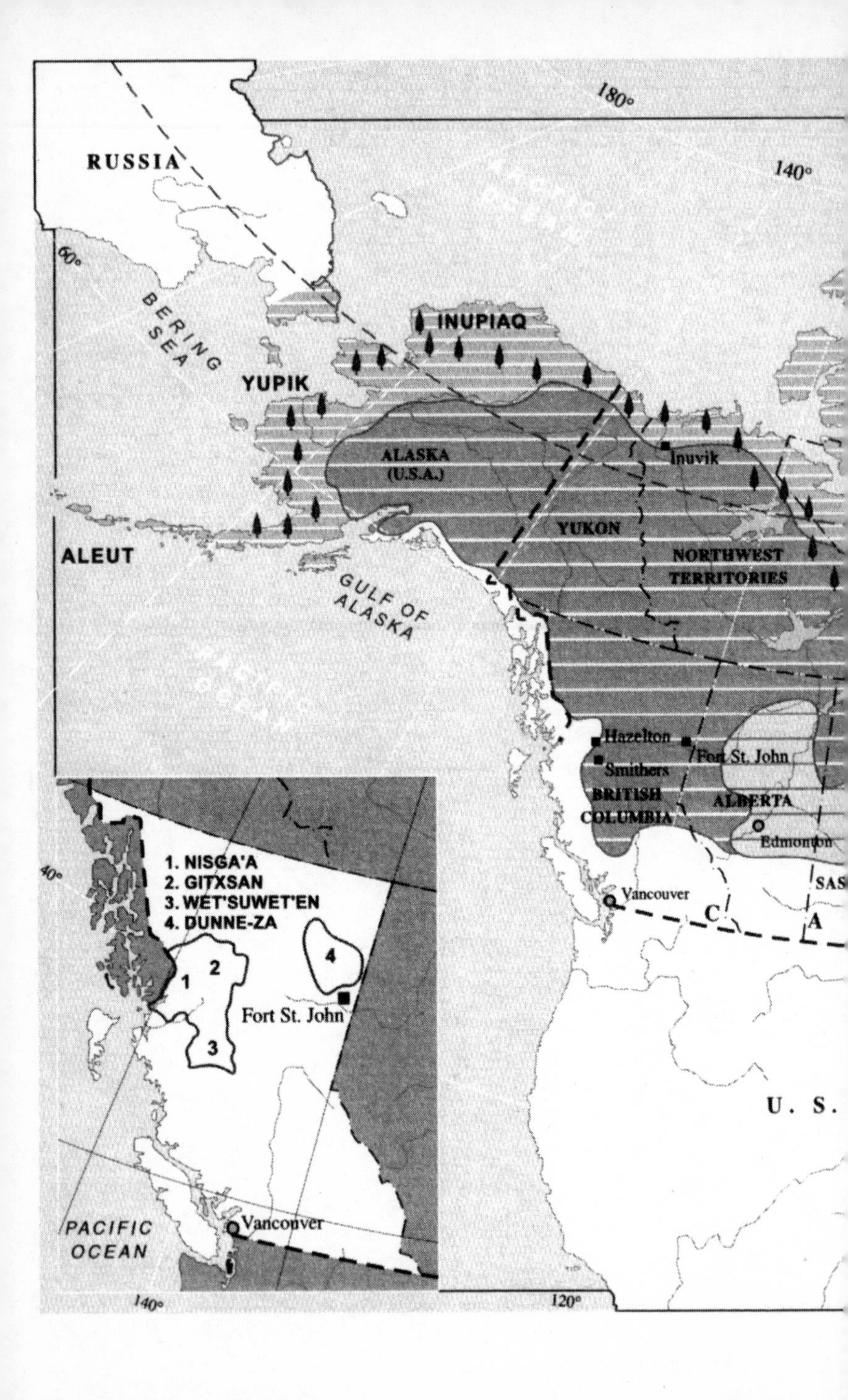

180°
140°
RUSSIA
60°
BERING SEA
INUPIAQ
YUPIK
ALASKA (U.S.A.)
YUKON
Inuvik
NORTHWEST TERRITORIES
ALEUT
GULF OF ALASKA
Hazelton
Smithers
Fort St. John
BRITISH COLUMBIA
ALBERTA
Edmonton
Vancouver
C
A
U. S.
40°
1. NISGA'A
2. GITXSAN
3. WET'SUWET'EN
4. DUNNE-ZA
1
2
3
4
Fort St. John
PACIFIC OCEAN
Vancouver
140°
120°

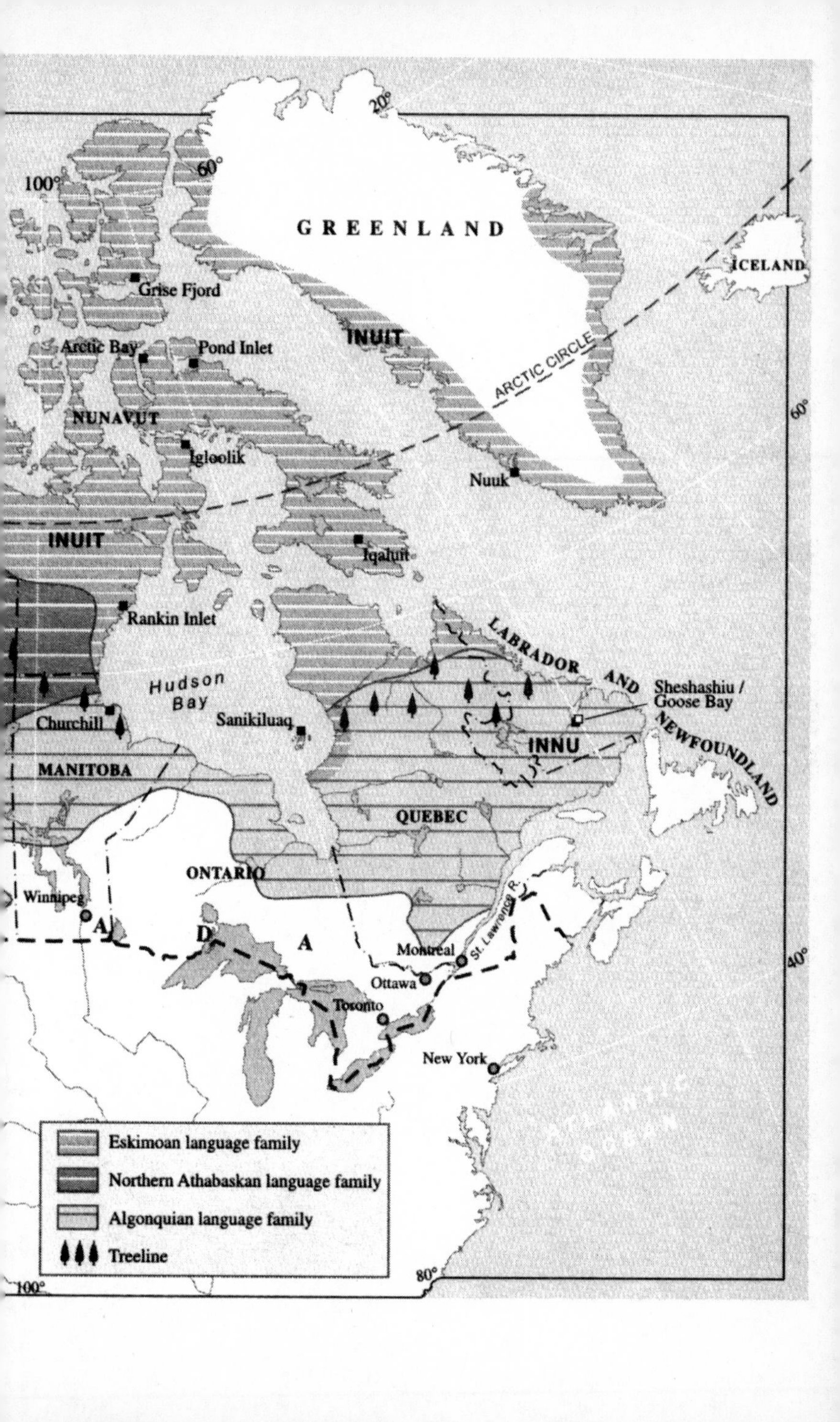
GREENLAND
ICELAND
Grise Fjord
Arctic Bay
Pond Inlet
INUIT
ARCTIC CIRCLE
NUNAVUT
Igloolik
Nuuk
INUIT
Iqaluit
Rankin Inlet
LABRADOR AND NEWFOUNDLAND
Hudson Bay
Sheshashiu / Goose Bay
Churchill
Sanikiluaq
INNU
MANITOBA
QUEBEC
ONTARIO
Winnipeg
St. Lawrence R.
Montreal
Ottawa
Toronto
New York
ATLANTIC OCEAN
Eskimoan language family
Northern Athabaskan language family
Algonquian language family
Treeline

THE OTHER SIDE OF EDEN

OPENING

Imagine the crystal darkness of an Arctic night. A canopy of stars and a glowing arc of aurora, the northern lights. A vast astral flickering and dancing; yet a sense of eternal, unmoving space. Under the moonlight, the surface of the world shines and fades into the distance. The sky is cloudless, open, with clearness that is like a sound, a crackling of frozen silence that many Arctic travellers claim to be able to hear. And there is a wind, strong enough to blow the snow across the ice.

I was travelling by dogteam with Paulussie Inukuluk. He was taking me to hunt seals at their breathing holes, in a favoured area beyond a headland that shapes the southeast corner of Bylot Island. We crossed the sound in front of Pond Inlet and followed the coast of Bylot Island, moving on bumpy sea ice. I was half running, half stumbling beside the sledge, while Paulussie ran alongside the dogs, urging them on. I remember a particular moment, close to the shore, when I lost my footing, almost fell, and stopped.

The entire surface of the world was flowing along at knee height. There were no features to the earth; the dogteam was half immersed in this strange current of snow. I stood long enough for the sledge and Paulussie to be no more than a blurred, grey movement at the edge of the light. I was encased in caribou skin clothing—parka, trousers, socks and boots. If I faced away from the wind, I felt nothing on my skin but the mix of my breath with the cold air. The world before me was as a vision, an unbelievable magnificence that filled me with awe, disbelief and, at the edge of my mind, real fear. I began to run, as fast as I could, on the uneven surface of the sea ice, my feet invisible in the layer of blowing snow, to catch up with Paulussie and his dogs.

This moment is held in my memory like a film clip—vivid, available for recall, but distanced by time and strangeness. There is a sense in me of a mystery. It comes to mind now, as I begin to write again about the words and people of the North.

In 1969 I spent five months living on the skid row of a Canadian city. Most of the people I met there had lived, or identified themselves and their ancestors, as members of "Indian" communities. I put quotes around "Indian" because it is a word Europeans brought to the Americas and used to categorise a huge range of peoples—none of whom, of course, had any connections with India. What the people I met had in common was that most of them had grown up in societies that were, or had been, dependent on hunting and gathering.

Two years later, in 1971, I first went to the Arctic. I lived and worked in several regions, encountering many people who had lived much of their lives as hunter-gatherers. From then until now, I have continued to work with hunter-gatherer communities, as anthropologist, land-claims researcher, filmmaker and, twice, as expert witness in land-rights court cases.

For all that I had written about hunter-gatherer societies, I was left with a deep conviction that I had yet to write about that which is most important. Something lay there that eluded not just me, but many who have experienced another way of life. We write about some facets of it, some surfaces, that we make our business. But the gold we find is transformed by the reverse alchemy of our journey, from there to here, into lead. Not into nothing, not into worthlessness, but into a substance that has more weight than light, more utility than beauty, is malleable rather than of great value. What is this reality that gets left behind? It is not simply some kind of otherness. In fact, anthropologists are often skillful at crossing divides between peoples in their field work, but clumsy when it comes to writing up the "findings." Perhaps the desire for the esteem of peers and critics leads to a tendency to make things unduly complicated or scholarly or heroic—depending on the audience we most need to impress.

This book draws on all parts of my work; it is rooted in my experience of hunter-gatherer ways of being in and knowing about the world. In many ways it is a personal account, with memories of people and places that influenced my life. The influence has been on how I see and understand both history and society, so this book is also about ideas. Several important points need to be made at the outset.

There are virtually no people in the world today who live purely as hunter-gatherers. Many kinds of colonial process have transformed peoples' economic lives, even in the remotest areas. Those who see themselves as hunter-gatherers, and are seen as such by their neighbours, may also be part-time labourers, do bits of farming, have domestic animals or rely on welfare payments and state pensions. Nonetheless, there are many individuals, families and societies for whom their way of raising children, using land and speaking of their culture is rooted in hunter-gatherer heritage. This is something about which people are often proud, and they do what they can to secure it against the incursions and criticisms of others, including the insistence by some anthropologists that hunter-gatherers themselves are a kind of myth. My book takes its inspiration from the courage and determination these people have brought to their struggle for survival, as well as from their skills and wisdom.

It must also be said that I have lived and worked in hunter-gatherer societies as a man; this places a limitation on what I have experienced. I learned far less about gathering than about hunting. I saw far less in the domestic sphere than I did on the land. In reality, the economic, social and political lives of the peoples I knew were as dependent on women as on men. Despite this imbalance in how I spent my time, I hope to pass on what hunter-gatherers can teach us not only about their own particular human genius but also about human history. I invite readers who are not experts in anthropology, archaeology or linguistics to come on an exploration that leads to wild places, harsh climates and concepts that may seem to lie beyond most people's actual and intellectual geography, but are, in reality, central to the history of all societies.

My work relies on that of many others. I have found information and inspiration in a wide variety of sources. There are also passages here that need some degree of qualification or explanation. To avoid burdening the text with too many references and refinements of argument, I have created a set of endnotes. They are—like much of the knowledge they refer to—a sort of shadow text.

It is not easy to write about other peoples without falling prey to conceptual and political misconceptions. The stereotypes that capture and tend to diminish tribal peoples in general, and hunter-gatherers in particular, are pervasive and powerful. I attempt to combat some of these stereotypes here. But I would like to establish, as a way of introducing these stories of exploration, three pivotal ideas.

First, hunter-gatherers live at what have become the margins of the "developed" world. Development means profitable farming and towns that exist thanks to the farms that feed them. Where farming is judged not to be possible or profitable, hunter-gatherers can sometimes continue to use and occupy their lands. At these margins, two ways of life meet and sometimes overlap. Yet there is a profound difference between these two ways of life, and an equally profound difference between the peoples who practise them. That difference is at the heart of what I have set out to explore.

Second, the difference between hunter-gatherers and farmers, or between hunter-gatherers and all other peoples, has nothing to do with evolution or with supposed levels of civilisation or development. Hunter-gatherers live at the margins of the farmer's world; farmers live at the margins of the hunter-gatherer's world. Each way of life is the centre of its own universe. This book places the sophistication of hunter-gatherers alongside the achievements of farmers.

Hunter-gatherers, like other peoples, use whatever technologies are available to them, including guns, engines and manufactured food, and they participate in national economic life insofar as they are able. We are all contemporaries, whatever lands we live on and whatever heritage we rely on to do so. All human beings have been evolving for the same length of time.

Third, a crucial difference between hunter-gatherers and farmers is that one society is highly mobile, with a strong tendency to both small- and large-scale nomadism, whereas the other is highly settled, tending to stay firmly in one particular area or territory. This difference is established in stereotypes of "nomadic" hunters and "settled" farmers. However, the stereotype has it the wrong way round. It is agricultural societies that tend to be on the move; hunting peoples are far more firmly settled. This fact is evident when we look at these two ways of being in the world over a long time span—when we screen the movie of human history, as it were, rather than relying on a photograph.

In one important way, hunters and farmers are not equals. Agricultural peoples, especially in the world's rich nation-states, are numerous, immensely rich, well armed and domineering. Hunter-gatherers are few in number, poor, self-effacing and possessed of little military strength. The farmers have it in their power to overwhelm hunter-gatherers, and they continue to do so in the few regions of the world where this domination is not already complete. Yet hunter-gatherers have experience and knowledge that must be recognised. Their genius is integral to human potential, their skills are appropriate to their lands, and their rights are no less because their numbers are small. Political inequality, hostile and racist stereotypes, and conflicts of interest over land have created incomprehension and suspicion of hunter-gatherers. The powerful find it difficult to listen. But listening is what must happen, somehow, on every frontier, for only if the powerful listen will the needs and rights of the vulnerable be respected.

I, as narrator, am in many of the episodes I use here to reveal peoples' needs and rights. At the same time, I reach for underlying ideas, linking societies across great distances and over immense spans of time. I therefore take a double risk: of being both too personal and too theoretical. I take this risk because I believe there are lessons to be learned from the hunter-gatherer world that go to the core of who we are as human beings. These are lessons about the nature of history, the way in which those who dominate the world have achieved their ends, and the extent to which language is inseparable from the identity and well-being of any people. There may also be lessons and explanations, of a kind, for some of the malaise and sense of inadequacy that afflict so many of us.

The journey from personal memories of the Arctic to these speculations is dependent on many trails. These trails come from places where I have worked as an anthropologist and filmmaker but lead into the meaning of myth, into issues of language and archaeology, into the history of many aspects of the human condition. The trails cross one another often, and are never far apart. When ideas and experience are set side by side, each makes more sense of the other. But the truth is that I begin with memories of the North, and move from there to what I have learned.

The Other Side of Eden explores an original and fundamental frontier, and therefore it is about the history that has shaped us. It is a search for what it has meant, and can mean, to be a human being. The book's title is intended to evoke somewhere not within the usual divides, somewhere not heaven or hell, not modern or ancient, not civilised or primitive, but a place where all human beings can be more fully themselves.

ONE: **INUKTITUT**

1

Imagine the darkness of the far north. Not as something in which the adventurous traveller moves in awe. But as a beginning, for those for whom the Arctic is home. Imagine the inside of a skin tent, or a snowhouse, or a government-regulation low-rental prefab. In this home, an Inuit baby girl wakes in the night. She is held, fed, cuddled—and talked to.

What words does she hear? The sounds of whoever is talking in the same space. The voice of her mother, encouraging her to eat. Words that tell the baby, over and over, that she can decide when to feed, when to stop feeding. Words of endorsement. After feeding, the baby girl dozes. With words of welcome, she is lifted into the *amautik,* the pouch shaped into the hood of her mother's parka, where she can lie curved against her mother's back. After a while, the baby begins to defecate. The mother, sensing the movements, lifts her out and holds her over the ground, murmuring encouragement. "*Unakuluk, annatiakulugit.*" "This sweet little one, have a lovely little shit." The mother wipes her baby's bottom, saying "*Kuinijuannu saluitutinnai.*" "Gorgeous and plump, aren't you nice and clean." The mother's father comes over to watch his granddaughter being wiped. He leans forward, his face close to the baby's, and talks to her softly: "*Nuliakuluga. Nuliagauvit? Ii, nuliaga una.*" "Sweet little wife. Are you my wife? Yes, this is my wife." The baby's mother smiles, holding her daughter for her father to adore, and says, "*Anaanangai. Ii, anaanagauvutit.*" "Mother? Yes, you're my mother."

In these words, the child is given the sounds of love, and can know that she is safe. Not safe just to feed, to sleep, but safe to do these things as and when she wants. For she is a baby who carries the *atiq*, the spirit and name, of her late grandmother. She is the adored baby; she is also her mother's mother, her grandfather's wife. Her grandmother is alive again in the baby. This means the baby is doubly and trebly loved. And she must be treated with respect. She can no more be denied food or refused the choicest morsels, be told to sleep when she wants to be awake or told to wake when she wants to be asleep, or be chided for being dirty than could her grandmother, were she still alive. But her grandmother *is* alive—in this baby who is also someone else. To her grandfather she will be "wife," and with this word, as well as all the pet names he used for his dead wife, he will call out to her. And the baby's mother will address her child as both "daughter" and "mother."

Imagine this little girl a year or so later, as she learns to speak. Like children in all societies, after making all possible human sounds, she learns to use the special consonants and vowels of her own language. Then some simple words. She begins to name things. Here, in the corner of her home where food is stored, is the body of *iqaluk*, an arctic char; the flipper of *qairulik*, a harp seal; the skin of *natia*, a juvenile ringed seal. Outside are the skins of *nanuq*, a polar bear, and several *tiriganiat*, arctic foxes. But there is no "fish," "seal" or "bear." In the Inuktitut the child learns, there are no such categories. It is the specifics of the natural world that are named. As the child gets older, she will learn to speak of *puijit*, the "breathers" that are the sea mammals; and of *uksuk* or *tunuk*, the fat of sea creatures or land creatures; and of *sijjarsiutit*, the "shoreline seekers," wading birds. But she will not hear generic words for mammal, fat or bird.

From the beginning of her life, the little girl will listen to stories. No one censors or limits that which is told. Her ability to make sense of what she hears is the only constraint. Her grandfather may give the details of the creation of sea mammals, at the earliest time of the world in which Inuit now hunt, with all its sexual and bloody details. He may tell a comic story about jealousy and fear. The small child listens for as long as she wishes—she is, after all, also her own grandmother. And she discovers that stories are always a mystery, for they have much that cannot be understood, and much that comes from knowledge and experience beyond understanding. There are words she knows, things she can make sense of; and these are both the border and the small gateways to an immense edifice of facts that she may not understand in any full way, but that creates questions, wonder and puzzlement.

As she gets older, the child recognises stories. Stories that are told many times. Details vary, but the same characters and principal events recur. This repetition is both the lessening and the maintaining of mystery. For the stories tell of events that are inexplicable and use words that are incomprehensible. No one would claim to understand every part of these stories, or to have a ready explanation for people, events or processes that are confusing and strange. These are stories that defy any complete understanding. To tell and to listen to them is to experience the delight and enigma of incomprehension. Mysteries are repeated, not explained. The ultimate wonder about the world remains.

As the little girl learns to be with her friends, in the community of children who roam on the tundra, play on the sea ice, share chores in one another's homes, or sit listening to the talk of adults, she hears the ways in which people deal with one another. She will notice the way in which individual choices are respected. More and more she discovers that she is embedded in a web of relationships that link her, through her *atiq*, to so many others. Her uncle calls her "mother" and she can call him "son." Some of her playmates are both cousin and nephew or niece. Others are her in-laws because they are her grandfather's siblings, or her sisters and brothers because they have the *atiq* of one or other of her grandmother's brothers and sisters.

The words with which the girl is addressed place her in a group of families, in a community. They also show that she is an individual—child and adult. She has a large, strong, unquestionable family, but she is expected to make her own judgements, take her own initiatives, be clear about her own needs and preferences. She is given a place in a system that is both communal and individualistic.

She hears men and women talk about where they have hunted, gathered and travelled, and she begins to learn the names of the land around her. She learns that many animals have to be given water when they are killed to ensure that some of their number will be willing to die again when she and her family need food. She discovers that animals and humans must be at peace with one another. Inuktitut has no words for "vermin" or "weed." There is no demarcation between the life of an animal and that of a human—no word for "it." There is no hierarchy of classes of people or, within her community, of rights to use land. Bit by bit, she will come to understand that the world around her is shared both among people themselves and between people and the other creatures that belong there.

She hears individuals referred to as the *miut* of particular places. They are "of" this or that hunting area, or a particular camp, or even of a country. She learns that she is a *miutaq* of both where she now lives and the place her family thinks of as home, an area where they lived when they were young. In these places *nunaqarpuq:* "she has land." Not land that she can buy and sell. Dealing of this kind is only in relation to whites and their trading posts or shops. Money, brought to the North by newcomers, is called *kiinaujaq,* "resembles a face"; its archetype is coins that showed the faces of monarchs and presidents. Beyond *kiinaujaq* there is no medium of exchange. Inuit did not have title deeds or contracts, prices or measurements of equivalent value. Inuktitut is without the categories and mathematics on which these depend. There are numbers for one to five, and words for ten and twenty, but no arithmetic system beyond these. The ways in which the girl's elders talk of belonging to or living in the places they have always known show her that the land all around her is irreducible, indivisible and inalienable.

This land to which she belongs is the subject of many kinds of stories. Stories about its creation, or the first appearance of various creatures. Stories about travelling on it and living from it. She listens to her elders describe ancient times and recent times, passing on their knowledge about what this place is, what inner meanings it may hold, how best to make use of its creatures. From stories of creation and the hunt the girl builds an image, or a set of images, of her world. As in all great narratives, history, geography, personal adventure and mysteries intertwine. There are misadventures, murder and starvation, to be sure, but spiritual powers and every kind of humour mean that even the worst is part of being in the best possible place, in one's own land. *Inuit nunangat,* "the people's land"—the expression Inuit use for referring to their part of the world—is an ideal. To change or abandon such a place, according to this world view, would be dangerous and foolish.

Thus is it possible to imagine this girl growing into the mind and land of Inuit culture, which is the northernmost example of hunter-gatherer societies. To go to the far north is to visit the most recent frontier between the languages of hunters and the languages of farmers—a place where it is possible to experience the divide between these two ways of being in the world. It is also the place where those who wish to describe this frontier, and the encounter between hunters and farmers, can experience some of the deepest difficulties with language—that of the hunters as well as their own.

2

The first Inuit village I ever saw was Rankin Inlet, about halfway up the west side of Hudson Bay. I flew there in 1971 via Fort Churchill, a supply post and military base at the northeast end of a Manitoba railway line. Churchill lies below the treeline, not yet in the Arctic. But a few minutes' flying time from the Churchill airport brought shock waves of amazement as the plane flew low over the land. It was March, early spring. I looked down on thousands and thousands of frozen lakes divided by endless, low ridges stretching across an infinity of tundra, and the spread of frozen sea with its long, long shoreline of heaved and broken ice.

I had read enough about the Arctic and the Inuit to be able to superimpose some elements of a way of life on the surface of this huge, white place. Hunting areas, fishing lakes, trails, groups of homes. Movements of families from one region to another. Camps at the ice edge, and places to intercept migrating herds of caribou. I knew a good deal about material culture, artifacts and clothing styles. I had learned something of the biology of the region, in particular about animals that northern hunter-gatherers hunted, fished for and trapped. And I knew a little about the plants people used. I had some sense of human links to this immense landscape, and did not see the North as intolerably cold or barren. In my mind's eye, I could picture societies and economic systems reaching across the entire, endless vista. I imagined people there, but in a form that was massively overwhelmed by nature.

I arrived in Rankin Inlet to take the Inuktitut language course offered by the Canadian government to teachers and other government employees who wanted to make a long-term commitment to working in the North. The course had been designed, and was run, by Mick Mallon, an expatriate Irishman with a genius for both Inuktitut grammar and teaching techniques. Mick met the plane at the Rankin Inlet runway, bade me a cheery greeting and took me to the trailer that served as classroom and residence.

For the next six weeks, five of us would live and take lessons here, separated by thin walls of insulated aluminium from Rankin Inlet itself. My fellow students were a teacher working in an Inuit community about three hundred miles to the west, an administrator from a settlement in south Hudson Bay, a social worker from Churchill and a young man who, like me, had come to the North for reasons that were a little unclear to him. Despite our varying experience of the Arctic, all of us struggled with the deep unfamiliarity of the language while living a life shaped and provisioned according to the habits and norms of the distant south. We existed in a cocoon—as, I was later to discover, did almost all white people who went to the North to do government jobs. A cocoon that protected us from discomforts of climate and isolation, but one that cut us off from most of the people for whom this was home.

Breakfast, lunch and supper punctuated our days. Lessons were divided into precise units. Fifteen minutes to grasp a new piece of grammar; fifteen minutes of practising this and other such bits; fifteen minutes of private study; fifteen minutes' break; then another new piece of grammar. And so on through the day. In the evening Mick often gave a sort of lecture: "Here, let me tell you in good old English about what you are learning." The pace was relentless as we struggled to understand, learn, remember. Meanwhile, we lived with a paradox, albeit one of which we were never conscious enough to speak: we occupied the institutions of one kind of world in order to learn the language of another.

The teaching programme, based on interpreters' courses, was designed to force the structures of the language into students' minds. At the centre of the method were drills. The students sat around a table with one or two Inuit participants from Rankin Inlet. An Inuk teacher or Mick Mallon or both stood at the front. The teacher asked a question, and one of the Inuit in the classroom answered. The question demonstrated some aspect of grammar. I remember one of the first:

Q: *Nanik nunaqarpit?*
A: *Qanallinirmik nunaqarpunga.*

Qanallinirq was the Inuktitut name for Rankin Inlet. The Inuk in the class then asked Mick Mallon:

Q: *Nanik nunaqarpit?*
A: *Ottawamik nunaqarpunga.*

And Mick asked each of us southern students in turn:

Q: *Nanik nunaqarpit?*
A: *Englandmik nunaqarpunga.* Or: *Belcher Islandsmik nunaqarpunga.* Or: *Churchillmik nunaqarpunga.*

Get it? Well, you do after a while. In the brain, in the strange way that translation is registered without anyone having to tell you, *nanik* becomes "where," and *nunaqarpit* becomes "do you live." This is a simple example. More complex items, built from many preceding bits and pieces, were presented in a sort of round-the-table game, which went like this:

Teacher, picking up a red brick: "*Aupaluktumik tigusigama, ilinnut tuniniarpara.*"

He hands the brick to an Inuk helper.

The helper, taking the red brick: "*Aupaluktumik tigusigavit, uvannut tunivait.*"

No translations.

Eventually we would grasp the meaning and grammar of the game. It translates as:

Teacher: "If I pick up red, I give it to you."

Respondent: "If you pick up red, you give it to me."

This goes on to:

"If he [or she—Inuktitut does not have genders] picks up red [or blue or black], he will give it to him."

And: "If he gives him black, then he'll give it to you [or me or someone else, depending]."

Inuktitut is more precise than English when it comes to the objects of verbs, making use of a fourth person. It is a language where the sentence "If he is not in a good mood, he will kiss his wife" is without ambiguity: the "his" signifies whether the wife is his own or someone else's.

The course moved at a predetermined rate, and concessions were not made (as far as I could tell) to those who were not keeping up. The assumption seemed to be that everyone, at some point, would fall by the wayside; it was a question of how long each of us could keep going. I like grammar and find a deep pleasure in the way it takes shape and becomes firm, predictable, usable in the brain. Mick's clarity made progress seem straightforward, even easy. He did not rely on a crude transforming of Inuktitut into the grammatical tables and lists of Latin or German that are part of so many schoolchildren's educational fare. Everything was alive, worked through dialogues and conversation in what educational theory defines as the direct method.

But the underlying strategy did have something important in common with the grammar learning I knew from my school days. At the heart of Mick Mallon's skill in making the language somehow familiar and manageable was an inevitable denial of subtlety and strangeness. An insistence on language as a set of rules, a system applicable by anyone to anything—the essential basis for teaching—obscures the difference between one kind of language and another. This is not to say that Mick himself was unaware of or unresponsive to Inuktitut poetry and mystery. Rather, his job was to get us students to acquire *some* Inuktitut basics. And we were, it must be said, a mixed bag of motives and abilities. For some of us, any kind of language learning lay somewhere between terrifying and unfathomable (we were deep in North America); for one student, absorbing Inuktitut appeared to be no more difficult than memorizing the lines of a rather odd play.

I have no doubt that Mick Mallon's method was the only way I could have grooved Inuktitut grammar into my adult brain in a six-week period. By the end of the course I could make myself understood in elementary but grammatical Inuktitut. I am still not sure what the alternative might have been, apart from attempting that which all small children achieve and few adults can bear—a total and random immersion into the flow of everyday talk, so that the language takes residence as both architect and worker. My guess is that I would never have managed it.

But the life we students led had consequences. When we were not engaged in the teaching sessions, we played cards or listened, I remember, to *Hockey Night in Canada.* I suppose the *struggle* with the language was, in its way, an encounter with the other world, but the mind-set and society beyond the aluminium walls and double-glazed windows of the trailer stayed out there, beyond the reach of what we were doing.

There were odd moments in the classroom when an Inuk helper brought the outside in. One morning, an elderly woman, Tautungnik, arrived with a story; something about the water supply and the taste of the water that a settlement tanker delivered to people's houses each day. With some difficulty, we were able to piece together what Tautungnik was saying. For a while, people had been noticing that the water tasted rather strange. Today, they had found the cause: a dead dog had been discovered in the lake, close to the pumping point. Very decomposed. It had been flavouring the water. No wonder she and others had noticed an odd taste. Retelling this story, making sure that we understood, she burst into fits of laughter. All these years later, I can see Tautungnik, as she stood in front of a large window in the trailer, describing the dead dog; I remember the light gleaming on a bank of snow through the glass behind her. And the laughter—that Inuit laughter, one hand to the face and the other to the midriff, tears running down and gestures of helplessness as the body becomes weak and bends at the middle.

The endeavour in the trailer, and in Mick Mallon's home, where he gave his evening talks, was to teach us the skeleton of a language, nothing more. There were no forays into ethnography, no discussion about the changes in the North that were bearing so hard, at that time, on every aspect of Inuit life. We did not think about the ways in which Inuktitut itself was being transformed, even displaced, by English. There was chat, of course, with its subtexts of expatriate gossip—an informal and rather subliminal recognition of the colonial setup.

About halfway through the course, I moved into the house of an Inuit family. This was the home of Karlik and Elisapik and their two young children, a two-bedroom government prefab with basic modern services that they infused with warmth, generosity and a great sense of peace. Karlik was a wonderful carver. He worked many hours every day shaping soapstone, the soft rock that the Inuit call *kullissak* or *kullissijak,* which translates as "material for lamps." As Karlik worked the stone, its dust filled the air around him, and he would cough over and over, deep in his chest—he had had a long struggle with tuberculosis. Yet he was always friendly, helpful, ready to answer my innumerable questions. Elisapik was shy and loving. They did everything to make me welcome, and in their house I learnt something of the ease of Inuit family life—and something of the confusion of life in a settlement that was designed and, in those days, run by southern officials.

We students at the Inuktitut course did get the occasional day off. One afternoon, two of us decided to go on a short ptarmigan hunt. In early spring, ptarmigan (the grouse of the far north) move in large flocks, white birds that stand out on the land but are invisible against snow. For a few hours we roamed on spring tundra, part snow, part slush, part explosion of grasses and ground squirrels. We came across one large group of ptarmigan, sitting on the side of a low ridge, and managed to shoot three. It was a small outing that made us eager to make a longer, more exciting expedition. So we were delighted when Mick arranged a two-day school camp, a miniature field trip, with all of us spending time on the tundra. We were to go out "onto the land." Nothing too alarming: we would be able to chat to one another in English, and would be sure to take with us an abundance of supplies.

This first journey beyond the settlement was startling. One of the men who helped Mick do classroom drills guided us, taking the easiest routes. We were encased in bulky quilted clothing, well defended by our outsider selves. But even a minimal escape from the trailer and safety of Rankin Inlet was an adventure. And knowing even a few phrases of Inuktitut created a small, perhaps inappropriate, sense of familiarity. Although I had never ridden a sledge, never been towed into the Barren Lands behind Rankin, never been among that great maze of frozen lakes and rivers, I could give names to things. As we bumped and lurched, the sharp chill of the Arctic on our faces, I recited Inuktitut words and phrases. In a small way I transformed the awesome, silent and immense "wilderness" into bits of language and culture, things I knew or could imagine getting to know. The ride from Rankin to our campsite took no more than two or three hours, and I had some difficulty learning to stay on the sledge, but I arrived elated.

My claim to knowledge was hubris; nemesis soon followed. Once we had pitched our tent and had a first mug of tea, I longed to be alone in the tundra. While everyone else was busy at the campsite, I set off once again to hunt ptarmigan. I had borrowed a .22 rifle from Karlik. It was a still, overcast day. The faster rivers were breaking through the winter ice. There were small areas of bare tundra. But the landscape as a whole spread into the distance as a white-and-blue patchwork. I walked over a low hill behind our tent and soon found a few ptarmigan. I shot one, picked it up and put it in my parka pocket; walked a little farther, and shot another. Seeing a small group of birds on another ridge, I made my way carefully towards them, skirting some rough ground, and shot a third. Then, for the first time, I looked beyond this preoccupied little hunt to see where I was. The campsite was no longer visible. I was surrounded by a landscape that repeated itself in every direction: flat, white, broken by many low ridges. I walked to the highest ridge I could find, no more than a quarter of a mile away, and climbed to the top expecting to see our tent. I saw more lakes, more wide expanses of snow-covered Barrens, and more ridges. No sign of a tent. I was lost.

I had read books about the North in which explorers stressed the importance of staying calm. Panic and haste cause the body to sweat; sweat makes clothing damp; wet clothes fail to insulate. I knew that I could not be more than a mile from the others. All I had to do was check each direction until I saw the camp. Or I would find my own tracks, then follow these back to our camp. I set off down the slope of the ridge and began to walk fast, then to jogtrot. This was crazy; I would get tired as well as hot. I was moving too fast, with too much anxiety, to be able to look for tracks. I struggled to slow down, to have some sense. But panic rose into every part of me; I began to shout, to scream. I dashed in one direction, then another. Everything looked the same—infinite, undifferentiated. I got very hot and became damp with sweat. I must have zigzagged at random for an hour, hurrying without any reason, before I managed to gain some self-control.

I knew that I must climb the ridges. I noticed one that was higher than the others, and I set off towards it. Some part of me was in the grip of a conviction that this was the end. I experienced waves of dismay: I had not even begun this journey, had learned so little. I was ashamed and angry. What conceit had made me think I could set off on my own across this huge and unfamiliar land? On my way to the high ridge, I saw many more birds; they sat motionless, relying on camouflage, allowing me to walk within a few metres of their black, staring eyes. I saw in those ptarmigan eyes a revenge. I said over and over again to myself: *Your first walk in the North and you die.* How absurd this may sound now, and overwrought. Yet in those few hours I struggled against many demons.

At length I came to the high point of the ridge. I could now see far into the distance. All around me were the Barrens. Mile upon mile of frozen lakes and tundra. No sign of a tent, and no clue as to which direction I should choose. I stood among some boulders, staring about me in disbelief. Then, as a quick flash in the distance, I saw a snowmobile appear and disappear between two ridges. It was moving along a lake. If I could get to its trail, I would have something to follow. With luck, it would lead to a camp. I saw that I could get to the lake the snowmobile had passed by following the ridge I was on, then walking along a small valley. Keeping my eyes fixed on the direction, I set off.

I came to the lake, found the snowmobile trail and followed the direction I thought it had been travelling—guessing that it was likely, at the end of a day, to be moving towards home. In fact, it was moving towards our camp, and I soon came to the tent where the others were sitting around eating. I had not been gone long enough for anyone to wonder if I were lost; and I was too ashamed to speak of what had happened. I handed over the three ptarmigan, then went and lay down. I was filled with a mixture of relief and shame. I had everything to learn.

A week or so after this chastening experience, Karlik took me out on a seal hunt. Then he suggested I come with his family for a long day's fishing through the ice at a lake. Watching, I felt like the youngest of the children. It was a wonderful time of year. Thousands upon thousands of birds were returning to their nesting grounds or arriving as migrants, waiting to fly farther north. Great flocks of ducks, geese and waders were feeding and resting on shallow water on the tundra. I learned as many of their Inuktitut names as I could, finding great joy when I was able to call out the words. *Ivugak,* referring to both mallards and pintails. *Kujjuit,* the swans. *Nirlik* and *qanguk,* Canada and snow goose. *Qadlulik* and *tudlik,* the black-throated diver and common loon. *Savvat,* the phalaropes. *Isungat,* the jaegers. And *tatigat,* the sandhill cranes that flew in slow loops, filling the air with their deep, rattling cries. Labelling what was around me connected me to where I was.

The spring thaw came to the land around Rankin Inlet. Grasses and flowers appeared everywhere on the tundra; the sea ice began to change colours. At the end of my course, Karlik and Elisapik told me that they would be going out onto the land for the summer, living in a tent, fishing, berry picking, looking for caribou. Would I like to come with them? The question filled me with unease. Here was an opportunity to spend time in the Inuit world. I would be moving far outside the confines and categories of the language school, far beyond the adventure of day-trips to fish or look for seals or hunt for ptarmigan. Karlik could not even say how long we would be gone—a few weeks, maybe more ... But I was not ready to take the plunge into so many unknowns. I had a plan to go to Ottawa, spend a bit of time there, then travel to Pond Inlet in the High Arctic. I needed these plans. The Inuktitut course had made it possible for an Inuit family to invite me out onto the tundra for the summer and for me to be able to say "no, thank you" in more or less coherent Inuktitut. But the school had not prepared me for living with the Inuit, in their world. I was not ready to live without my language.

3

Back in Ottawa, I learned that my contract with the Northern Science Research Group, a division of the Canadian Department of Indian and Northern Affairs, could be converted into a full-time position. So I became a civil servant. It was a position with unusual terms of reference: I was to continue learning Inuktitut and living in the North long enough to be able to give informed advice to Canadian policy makers. As planned, I would go to Pond Inlet, an Inuit community at the northern tip of Baffin Island.

I flew from Montreal to Frobisher Bay, on south Baffin Island, and from there to Resolute Bay, a government and military base on Strathcona Island, five hundred miles north of the Arctic Circle. In those days there were no scheduled flights to the settlements of the far north; passengers hung around Resolute waiting for a ride with one of the charter companies that carried in government officials, mail and supplies. For two days I was stuck at Resolute, confined in buildings that were part hotel, part military barracks, in a landscape that was bleak and bitter. Then I hitched a ride in an Atlas Aviation Twin Otter taking the post to Pond.

I was the only passenger. The sky was clear, and I sat by a window at the front of the plane, staring with a mixture of fear and incredulity at immense fjords, ranges of mountains and the edges of the sea ice. We covered about 350 air miles, a tiny fraction of the region, but a step into the most awesome of landscapes.

It was the middle of June; spring in the High Arctic was at its most magnificent, most intoxicating. When I landed at Pond I had no school to go to, no work plan that was anyone else's responsibility, nowhere to turn for the security of southern schedules. I had arrived only in order to be there.

Pond Inlet, like all of Canada's Arctic communities, is a "settlement," a place where "native people" are expected to settle—that is, to come in from their hunting "camps" and begin to receive modern housing and services. Settlements were designed by government to offer something between aboriginal simplicity and modern complexity, providing what planners believed would be a halfway stage along the development road. By the early 1970s almost all Inuit in Canada lived in such settlements, with rows of prefabricated low-rental housing, basic amenities, a day school and a nursing station.

Each settlement had a "settlement manager," the government-appointed person who acted as general overseer of life. The Inuktitut name for this role was revealing: *inulirijik,* the one who fixes up the Inuit. The Pond Inlet settlement manager was John Scullion, a friendly man who came to the airstrip to meet my plane. He showed me to the Pond Inlet Transient Centre, something between a hotel and a bunkhouse. It was one of two identical buildings that sat side by side on the edge of a steep slope leading down to the shore. The buildings had pointed roofs, from which they got their Inuktitut name, *nuvulik,* a peak. They had been designed for children brought in from hunting camps to go to the new school. The transient centre had been a dormitory; its neighbour had been a mixture of dormitory and schoolroom. Now that all of their parents had begun to live more or less full-time in Pond, the children went to day school. The building I was in had become the place where migrant workers (important for house construction) and other visitors were given a place to sleep, shower and cook for themselves. The other of the two buildings became the "adult education centre," where the settlement manager arranged for cooking, sewing and English language lessons.

The *nuvulik* sitting area looked out, through a large plate-glass window, onto the sound between Baffin Island and Bylot Island. Daylight lasted twenty-four hours, and most of the time the sun shone on a twelve-mile width of sea ice, meltwater and tidal cracks. Beyond this rose the immense mountains of Bylot, with snow-capped peaks and many glaciers. On the shoreline below the buildings and out onto the edge of the sea ice spread a muddle of skidoos, sledges, fuel drums, boxes of camping gear and small sheds where families could store equipment and seal carcasses. At that time there were only two or three dogteams left in Pond Inlet, but the animals were tied up in lines on the shore; periodic howls reminded everyone of their dismal conditions.

For the first few days I spent much of my time in the transient centre, staring out at the panorama, watching people come and go on the ice, listening to those dogs. I was holed up. A sense of being incompetent and foreign created a rather absurd but paralysing shyness. I had no defined role, nothing with which to disguise my awkwardness.

When I did venture out, the few southerners I met were friendly and helpful; traditions of frontier or northern hospitality were strong. There were several teachers, two missionaries, two nurses and the manager of the Hudson's Bay Company store, as well as the settlement manager and his wife. I began to settle into the compromise between being there and avoiding being there that comes from accepting the generosity of these homes. Good southern food and easy conversation were on offer; and several of the community's white men and women had much to say about their experiences of the North, in some cases reaching back decades and across many Arctic communities. Here was an alternative to my nervous, incompetent attempts to get to know the Inuit of that time and place. Yet I knew I needed to be more at a loss, not less.

4

One of the few Pond Inlet men who spoke confident English was Elijah Irkloo. Elijah agreed to work for a part of each day as my social guide and interpreter, making sure that I got to know as many people as possible and heard as much oral history as I could. He took me into many homes, where we sat and listened to whatever people wanted to tell us. I heard again and again about the extent to which life on the land had given way to life in the settlement. Policemen had come in the 1930s, then missionaries and Hudson's Bay Company traders, white men with many powers and purposes. Some were fun, others were frightening. Under their supervision, the "old days" changed. A mix began to take shape, with life on the land yielding to various forms of modern poverty. Inuit women became servants in the white officials' homes; men had opportunities for occasional wage labour. This was work that yielded small amounts of money or credit at the store. It also meant a shift towards store-bought food and clothes, and some degree of dependence on newcomers who were anything but generous. Inuit families began to spend far more time in settlements, making do in shacks built with whatever materials came to hand, from packing cases to sheets of plywood. The balance of this mix shifted and changed, as it has continued to do. The effect is a blend of wage labour and hunter-gatherer resources.

I saw the broken-down look of so many of the houses and the paucity of everyday goods and services. Many people told me about the confinement and inactivity that came with living in a settlement. They could not hunt or fish or trap without making a journey that was almost from town to country, from one way of life to another. In house after house I saw the results of modern northern planning—low-cost housing, economies of administrative scale, reliance upon wage employment (and therefore the creation of unemployment), schooling, illnesses that come with settlement life, bottle-feeding of babies, a nursing station to meet the medical problems—and from person after person I heard quiet, understated dismay.

At the same time, there were also signs of a life that was independent of, or at least not obscured by, southern plans and administration or the everyday customs of southerners. No one expected Elijah and me to make an appointment, and no visit ended before it had run its course: since there was no right time for beginning, there was no time by which we could expect to finish. We would sit at kitchen tables as life in the houses went on around us. Children came and went, toddlers often clambering onto mothers' and fathers' shoulders as their parents talked. These children were never chided but, instead, were helped on whatever journey they had chosen, from back to shoulder or down again. This gentle and automatic support for the child left parents free to talk at great length and with intense concentration.

The conversations were in Inuktitut. Every adult spoke and lived in the language. Yet English was the language of the school, and many adults said how much they wanted to be able to speak and understand it. Their desire was animated by a belief that jobs in the future would go to those who spoke English, and by a sense that, without English, social or political effectiveness was limited. At the same time, many parents complained that their children spoke too much English at home and were not gaining a rich enough fluency in their own language. People were proud of their heritage and they delighted in their language; yet they had been persuaded, by both intangible historical forces and the activities of southerners, of the need for Christianity, Canadian law, compulsory schooling, southern-style political organisation and even Canadian cooking—all of which were tied to the power of English.

The tension between an Inuit sense of culture and the colonial enterprise was demonstrated by a curious argument between Pond Inlet parents and the southern educational project. The long season of continual daylight had begun; and as the sun warms the land and the Arctic bursts forth in full brilliance, the Inuit day shifts towards night. People stay up late, often until morning, using the midnight in preference to the midday sun. Children play games until breakfast time, then go home to bed. School, which had *not* changed its timetable from day to night, was something of a problem. For the past month, children had been falling asleep in the classroom; others were failing to get to school at all. It happened each spring.

Teachers asked to meet with a group of parents to explain the nature of the problem. Something must be done. Ah yes, replied the parents, they quite agreed. The children must go to bed at night and be able to go to school, and stay awake there, in the morning. It was very important that they not miss this chance for an education; above all, they needed to learn English.

So, what was to be done?

Well, said the parents, it was obvious enough: the children must go to bed.

You parents, said the teachers, will you make sure the children go to bed?

No, no, said the parents, you teachers must do that. You are the ones who run the school.

And how can we do that?

You must come round to our houses, said the parents, and make sure our children go to bed.

The teachers were dismayed and bewildered. They could not see themselves going from house to house in the middle of the night, even under the blaze of the midnight sun, persuading children to stop playing their baseball games, to come off the ice, to put down their bicycles and go home to bed. This was absurd. It all amounted, in the teachers' minds, to a failure of parenting. So nothing was done.

The Inuit way is without authoritarianism; parents are inclined to trust children to know what they need. Individuals have to be left to make decisions for themselves; and children are individuals just as adults are, since they carry the names—for which we may say souls—of their late and much admired relatives. This belief is fundamental to the Inuit way of being in the world, to their culture, and to hunter-gatherer cultures more generally. School might be important, but its disciplines, its theories of need, had to make concessions to Inuit customs. The needs that school was supposed to meet might be a matter of agreement, with Inuit and teachers of one mind about the importance of southern kinds of education; but the imposition of the routines and authority of school was still alien. To suggest that parents should impose authority, should defy the respected elder who lives in the core of the child, was not acceptable to the parents in Pond Inlet. If there was a clash of schedules, let the Qallunaat, the southerners, in whom these clashes originated and in whose role was their orchestration, do the job. Qallunaat had designed, built and even furnished Inuit homes, so why should they not go there and organise bedtime too?

The tenacity with which Inuit held on to the core of their culture was both startling and delicious. It offset the troubling look of the place, so much a southern plan for a settlement, so little an Inuit village in design. It showed what it means for a people to insist upon that which defines them. Refusing to force bedtime on their children in an Arctic June sustained a heartland of Inuit life.

5

I longed to get out of Pond, onto the land, into the places where people hunted, trapped, fished and said that they lived most fully as themselves. Through the windows of the transient centre, I watched skidoos setting off onto the ice, towing families and supplies out into the distance, and others returning to shore, loaded with the bodies of seals. Occasionally I saw a dogteam, harnessed in a fan with a figure running alongside waving an immense whip. I became convinced that I must make a journey out there, and that it should be by dogteam.

With the help of one of the settlement's schoolteachers, I met Inugu, a man of seventy-five who still hunted. He was going fishing with one of his children; they would be travelling by dogteam. He sent a message that he would be happy to take me along. We should meet in the Hudson's Bay Company store, he said, where we could buy whatever we needed for the trip.

I found Inugu in one of the aisles of the store, surrounded by shelves loaded with tinned meat and fruit and many kinds of biscuit. He was old, but every part of him seemed strong and clear. He stood upright and walked with definite movements. Although he hesitated before he spoke, he was decisive and clear in all he said. There was no doubt about what took place in the store, among all those supplies. I would point to something I thought we might want, and he would pause for a moment, consider, and say no. No, we did not need bread, or butter, or jam, or packaged meat, or coffee, or beans. Nor did we need fishing line or lures. Nor would we take tinned fruit. He conveyed to me that these things were not part of his way of being on the land. He led me to the pilot biscuits—packets of cheap, hard, round, desiccated, more or less indestructible things. This was the carbohydrate staple of Inuit hunting trips. We bought several packets. Inugu did agree that we should take a bag of sugar and a packet of tea bags. He also suggested that I buy some naphtha, the fuel that Coleman stoves depend on. And a box of .303 rifle ammunition.

As we left the store, carrying our small load of supplies, I asked Inugu how long we would be away. He said he did not know. Maybe one week, maybe more. I asked if it might be a good idea to bring a fishing rod, since I liked spinning for trout and char. He looked confused. Perhaps he did not understand the word I used for rod. I tried to explain what I meant. He made the movement of plunging a harpoon and said *kakivak,* a word I knew to mean fish spear. I deduced from all this that Inugu wanted our trip to be, as it were, traditional.

I was thrilled. After he left, I slipped back into the store and bought a supply of chocolate bars and a few spinners. I decided that I would hide them, along with my spinning rod and reel, in my sleeping bag.

We set off late the next day. Inugu brought his youngest son, Willie, aged fourteen. The two of them loaded the sledge while I stood and watched. The box of supplies included no more than the few things we had bought at the store. But Inugu had brought a tent, tent posts, a rifle in a sealskin gun cover, and his *kakivak*. The three of us and fourteen dogs set out across the sea ice.

It took almost three days to reach the fishing place. Amazing, beautiful, alarming days—or nights. The sun circled above us, with no cloud in the sky for all the time of our journey. A drop in temperature separated day from night. By moving when it was coldest, Inugu explained, we would suffer least from thawing ice and excessive meltwater. So we travelled through the Arctic night and camped in the daytime.

I was wonderfully disoriented. We had no watch, and I did not yet know which angle of the sun meant what time of day. For me, time was seamless. We would arrive at a camping site, pitch the tent, find fresh water, roam about on the land nearby, go to sleep, wake up, set off again. My ignorance allowed me to be lost, at last, in just the right way.

An hour or two after leaving Pond, we had come across some seals basking on the ice. Inugu stopped the sledge and set off to stalk them. With immense care he made his way towards the animals. When they had their heads down, he moved with quick little steps, crouching low; when one or another of the seals raised its head to peer and listen for danger, he stopped moving and crouched even lower; once the seals appeared to doze off, Inugu stepped forward again. Taking all the time he needed to move unnoticed, he came to within range of the seals. He squatted, raised his rifle, took aim, fired, and missed. With a quick lurch, the seals were gone from the ice. Inugu stood up, walked to the breathing hole of the seal he had hoped to kill, looked at the surrounding ice, and made his way back to us. This was repeated two or three times before we made our first camp. After the second miss, Inugu told me that the sights of his gun were faulty.

We had our pilot biscuits to eat and my few bars of chocolate. We also had some glaucous gulls' eggs, which Willie had collected from a small cliff face not far from our campsite. We ate them boiled. They were delicious—as, indeed, were the pilot biscuits and sweet tea that went with them. By putting the eggs in a pan of cold water, I had sorted out the ones that were fresh-laid, marking them with a pencilled cross. (Fresh-laid eggs sink and lie on their sides at the bottom of the pan, whereas those with a developing embryo half-float, and those about to hatch float high up on the surface.) Inugu and Willie enjoyed the mixture of unhatched chick and egg white.

Thus we travelled for two more days, stopping for Inugu or Willie to stalk seals, all of which they missed, and living off pilot biscuits, sugary tea, bits of chocolate and a few gulls' eggs. The dogs had no food at all. We crossed a wide sound to reach the entry to long fjords that cut deep into the broken shores of Baffin Island. We headed down one of these fjords, sometimes in its shade, with cliffs towering above us, but often under full sun. We circumnavigated cracks in the ice that were too wide for the sledge. Hour upon hour we sat on the sledge or jogged alongside. Inugu taught me vocabulary, pointed out landmarks, but for the most part we moved in a quiet and perfect peace. At one place, Inugu took me with him for a walk up a small hillside that sloped to the shore of the fjord. As we came to its peak, I noticed many bones scattered there.

"*Tuttu,*" I said. "Caribou."

"*Aakka, Inuit,*" said Inugu. "No, people."

He was showing me a place where some of his relatives and ancestors had been buried under cairns of rock. Later he pointed out a grave that was only partially scattered: a pile of rocks among which it was possible to see the long bones and skull of a human body.

On the third night of our journey, Inugu chose to pitch our tent at the mouth of a tiny stream that flowed down the steep side of the fjord. The valley was narrow and the campsite an area of tussocky grass no larger than a tennis court. The slopes were steep and thick with boulders and rock. After our usual meal of tea and biscuits, I climbed the slope behind and reached the peak of a ridge. In front of me, in an immense panorama of peaks and valleys, range upon range, the mountains of north Baffin Island stretched to a far horizon. Behind me was the long fjord down which we had come.

We had seen no other travellers or hunters since we set off and expected no one else to be at the fishing place. In one sense, this whole landscape was "empty," a "wilderness." But all of it had been given a set of complete human shapes—names and purposes and meanings—by Inuktitut. Inugu and Willie knew this land. They could navigate, select routes that took them over hundreds of square miles of sea ice. As we travelled, they named each bluff and headland, identified bays and fjords. They gave the names of every river, and told me the names of many inland lakes. Again and again Inugu took me to places where we could see far into the distance, and there he would point and name and take delight.

To move around with safety, to hunt with success, to make the land's resources available and nourishing, the hunter works with a mass of details and the names of many, many places. Nothing could be better, for there could be no alternative: to know this particular territory is to prosper; neither the land nor the knowledge of the land can be replaced. A territory is made perfect by knowledge. Inugu was revealing his profound conviction that this was his only imaginable home.

6

At last we arrived at Iqaluit, the river where Inugu wanted to fish. *Iqaluit,* the plural of *iqaluq,* means arctic char, the species of fish that spends part of each year in lakes, migrating to the sea to feed when the ice breaks in the spring. I was to discover that great fishing places are often named for the species caught there.

We had arrived at the mouth of the Iqaluit River, where it widened in a small estuary at the end of a long fjord. Inugu was worried, he said, by the lack of snow on the land. Everywhere around the estuary the tundra was bare, and the river itself ran more or less free of ice, spreading out into an open bay. At first I did not understand the problem. Then I realized: to use a *kakivak,* the fisherman had to be able to stand on ice, looking down into clear water. If the ice was gone, then there were shelving banks with either no ice at all, or ice that was too broken or rotten to stand on.

Inugu chose a campsite near the shore. He took his *kakivak* and set off to look for a place where it might be possible to spear fish. I took my fishing rod, reel and bag of spinners from inside my sleeping bag. Willie and I walked to the last stretch of river before it widened into its estuary and joined the fjord. In the distance, far from our campsite, Inugu had pointed out that the river was full of fish—we could see swirls and small splashes on the surface. As soon as we reached the river, I rigged up the rod and cast into the water. A char grabbed the lure. We landed it and I cast again. Another char was hooked. And another. Every cast produced a fish. After a while, I pulled some line off the spool and gave the rod and reel to Willie. I think it was the first time he had used one, and he was keen to have a go. I took my line, knotted a lure to it, waded knee-deep into the river and flung the line out into the current, using myself as rod and reel. Every few times I cast, I pulled out a char. They varied in size from two to seven or eight pounds.

Inugu came back from the estuary ice. The ice nearby was too broken, he said; the fish could not be lured to the spear. He would have to move farther around the shore, out into the estuary where the ice was solid, to find a place where the *kakivak* method would work. But Willie and I, with our rod and line, or line without rod, could fish here. Inugu laughed at this newfangled way of catching arctic char, but he welcomed its success. After all, we could now avoid moving camp. Squatting on the ground, Inugu killed and cleaned the fish as we caught them, sorting them into those we could feed to the hungry dogs and those we would load onto the sledge and take back to Pond.

We spent two days fishing, cleaning fish, and taking time to roam on the tundra. Inugu taught me the names of the birds there—greater snow geese, semipalmated plovers, northern phalaropes and two species of jaeger that dive-bombed us for getting too close to their eggs or chicks.

Willie and I spent hours searching for nests. As we walked, he showed me the saxifrages, tundra flowers, grasses and lichens that burst into life and colour once the snow has gone. The abundance of life in the High Arctic is the region's deepest surprise of all. To be able to speak of it, even a little, in Inuktitut gave me in some essential way both the glory and the detail of the land.

Three days after arriving at the river, Inugu decided we had enough fish and should begin the journey back to Pond Inlet. He laid the fish along the cross-slats of the sledge; there were enough to cover its full length. Then he arranged our bedding and gear on top. We travelled back with a sledge that was piled high and heavy with char. The journey was slow, the dogs struggling to pull the sledge's weight through ice that was more brittle, fragmented and watery than ever. Whenever we stopped, we ate boiled fish.

A year or two later, when Inugu and I were reminiscing about the trip to Iqaluit, he laughed at the time and effort he and Willie had taken to cook our chunks of char. They had feared I would be shocked if they did otherwise: they knew that many Qallunaat were disgusted by raw fish. Had I not been there, or had they known me better, they would have boiled the char only when we camped. Maybe not even then. We were all engaged, in our own ways, with the cultural divide between us.

7

Once we got back to Pond, my life there changed. Simon Anaviapik, a man of sixty who had always wanted to teach Inuktitut to a

Qallunaak, offered to give me lessons. He arranged for me to live in the house of one of his nieces, Inuja, along with her husband, Inuk, her two sons and her daughter-in-law.

Most of my Inuktitut lessons took place in Anaviapik's house, which was two or three doors up the street from Inuk and Inuja's and, like theirs, was built according to government-regulation northern housing design. Prefab sections made a largish living room/kitchen and three small bedrooms. There was a bathroom, with toilet and bath but no running water. Cooking and heating were done with the help of an oil-fired stove and space heater that kept the inside temperature at about 80° Fahrenheit. In all but the warmest weather, meat was stored either on the roof or in an unheated porch, along with skin clothing and hunting gear. Pieces of caribou, seal and fish ready to be eaten were laid on a piece of cardboard or plywood to the side of the stove, where the family and guests could squat and eat when they felt like it. In many ways, these houses were as much campsites as modern homes. The Inuit made them their own with an ease of manner and a quiet indifference to the images of everyday life firmly fixed in the minds of those who had designed them.

Anaviapik's household included his wife, Ulajuk, their daughter Rebecca and their son Jake. There were visits for much of every day from his daughter Daniki, her children, and many other relatives. Anaviapik's close companion and cousin Arnatsiak also spent part of each day there. The younger children spoke good English, and sometimes preferred to use it with their friends, thereby causing some distress to their parents and elders. But the language of the home was Inuktitut. Nobody in the family over the age of twenty spoke anything else.

Anaviapik was eager to teach me. As we sat for the first time at his kitchen table, he told me he had always wanted to teach the language to a southerner. But he wanted to do it well. Was I willing to do it well? I did not understand. Of course I hoped to be able to speak as clearly as possible, and to understand anything people said to me. I asked if he meant something other than this. Yes, said Anaviapik, he meant learning Inummarittitut, the "real" Inuktitut. I told him I would be very happy to try to learn Inummarittitut, though I was not sure what it was.

He explained to me: White people come to the North. They say they like it, and seem to have a good time. Some of them do good work, and the Inuit get to like them. Some are not so good. But just about all of them, after a while, go south again. Whatever they say, they do not come to the Arctic to live. He was very happy to teach Inuktitut to a southerner. But it must be the real Inuktitut, not the bits and pieces that traders or policemen learn. It must be the Inuktitut of the older people. It was important that their Inuktitut be learned. Many young people were not interested in learning. So that was what he would teach. But to do that would take a lot of time. We would have to spend a lot of time together. Every day. I would have to come to him every day, and maybe we would go out hunting too. We would get to know each other well. More than that: we would get to like one another. Then the time would come when I went back to the south. The way all southerners do. That would be very sad. So we must be able to write to one another. In order to write, I would also have to learn the Inuit way of writing.

Thus Anaviapik set out the conditions for his becoming my teacher. In the coming months, he stuck by them with great determination. I did go to his house every day, usually for several hours. He would have the lessons prepared in advance. Exercises, questions, and small outings onto the tundra or along the shoreline. There was much laughter, often on the part of those members of the family who sat and listened or joined in.

Learning the writing system was not difficult. It used a set of syllabic symbols that missionaries had developed in the nineteenth century to create a written form of a northern Cree language. Inuit, whose territories in southern Hudson Bay bordered those of the

Cree, picked up syllabic writing and used it for Inuktitut. It worked well and soon spread throughout the eastern Arctic. By the early part of the twentieth century, it had become part of the region's traditional culture, something elders were proud of and used often. Mick Mallon taught all his students by syllabic symbols. They are quite easy to write. Each represents one syllable: ᐱ (pi), ᐸ (pa), ᐳ (pu); ᒥ (mi), ᒪ (ma), ᒧ (mu); ᑎ (ti), ᑕ (ta), ᑐ (tu), and so on. These are only approximations of some sounds, and they fail to make some important distinctions. As you write, you do not have to be sure whether a consonant is single or double, or whether a sound is made in the throat or at the back of the mouth. But when it comes to reading, critical meanings come from context rather than the actual symbols on the page. Soon I could write letters to Anaviapik; it was quite some time before I could read most of what he wrote to me.

Usually our lessons began with a very simple written question that introduced new words and new ways of using them. Anaviapik, like many Inuit I knew over extended periods of time, worked within the vocabulary and language skills that he knew I could manage. Even after a year of working together, he would notice if he used a word or some piece of grammar that was new to me. This was an aspect of Inuit thoughtfulness and hospitality; it also showed a remarkable awareness of, and extensive memory for, language use, which may well be a feature of a primarily oral culture. So our lessons were very precise, at least as they began. We focussed on language, but I was also instructed when to be quiet and how to wait for others to ask questions. Anaviapik showed me how to come into my house when returning from a hunt. How to ask for tea and some food but to say nothing of success or failure, waiting for those in the house to ask me about what had happened. How to answer these questions with the right kind of modesty, with precision, leaving it to others to ask more questions before I gave more detail.

Sometime in July, when the ice had gone and Pond looked out on a wide stretch of open water, Anaviapik began a lesson with the written question "Do you know how to hunt seals from the shoreline?" This exercise led, first, to my learning how to say that, since it was summer and I had no boat, I was going to hunt seals from the land. Much care was taken to ensure that I did not claim I was going to be successful (I had to keep an infix in place that showed I would be making an attempt) and to ensure that I conveyed a suitable degree of uncertainty (I had to use an affix that indicated shoreline seal hunting was probably, but not definitely, what I was going to be doing). Then Anaviapik said we must now go and look for seals. He gave me a gun from the porch and a few shells, and we walked down from his house to the edge of the sea, then along the shore to a place where he knew seals often fed close by. As we sat and waited (unsuccessfully) for a seal to surface within range, I had lessons in how to sit and look out at the water, as well as in how to name different sea mammals. When we got back to Anaviapik's house, there was a concluding session in which I was told how to talk about the hunting I had just done.

This daily reporting of events included a somewhat alarming learning game: Anaviapik liked to send me to visit someone who could be relied upon to ask me what I had been doing. Since I had been instructed in what might be called the classical forms, my mixture of very proper manners and what may well have been archaic vocabulary caused great hilarity. I suspect I was taught to behave in imitation of the most correct elder and to use words that matched, rather as a butler might be schooled to ask an employer if he would care to imbibe a glass of a particularly delectable Sauterne, or a pompous schoolteacher tell a pupil that his homework had, atypically, been without its habitual blunders. These comparisons rely on social inequalities and may therefore be misleading. Even though Inuktitut is a language that makes little, if any, use of terms to show status, it does have levels of sophistication, including especially elegant and polite ways of saying things. My use of these must often have seemed ludicrous.

My attempts to describe events caused much laughter. So did my failures to get words right. Inuktitut is full of possibilities for embarrassment or comical nonsense: *uttuq* is a seal sleeping on the ice, *utsuk* a vagina; *ujjuk* is a bearded seal, *ujuk* is soup; *uksuq* is the fat of sea mammals, *usuk* is a penis.

Anaviapik did his best to make sure the mirth was not always at my expense. He also made a point of showing me how to tease others in Inuit fashion. For the most part, this teasing was directed at the one or two older men who were Anaviapik's best friends and former hunting partners. The jokes were bold. One lesson was spent teaching me how to tease men who had wives much younger than themselves. This lesson led to my being urged to say to one of Anaviapik's friends: "Well, things must be very tiring for you since your wife is so youthful." Everyone thought this immensely funny, though I do not believe I ever agreed to try it out as required. But one day we were talking about another elder with whom I had spent some time, and who had a wife some thirty years younger than himself. I said, by way of continuing the joke of that earlier lesson, that next time I visited this elder I would of course ask him if things were tiring for him, since his wife was so young. Anaviapik was aghast. No, no, I must do no such thing. My potential foolhardiness led to a lesson about *ilira*.

8

Inuktitut has several words for fear. Each is a root, to which the infix -*suk*- is added to show a feeling or -*na*- to show a circumstance. Many words for danger are based on the root *kappia*. Hence fear of danger, *kappiasuk-,* and something being frightening, *kappiana-*. Another, less common word for fear is based on the root *irksi,* which denotes a source of terror. Polar bears are sometimes said to be *irksina-,* terrifying.

During one of our lessons, Anaviapik talked about white people who came from the south and bossed Inuit around. He gave the example of a policeman who was especially domineering, who gave orders that resulted in men working intolerable hours, and who had sexual liaisons with women who did not like him. This example led to our talking about why Inuit had, at times, done things that were not in their own interest. "*Ilirasulaurpugut,*" Anaviapik said. "We felt *ilira.*" I did not know the word. He began to explain it to me. "*Iliranalaurtut.*" "They were *ilira* causing."

So *ilira* is to do with being afraid? I asked. Like *kappia?* No, not that kind of fear. And not the *irksi* kind of fear either. Anaviapik gave examples of what might make you feel *ilira*: ghosts, domineering and unkind fathers, people who are strong but unreasonable, whites from the south. What is it that these have in common? They are people or things that have power over you and can be neither controlled nor predicted. People or things that make you feel vulnerable, and to which you *are* vulnerable.

Anaviapik explained further: when southerners told Inuit to do things that were against Inuit tradition, or related to the things that Qallunaat wanted from the North, the Inuit felt that they had to say yes. They felt too much *ilira* to say no. There was danger—not of a kind that was easy to describe, but real enough. A possibility of danger. White people had things that Inuit needed: guns, ammunition, tobacco, tea, flour, cloth. They also were quick to lose their tempers, and seemed to have feelings that went out of control for no evident reason. They had power, and there was no equality. These circumstances inspired *ilira.* I asked if this still held true, in the Arctic of the 1970s. Yes, Anaviapik said, for the most part; there was some *ilira*—not with every southerner, but with most. They had the power, and they were not like Inuit.

The word *ilira* goes to the heart of colonial relationships, and it helps to explain the many times that Inuit, and so many other peoples, say yes when they want to say no, or say yes and then reveal, later, that they never meant it at all. *Ilira* is a word that speaks to the subtle but pervasive results of inequality. Through the inequality it reveals, the word shapes the whole tenor of interpersonal behaviour, creating many forms of misunderstanding, mistrust and bad faith. It is the fear that colonialism instils and evokes, which then distorts meanings, social life and politics. The power of colonial masters is indeed like that of ghosts—appearing from nowhere, seemingly supernatural and non-negotiable.

In our lesson about *ilira* and the limits to jokes about young wives, Anaviapik explained to me, with great care, that not all people could be talked to in the same way. Inuit elders who inspired *ilira,* though obviously not part of the colonial system, were often men and women of immense importance, with power that was beyond the norm. Anaviapik hinted that this may have been to do with shamanic skills or qualities that evoked equivalent kinds of awe. But it was Qallunaat who gave rise to the greatest and most troubling degree of *ilira.* The core of the lesson was: you cannot talk with freedom and humour to those who inspire *ilira,* whereas it is possible to say anything at all to those who do not. In this way, to be talked about as someone who does not cause *ilira* is to be paid a deep compliment.

9

Anaviapik never spoke a word of English, not even "yes" and "no." He did not use everyday Anglicisms that were common in Pond, things like "hello," "bye-bye," "good morning" and "thank you." In our lessons, he conveyed new meanings through examples, relying on vocabulary and grammar I did know to get me to understand bits that I did not. He acted things out, clowned and, in desperation, sometimes called on Rebecca or Jake, who went to school, to translate for me. He and Ulajuk insisted that they did not understand a word of Qallunaatitut, the language of the southerners.

But one must be careful about judging language skills. Shyness, modesty and cultural pride can cause many older men and women to hide their English skills. Some Inuit had close and long-term dealings with traders, policemen and missionaries, going back to the 1930s and '40s. Those who had worked and lived most closely with outsiders heard, and no doubt learned, at least bits of the newcomers' language. Contact between Inuit and Europeans in the Pond Inlet area reached even further back. On our fishing trip, Inugu had explained to me that he was part Qallunaak, for his grandfather had been the captain of a whaling ship. Inugu, however, had never learned any English, for he had always lived far away from trading posts, in an area some two days by dogteam from Pond Inlet.

I did once get a surprise from Anaviapik. It was February, seven months after I had first stayed in Inuja's house. Anaviapik and I were returning from a caribou hunt. He was driving a skidoo, while I rode on the sledge. We had encountered hard weather. The temperature for the two or three days we were out could not have risen above –40° (at which number Celsius and Fahrenheit coincide). The day before there had been a fierce storm, and as we headed back to Pond, the wind still bit into our faces. This was the occasion on which I first experienced my eyes freezing shut: as they watered in the wind, the tears turned to frost that was firm enough to stick upper and lower eyelashes together. Although this was alarming when it first happened, I found that it was easy to fix with a brush of the back of a glove against the closed eyes. In fact, if you are dressed in caribou clothes, extreme cold is more startling than uncomfortable: the fur creates almost complete insulation.

The skidoo, however, could not keep warm. The engine kept cutting out. This had been happening throughout our return journey. Fuel would not keep flowing, perhaps as a result of minuscule droplets of water in the gasoline that were freezing in the carburettor jets. Each time, with a little difficulty, we got the engine going again. But about thirty miles from Pond, the engine stopped and would not restart. We were far from any other hunters. Anaviapik was no mechanic; nor was I. We were eager to be home. The weather was getting worse, and I think that Anaviapik was physically tired. (A thing he would never acknowledge, speaking instead of being sleepy; lack of sleep is admissible, but failure of the body's muscles is not.) He climbed off his dead machine. I walked to it from the sledge. We stood there, side by side, peering at the engine. Then Anaviapik said, in a quiet voice that was firm but without real anger: "Shit, bastard, fuck."

I stared at him in exaggerated amazement. And made a joke: "Ah, so you speak English after all."

"Yes," he said in Inuktitut, "I speak the English my father learned when he worked for the cook on a ship. He knew the words for being angry. Shit, bastard. There were others, but I have forgotten them."

Inuktitut is a language without swearing or cursing. Words for sexual actions, genitals, or excrement and urine are not used to express dismay or insult. Nor is there an invoking of the supernatural to express surprise or hope. Inuktitut is also remarkable for being free of the language that expresses social difference—the "please," "thank you" and "by your leave" that, along with cursing, can indicate so much about status. The equality and forthrightness of Inuit everyday life are well marked by this absence of words of polite obeisance. Seventeenth-century Quakers would have been at ease with the Inuit non-use of language forms that speak to bad faith, obsequiousness or hierarchy. In the course of colonial changes to the North, however, Inuit adopted or created words for "hello," "good-bye," "thank you," "you're welcome" and many others. These were needed to satisfy the newcomers' demand for verbal politeness and the Inuit wish to appear, at least, to be cooperative. Where the Qallunaat inspired feelings of *ilira,* they elicited a new vocabulary of deference and respect.

Anaviapik told me he did not know what those anger words he had learnt as a child referred to. With much laughter, we translated them into Inuktitut. Anaviapik wondered why Qallunaat would want to shout about shit or sex when they were angry.

My attempt to translate the words and the ensuing conversation led to a lesson, a few days later, in which we explored the difference between Qallunaat and Inuit expressions of feeling. In the course of this, Anaviapik told me, as other Inuit elders had done before, that the strong must conceal bad feelings, must be angry without anger. To make the sounds that disclose inner rage is, he pointed out, to be like a small child. It indicates a failure of *isuma*—that capacity for sense and reason which grows as part of becoming an adult. Anaviapik had often seen Qallunaat reveal their anger, not only in words but through the changing colour of their faces and their aggressive gestures. He had felt sorry for them, he said, and supposed that this want of *isuma,* this childishness, was a result of their being so far from home, in a strange place, and frustrated by the Inuit.

10

As time went by, Anaviapik and I did more and more things together on the land. When it came to travelling in hard conditions, he sometimes preferred to hand me over to younger hunters. But he took me for day-trips to check fox traps and on at least one caribou hunt in extreme weather. And in late April of my first year in the North, a chance came for us to make a long journey. A doctor who did occasional clinics in Pond asked if I could arrange for him to travel by dogteam. The doctor was based in Frobisher Bay, the region's centre far to the south. Before returning there, he was scheduled to go to Arctic Bay, the nearest community to Pond Inlet, 150 miles away across the north tip of Baffin Island. He wondered if it would be possible to use this as his chance for a dogteam adventure. When I asked Anaviapik who he thought might like to take up the request, he said he would go. He would take his son Inukuluk's dogs and get Muckpah, a younger hunter with a good team, to go as well. Our doctor friend could ride with Muckpah; I could go with Anaviapik. When the idea was put to Muckpah, he was eager to join in. Everything was soon arranged. Caribou clothes were found for the doctor, and there was much talk of what supplies we would need.

There was also much talk about the route to be followed. Anaviapik proposed that we go there across land and come back on the sea ice. He thought it would take six or seven days in each direction, and we would build snowhouses for shelter. The overland journey was through mountain passes and, in late April, would not create chances for hunting. Going out, we would carry all our food with us. The return, however, was round the northern point of Baffin, an area known to be good for seals and polar bears. We would hunt our way home.

11

The Inuit of the eastern Arctic are supreme builders of *illuvigat*, snowhouses. The root *illu* means an interior or home of any kind, and is the origin for the English word "igloo." Before the use of modern houses, the large family snowhouse was built in late autumn and designed to last several months. When Inuit travel, they still make small, overnight snowhouses. A single hunter will do this by building on a snowbank, cutting blocks out of the snow where he stands and using these to form walls and a roof around him. On sea ice, where snow is relatively shallow, the blocks have to be cut first and then moved to the site chosen for the house. This is slower work, and best done by two or three hunters who share the tasks. One person cuts blocks, using a medium-sized carpenter's saw, another carries the blocks to the building site, and the third builds the house around him. Once a layer of good, wind-packed snow is found, three people can build a shelter large enough for them all in as little as twenty minutes.

Expertise in building a snowhouse depends on close observation and knowledge of snow, in all its many manifestations. This knowledge is evident in the detailed ways in which Inuit categorise different kinds of snow—which brings us to the famous stereotype surrounding Inuit words for snow.

The question about snow is, or has become, one of phenomenology rather than ethnography. An ethnographer can explain the ways in which a particular person or group of people describes and responds to and manipulates the world. A broad humanistic assumption stands behind such work, namely that all people are using the same kind of brain to achieve their particular version of the human task, albeit in varying circumstances. Pastoralists in the Arabian desert, farmers in the west of Ireland, and Inuit in the High Arctic live in very different circumstances. They have very different ways of talking about the world. But according to the ethnographer, if they make the necessary effort, people in each of these societies can learn the language of the others. In this view, all languages are intertranslatable, and the meanings that specific circumstances give to words are also communicable. So we can say that the Inuktitut word for the sea bird *qaqudluk* translates into English as "fulmar"; and we can explain that the Inuit have built into their word the sound a fulmar makes *(qaqu)* and an infix that signifies wrongness or unpleasantness *(dlu),* since the fulmar has an unpleasant-smelling gland at the base of its bill that makes it a bird one eats, if at all, only after some careful preparation. This is a simple example, but a difficult one would be a matter of degree, not kind. Many words may be necessary to achieve a good translation, but it usually can be done.

Those who challenge this belief in the intertranslatability of languages and cultures often look to the Inuktitut words for snow to argue that the way the world is known in language determines the speaker's reality. According to this view, the words of the Inuit create the world as well as describe it. That is to say, those who are not Inuit (or have not been brought up in the language and environment of the Inuit) are unable to know or actually "see" the world that the Inuit know and see. Another way that this point has been made (by Wittgenstein, for example) is in relation to the nature of language itself: a person can explain how a word is used and what it refers to, but the word's *meaning* depends on knowing a web of contexts and concealed related meanings. A good example is the word "worship"—how can anyone who has not lived in a society that practises some form of religious worship understand what the word really implies? Therefore, it is held, the language of the Inuit cannot be translated into the language of the Qallunaat. The multiplicity of Inuktitut words for snow gets cited so often that it has become an everyday dictum about both the Inuit and the connection between language and reality.

There are indeed many Inuktitut words or terms for different forms and conditions of snow. These include snow that is falling, fine snow in good weather, freshly fallen snow, snow cover, soft snow that makes walking difficult, soft snowbank, hard and crystalline snow, snow that has thawed and refrozen, snow that has been rained on, powdery snow, windblown snow, fine snow with which the wind has covered an object, hard snow that yields to the weight of footsteps, snow that is being melted to make drinking water, a mix of snow and water for glazing sledge runners, wet snow that is falling, snow that is drifting and snow that is right for snowhouse building. Also, Inuktitut has a number of verbs that have snow as their root, including picking up snow on one's clothes, working snow with an implement of any kind, bringing snow to someone, covering with snow, living on snow-covered ground, and putting snow in a hot drink to cool it down.

The linguistic roots of these terms are varied. Some are specific terms, as in the case of blowing snow (*pirqsirq*), or soft snow that is hard to travel on (*mau*), or freshly falling snow (*apu*). These words are very different from one another. There is no root term, no category that is equivalent to the English "snow," which then repeats with modifications to refer to all the different kinds and conditions of this one thing, snow. Some of the Inuktitut roots do not relate to snow at all, as, for example, snow that is right for snowhouse building (*illuvigassak*), the root of which is *illu,* to which is added an infix to signal that it is a home made from snow, and then an affix that refers to it being good material. Similarly, snow that has been rained on (*kavisirdlak*) is based on a word for fish scales (*kavisiq*), referring to the hard and crystalline look to the rain when it blends with snow and freezes. The word for snow that is hard as a result of powdery snow being compressed into a drift, *sitidluqaaq,* is based on the word for hard (*sitik*). Snow that yields underfoot (*kataktanaq*) is based on the root *katak,* used for many words that refer to falling.

In the course of an Arctic spring there are many kinds of snow and ice. The weather can be wintry, with temperatures staying below

−10° Fahrenheit for days on end and the chance of some of the year's most fierce snowstorms. Yet by late April, the days in north Baffin are twenty-four hours long, and if the sky is not darkened by snow clouds and the air pallid with blizzards, there can be hour upon hour of warm sun. Snow and ice shift from hard to soft to hard again. This variety of temperatures means that spring is also the time of year when the sea ice begins to show its greatest beauty, for there are areas of both liquid and frozen meltwater, patches of old and new snow. Leads in the ice—long fractures where icefields join in winter and now begin to separate—open and widen, creating dark lines of sea. At the same time, seals' breathing holes widen under the sun, and the tides push salt water up onto the ice. Each kind of snow and water is a different colour, and when the sun shines, the sea ice can gleam with a mixture of blues and greens and whites that must be one of the great aesthetic wonders of the natural world.

These varieties of snow and ice are things that Inuit differentiate and talk about. People must choose sledge routes, find water or snow for drinking, select places where they can make a house, consider which surfaces are safe to walk or sledge across, decide where to stand beside a breathing hole without making sounds that will reach the sensitive hearing of seals, and cut holes through snow and ice to fish. They also must predict the weather, then accommodate to its vagaries. The language for snow is integral to making decisions that will determine the success or failure of hunting, and has vital importance in assessing the probable degree of comfort and discomfort, as well as the dangers, of even a short journey. There is nothing surprising about the richness of Inuktitut when it comes to snow.

Yet the attempt to translate all the Inuktitut words for snow reveals just how many terms there are in English for types of rain or snow, or for winter and spring conditions to which Inuit also refer. Many of the English phrases or sentences that explain what a specialized Inuktitut word is demarcating are straightforward. Once Anaviapik had pointed out and explained a particular type of snow or ice to me, and had told me the words for it, I had no more than the (considerable) difficulty of remembering them. I could see or feel or make sense of all the things he named. There are grammatical forms in Athabaskan languages, notably to do with motion and time, or in Algonquian, to do with the animate and the inanimate, that are indeed difficult for a speaker of Indo-European languages to grasp. Grammatical categories in these other hunter-gatherer languages of the North are deeply unfamiliar to most other peoples in the world. Yet even in these cases, the difficulty of translation relates to unfamiliarity, not to any seeming intrinsic incomprehensibility. I can explain the grammatical principle at issue, even though I may have immense difficulty when it comes to using and applying this principle in the new language. I can set out what a grammatical distinction is doing, even though I may not be able to reproduce that distinction in ordinary English grammar. Learning to use words and grammar presents one kind of problem; learning the meanings of words and the intentions of grammatical devices presents another. Insofar as one can learn the latter, the ethnographic assumption about the intertranslatability of all languages would appear to me to be sound.

No one is surprised that experts of all kinds differentiate the things they know, use and work with. Doctors have many words for illnesses; carpenters identify a wide range of tools and types of wood; gardeners discriminate between many kinds and qualities of soil; and so on. Differentiation of this sort is integral to human skills in all societies. It does not follow from differentiation that there will be a problem of translation. But by the same token, intertranslatability does not mean that language does not reflect profound differences in how people live and see the world. My speculation is that the persistence of the stereotype regarding the "mysterious" Inuit range of words for snow, overstated or misrepresented as it is, symbolises a divide, and therefore a *sense* of this divide, between agriculturalists and hunter-gatherers.

In this debate about whether language creates reality or reality creates language, I am inclined to seek to have my cake and eat it too. There *are* profound differences between hunter-gatherers and other peoples, and these differences are going to be evidenced in language. On the other hand, languages are for the most part intertranslatable. Wittgenstein and Anaviapik are much in agreement: it is possible to learn another language, but there are limits to what translation can achieve. The Inuktitut words for snow, however, are not a good example of these limits.

12

The route from Pond Inlet to Arctic Bay took us from sea ice that was sometimes bare and crystalline, sometimes covered with a layer of hard snow, to wind-packed drifts along the slopes and valleys of mountain passes. The journey was a joy. We made our way up narrow gulleys, reaching high passes, then hurtled down the other side, often onto gleaming lakes. At the end of each night's travel, Anaviapik and Muckpah checked for the right quality of snow, cut building blocks, and made our snowhouses. We slept through the warmest part of the day. We ate our way through our supplies—tea, sugar, biscuits, soups, caribou and seal for us, frozen seal for the dogs.

On the third day, high in the mountain range that divides the hinterlands of the two communities, we stopped in a wide pass. In every direction the mountains were huge and jagged; it was hard to believe that we had found such an easy way into their very peaks. But Anaviapik and Muckpah seemed unsure about the next part of the route. They stood and looked around them, peering as if to see some road, some feature that would act as a signpost or beacon. "*Naukkut?*" said Anaviapik, turning towards different possible passes in the mountains. "Which way?" Then he took a knife from the box on our sledge and began to draw a map in the snow. I could not understand what the lines referred to, but they seemed to lead to a decision. Anaviapik pointed to the snow map, then pointed out the route we should take. We set off again.

Two days later we arrived at a steep coastline with a round and sheltered bay that gives Arctic Bay its Inuktitut name, Ikpiarjuk, "a pocket." Anaviapik and I were some way in front of Muckpah and the doctor. As we stopped to wait for them, we untangled the dogs' traces, tidied our sledge and spruced ourselves up. We wanted to arrive in Arctic Bay looking our best. Anaviapik was full of excitement. I knew that he had many close relatives there.

"When do you think I last came here?" he asked.

I guessed. Three or four years?

"No," he said. "Not three or four years ago. I was last here in 1938."

"So how did you know your way through the mountains?" I asked.

"Because Inuit cannot get lost in our own land. If we have done a journey once, then we can always do it again."

In Arctic Bay, Anaviapik arranged for me to stay with his wife's cousin's family. He took me around to visit many households, encouraging me to speak in the ways I had been instructed. During these visits we met two men from Igloolik, the first Inuit community to the southwest of Arctic Bay, though the distance between the two settlements is about three hundred miles. These men planned to leave Arctic Bay soon. They would begin their journey back to Igloolik by following the sea ice around the headland, in the Pond Inlet direction, before turning west again towards Igloolik. They were going to hunt for bears. Anaviapik proposed that the five of us should travel in a group for the first two days. Our doctor friend had already taken his plane south. The Igloolik hunters welcomed the idea. A few days later the five of us set out, the four Inuit each with a team of between twelve and fifteen dogs and a minimum of supplies.

Igloolik dogteams are famous for their strength and speed, and remarkable for their use of northern rather than imported materials. One of the sledges had long whalebone runners, and both used only sealskin lashings and lines. All toggles for fixing lines to the sledge or the dogs' harnesses were carved from walrus ivory. The harnesses themselves were cut from skins. And the Igloolik men's sleeping bags were made from caribou hides. We, on the other hand, were using nylon ropes, canvas harnesses, metal toggles and down-filled sleeping bags. Whatever the Pond Inlet hunters' reliance on trade goods, though, I was very aware that none of the men I was with spoke English. They took pleasure and pride in their language, as well as in being deep in their land. I was a visitor in a world celebrated, puzzled over and given shape by words, anecdotes and jokes in Inuktitut.

We were going to find a cache of seal meat that would serve as dog food for a day or two, and then we would be hunting. The cache was soon discovered, and three seals were loaded onto the sledges. We travelled fast through that night and stopped, full of joyous confidence, in the morning. The Igloolik men built a snowhouse for us and, alongside, another smaller one for our supplies. Then one of the men cut a seal into chunks with an axe, while another kept the fifty-five dogs at bay by cracking his whip into the noses of those that pressed too close. They scattered the chunks among the dogs, making sure that each of them got something to eat. As we drifted off to sleep, I can remember thinking that this was the first time on our journey that the dogs had been left untethered.

When we woke, we found that the dogs had broken into the supply house. Everything was eaten or destroyed. Bits of packaging, cigarettes, chocolate and biscuits were scattered all over the ice. Nothing could be salvaged of either the dogs' or our own food. We had with us, in the snowhouse, a large piece of seal meat and packets of tea, sugar and biscuits—enough to keep us going for a day or so. We divided it among us.

No one seemed to be worried. We would be hunting later that day, and we had enough tea and sugar to go with the seals we would kill. I felt a flutter of excitement: now we were living as hunters must often have lived, reliant on judgements about place, weather, animals, with minimal margins of safety. Our decisions about where to hunt would have to be made with great care and accuracy.

Towards the end of the first night's shared travel, the Igloolik men, seeing no sign of bears, decided to turn farther to the north and see if they might do better closer to the floe edge. Anaviapik and Muckpah thought we should keep to a route that followed the shoreline; this was the direction, they said, in which we would find the best seal hunting places. They based this on a knowledge of leads in the icefields that spread out from the tip of the headland in front of us. These leads formed each year in the same places and created concentrations of seals. At this time of year the leads would still be frozen and covered with some snow. But there would be many breathing holes; we would spread out and have a good chance of making some kills.

But in this direction the ice was rough. The force of tides and currents around the headland, along with autumn storms, had broken the sea ice as it formed. Huge slabs and fractured ice boulders had then been heaped onto one another, to be frozen solid, as winter came, into a vast obstacle course. For as far as we could see, the surface of the ice was a litter of jagged hillocks of ice. Many of these rose ten or fifteen feet into the air; in places they created miniature mountain ranges with steep, sharp peaks. We climbed and fell and lurched, the sledges again and again becoming trapped, and often capsizing, as the dogs tugged sideways around some tight corner. Traces snagged on jagged ice, causing the sledges to skew and then to stop. We struggled up steep places and jarred shockingly down the other side. Progress was slow and hard. Within a few hours, only a few miles into this crazy field of ice, we were exhausted. Several slats on one of the sledges were broken and had to be repaired. Anaviapik found a stretch of snow that would be good for a snowhouse; he decided that we should stop.

We ate the last of our seal meat, tied the dogs in a circle around the snowhouse to ensure that any approach by bears would be barred by snarls and howls, and went to sleep.

We woke to the sounds of a storm. Outside, the snow swirled

around us, and the dogs were already partially covered by small individual drifts. We could see that the weather was going to get worse. We set off, continuing to push our way through the pressure ice. As the combination of wind and icefield made our journey harder and harder, Anaviapik and Muckpah agreed that we should change direction and head towards land. From there, we would be able to use the ice when it was passable and travel on the snow along the shoreline when it was not. By the time we reached the shore, the wind was howling, and we stopped as soon as we found a drift that would provide blocks for another snowhouse. We built in haste: the weather was becoming ferocious. Inside, we were warmed by the Coleman stove and sugary tea. There was no meat.

The storm continued for another day. We stayed inside the snowhouse. Muckpah scraped all the bits and pieces of spilled soups and grease from around the burners of the Coleman stove and made them into a stew. We were surprised by its sweetness; no doubt many pans of tea had been sugared without much care. We ate the last biscuits.

The next day the storm subsided enough for us to continue the journey. There was still a wind that moved the surface of the snow. The Inuktitut word for this is *pirqsirtuq,* from *pirqsirq,* the powdery snow that carries in the wind. When *pirqsirtualuq*—"it *pirqsirqs* a great deal"—the result is *maujualuq,* a thick deposit of *mau,* soft snow that makes it difficult to travel. The storm had caused a layer of snow some six or seven inches deep to be spread everywhere.

Our first task was to find the dogs. Anaviapik and Muckpah walked around thwacking each high undulation in the snow with their whips. The dogs did not seem eager to resume hauling our sledges, but one by one they were discovered and harnessed; the sledges were loaded, and we set off.

The softness and depth of the windblown snow was a sort of torture. The sledges tended to stick, and every footstep was an effort. For the first few hours, the sticking of the runners and the effort of walking were relatively small problems. The dogs had energy enough to keep us moving, and we had strength enough to walk or jogtrot alongside. But as the night wore on, and the wind continued, we slowed and then began to stop altogether. The dogs would give up pulling, and then the inertia of the sledges, sunk deep in the soft snow, had to be overcome. The trick was for one of us to catch hold of all the dogs' traces and then haul as hard as possible, dragging the dogs backwards. In their intense resistance to this pulling, the dogs strained forwards. As the man holding the traces felt this energy building to a peak, he would let go of the traces and jump out of the way. Thus the dogs found themselves rushing ahead with enough momentum to get the sledge moving. The job then was to *keep* it moving—with urging, use of the whip, and even a glove thrown ahead of the dogs, leading them to believe that they could, if they hurried, grab an illicit mouthful of caribou skin. Their dismay when the glove was snatched away from in front of their noses no doubt gave them some further impetus.

At the beginning of that night's travel, we headed out from the shore and across the sea ice again. We had now travelled beyond the worst of the pressure ice and could find routes along a level, if snow-covered, surface. Anaviapik and Muckpah often ran out to the sides of our route, searching for breathing holes, looking for signs of one of the predicted leads. But the depth of the blowing snow frustrated them: no breathing holes were found. From time to time we paused to brew tea.

After moving in this way for eight or nine hours, we stopped and built another snowhouse. The dogs were again tied in a ring around us. Muckpah took all the caribou skins we had with us as floor-covers and mattresses and proceeded to dig small pieces of fat from the ears. We ate these, with much laughter. There were jokes about whose clothes we would boil up first.

When we woke, the wind had dropped, and Muckpah proposed that we change direction and head out towards some islands to the north. We would be heading away from Pond Inlet, but it was an area that could be relied on for its seal hunting. Anaviapik agreed. So we swung left, moving towards a horizon that was without mountains. We were looking out to the open sea, which must have been some fifteen or twenty miles ahead of us. The going was hard—the snow soft, the dogs weary. We moved for a mile or so, then stopped. The dogs had to be tricked and forced to move again. Three of them collapsed in their traces, no longer able to pull. They were set loose and left behind, lying in the snow. We found no breathing holes, no seals. We made another snowhouse, had another sleep, then turned back towards Pond Inlet.

Anaviapik said that we must ration our sugar. We were careful to have only one spoon each per cup of tea, instead of the usual two or three. During one pause for a drink and a rest, Anaviapik, standing very close to me, looked into my face and said, in a tone of slight surprise: "*Saluttualuvutit.*" "You're so thin." I looked at him and noticed that his face had become all cheekbones and jawline. His eyes seemed very large. Then he said: "But our companion [meaning Muckpah] is not thin at all. He must be eating something we don't know about." And he laughed. This was a joke Anaviapik had made to me before, comparing his own strength teasingly with the strength of young Inuit men.

We passed a difficult few hours. By now the dogs were weak, and conditions did not improve. The new cover of snow had not been wind-packed enough to make a firm surface on which we could travel well. We struggled on, making our way across the fjord. We kept going for seventeen hours, covering what must have been no more than twenty miles. (By comparison, on a good day, we were able to travel thirty or forty miles in ten hours.) During one of many tea breaks, Anaviapik turned to me and said, in a matter-of-fact way: "*Mitimatalingmut tikijjajunirpugutqai.*" "Probably we'll never make it back to Pond Inlet." I have a vivid memory of the shock of fear that went through me. Although I had known this was a journey full of difficulty, and while some of the dogs were starving and we were very hungry, I had never for a moment supposed that we were at risk.

How could I have felt so little apprehension? We had gone for five days with little or no food, and had seen the chances of killing a seal fade to hopelessness; several of the dogs were too weak to be harnessed, and the distance to Pond Inlet stretched far away ahead of us. Yet until that startling remark about not making it home, neither Anaviapik nor Muckpah had given a hint of worry. Despite exhaustion, hunger and disappointments, they had expressed neither disingenuous optimism nor disgruntled pessimism. The mood never changed. The same balance of quiet conversation, jokes and friendly silence had continued each day.

Inuit elders are remarkable for their equanimity. Smiles and laughter are used to deal with many forms of disquiet, and a judicious withdrawal is the proper way of responding to conflict. Anger was understandable in children but quite unacceptable in adults. In my language lessons, the question "*Ningngarpit?*"—"Are you angry?"—was somewhere between a tease and a reproach. If I was a little withdrawn on a particular day, or chose to go for a walk on my own on the tundra, my lessons would be sure to include words for "depressed" or "unhappy," often with an affectionate or consoling set of infixes: some such question as "*Numasuktukulungulitainnarpit?*" "Are you at last a wee unhappy one?"

To laugh, to be happy, to feel welcome or welcoming, to experience shyness, to be nervous about dangers in the world or society, and to feel *ilira,* the mix of apprehension and fear that causes a suppression of opinion and voice: all these states of mind are spoken of and exhibited with real freedom. They have helped to shape the stereotype of the Inuit as a people of unfailing goodwill, good humour and generosity. Yet the intensity of feeling that can exist within this restraint and dignified self-control, breaking through at times of extreme difficulty, is also remarkable. When one of Anaviapik's grandchildren was killed in a fire, the family's grief was fierce. The boy's father spoke to me of having been insane with grief—meaning that he had not been able to check outbursts of crying or periods of open despair. By European standards, this would appear to be moderate; by Inuit standards, it was extreme. Similarly, expressions of real anger have a smouldering, sharp quality that gives them more menace than mere shouting and yelling would convey.

Our hardships on the journey from Arctic Bay to Pond Inlet would not have warranted any extremes of emotion. I am sure that neither Anaviapik nor Muckpah anticipated catastrophe. Also, no one was blamed for things that had gone wrong. The real source of problems was ice and weather, not the break-in by the dogs. So there was no reason, in their terms, for expressions of fear or dismay or anger. Yet we had gone for a long time without adequate food, walking and pushing and working in cold weather for many, many hours each day. And it was not at all clear how we would escape without worse problems. In which other societies would circumstances of this sort have been met with such equanimity?

As we continued to struggle towards Pond, Muckpah suggested that we make a detour to the south and head for a campsite where he and others had spent part of the previous summer. He knew that a bearded seal had been cached there, in a stone-covered depression on the beach. It could well still be there, if no one else had needed it and bears had not managed to pull away the stones of the cache. We turned towards the shore again.

As we came, at last, to the beach, the dogs got the scent of something they could eat, and for the first time in several days they rushed ahead of their own accord. We hurtled into a line of rough ice, where the traces tangled and the sledge runners jammed. The dogs were held there, straining forwards. We left them and walked up onto the land. And there, sure enough, lying on top of an empty forty-gallon oil drum, was a large piece of *muttuk:* the skin, with a thin layer of subcutaneous fat, of a narwhal. The *muttuk* must have been there for several months, probably since the open-water hunting of the previous fall, and it was no doubt all that remained of a much larger pile of narwhal skin and meat that would have been devoured by passing foxes, gulls and ravens. This *muttuk* was now what is categorised in Inuktitut as *igunaaq,* meaning meat that has been transformed by slow decomposition. This is the strong cheese of the Inuit diet, and a great delicacy. Anaviapik and Muckpah pulled out knives, cut off chunks, offered some to me, and ate.

I had not had a solid meal of any kind for five days. Yet when I took a bite of the *igunaaq* I was instantly without appetite. I swallowed a tiny amount, then pinned my hopes on the cached seal, which was in due course discovered. Its meat carried the strong flavour of several months in cold storage under rocks on the beach. Cut into small chunks and boiled, however, it was tasty and meant an end to being hungry.

This was not the last of our troubles. After twenty-four hours at the camp and cache site, Anaviapik announced that he wanted to begin the last leg of our journey back to Pond Inlet. Muckpah demurred, saying that the dogs needed time to recover their strength, and that they would be gorged with overeating (they had been devouring large amounts of the cached bearded seal). But Anaviapik insisted. There was adamance on both sides, resolved in Inuit fashion by each doing what he thought best. So Anaviapik and I set off alone, to cross the fjord yet again. We planned to reach Bylot Island, where there was an old cabin we had stayed in before, and which marked an approximate halfway point between us and Pond.

But Muckpah had been right about the dogs. Once again we had to take turns to catch the traces in our hands, haul backwards, and wait for the dogs to pull with some real energy in the direction we wanted to move. We spent much time throwing gloves or pieces of seal meat in front of them to cause some bursts of determination. There was still a cover of soft snow—no longer so deep, but enough to cause the sledge runners to sink and stick. As we came to the Bylot Island shore, Anaviapik declared that we could go no farther. It had taken us twelve hours to get to where we were. I was relieved.

"So we'll build a snowhouse," I said.

"No," said Anaviapik. "There is no good snow. We can sleep here, out in the open."

I was alarmed. The temperature was no more than −10° Fahrenheit. We would have been able to lie side by side, in our sleeping bags, on some caribou hides. But I urged a final effort to get onto Bylot and then push our way along those last few miles to the hut. Anaviapik indulged me and agreed.

It must have taken us two hours to cover those last six or seven miles. The snow was deep and soft. We struggled to keep going. Chunk after chunk of bearded seal meat was thrown over the dogs' heads, causing them to dash forward as Anaviapik and I pushed and tugged the sledge. It was the one time on the long journey from Arctic Bay that exhaustion felt like a threat, a dangerous obscuring and distorting of the world. I remember that Anaviapik did not talk at all for the last stretch; no sharing of words for things we saw around us, no teaching, no banter. Yet there was nothing of what English speakers refer to as bad mood or strain, no sense of hostility or explosive frustration. Just silence. We walked and worked with our eyes always fixed on the snow into which we sank, or on the traces that we clutched and tugged, or on the ridge just ahead of us.

At last, of course, we arrived. We came over—in part through—a bank of snow that lay alongside one wall of the hut. We stopped. For a moment it looked as though the entrance were buried. We walked around the rim of a huge drift and found that there was a space between the snow and the door. We clambered down and let ourselves in.

We were quick to unload everything from the sledge, unharness the dogs, spread caribou skins and sleeping bags onto the sleeping areas inside the hut, and get a pot of meat onto the Coleman stove. Anaviapik found a checkers set some hunters had left in the hut and suggested I set it up for a game. We sat down opposite each other, about to play. He arranged the pieces, then looked at me. I waited for the usual series of jokes about who was sure to win, about why he was so nervous of playing when he knew he'd lose, or perhaps about how I would be very sad if I lost—the ritual joking that acknowledged we were going to compete and defused with laughter any inappropriate competitiveness. Instead of making these jokes and teases, however, Anaviapik was rather serious. He said: "I am going to write to Ottawa, to the government, to the man there who sent you to the North. I want to tell him that now you have learned Inuktitut. You have seen how we lived, in the old days. Journeys were often hard."

I had learned some Inuktitut. But these past few days had not been about words and language. I had not been making vocabulary lists and working out new pieces of grammar. I could see now that there had been a misunderstanding, something I had sensed but never named. Again and again a lesson that I had expected to be about language had also been, or become, a lesson about other things—how to hunt, how to behave when talking, how to use the telephone, how to walk, how to sit, how to make jokes, how not to make jokes, how to play checkers. When I had asked Anaviapik to teach me Inuktitut, and when he had said he was eager to do so, I had thought we were talking about words and grammar, about speaking, while he had supposed we were talking about a way of being. He had embarked upon the task of teaching me how to do and to be *Inuk-titut,* "in the manner of an Inuk." Anaviapik had always known what it would mean to learn his language.

TWO: **CREATION**

1

I first discovered wild landscapes in the north of England, on the Pennine hills between Yorkshire and Derbyshire. My parents took my brother and me for walks, then left us free to roam among bracken-covered slopes, on crags of granite boulders, beside peaty streams. I learned to recognise the birds of these high places—curlews and golden plover, ouzels and snipe. Even as a small child I found some kind of escape into large spaces and cold, strong air. As I grew older and heard about Lapland, the Russian steppe, Canada, I realised that the vast, wild Pennines just beyond the suburbs of Sheffield were neither very large nor very wild. I began to associate freedom, and the future, with travel much farther north. Later still, I learned about some of the peoples who lived as hunters, fishermen and trappers in those northern expanses; and I began to associate their lives with my own imagined freedom—an escape, perhaps, from this England that was turning out to be so much smaller than it had once seemed. My journeys to the Arctic, at least to this extent, may have been a form of personal destiny.

Childhood in Sheffield, however, was more than long walks or searches for birds' eggs. My parents were Jewish. My father came from an Orthodox and rather inward-looking family, with roots in Russia and the Polish Ukraine. My mother had been born in Vienna and brought up in the aftermath of the Austro-Hungarian Empire; her mother spoke Polish and French as well as German, and belonged to the intellectual, emancipated elite of Central European Jews who, as they have said with such retrospective anguish, thought they were Austrians. My mother and her mother arrived in England as refugees after the Nazi occupation of Austria, escaping the Holocaust by only a few months. Many of the family did not escape. The concerns of our home were shaped by a father for whom Sheffield, where he had been born and educated, was the only possible place to live, and a mother for whom it was a refuge. Both my parents hoped that their children would be able to find a life in England, become members of this new and old society, taking advantage of Englishness in ways that were never available to them.

It was the late 1940s, the early 1950s. We were sent to private Church of England schools. We were also sent, three times each week, to *cheder,* the classroom, to learn Hebrew. The teachers of these lessons would have said they were making sure we learned the stories and rituals of the Jewish people. In fact, we were schooled in a mixture of elementary Talmudic scholarship and the rudiments of Zionism.

We were not far from the events of the Second World War: as well as Hebrew and Talmud, we were learning how to take our places in the community of puzzled immigrants. Many of our parents had had to re-create themselves as English, as professionals, as human beings with a right to life. And a part of this was to make sure that their children, speaking English and going to English schools, achieved every kind of Englishness, from a taste for fried breakfasts to an appreciation of cricket to a reverence for Shakespeare.

The *cheder* was a jumble of buildings and rooms that included a hall large enough to hold the Saturday morning children's service and, on other days, a ping-pong table. There was a tiny corridor with a hatch where we bought a pink, fizzy drink called Tizer and small, homemade buns. There were two or three classrooms where we learned Hebrew letters and recited, in Hebrew and English, sections of the Old Testament.

The teachers were volunteers, members of the community ready to give their spare time to this work; but they were dour and pedantic, requiring the children to repeat and learn by the poorest forms of rote. We were given lines of Hebrew, then the translations into lines of English. We recited them, one after the other, over and over. Our teachers did their best to explain words, and I remember them explaining, with as much weary repetition as we brought to our recitations, the significance of sacred texts. Yet the meanings escaped me. The Hebrew seemed to have so little to do with the English: I failed to grasp that each bit of the one language had a translation in the other.

Hebrew, unlike English grammar or the Latin and French we learned at ordinary school, did not appear to me to have verbs or nouns, did not tell me anything that I could report or write down. But when I think back now to those classes, I remember with surprising intensity the Hebrew letters and words. They had their own magic, as images and sounds rather than as mere meaning. They caught something deep inside me—a set of moods, a kind of feeling, perhaps a new relationship to the outside world. Even in that drab classroom with those tired teachers, I came to love the letters, both the sounds they stood for and their shapes. The strangeness of the script, with vowels that could be left out, like a secret code; the possibility of writing the same sound with quite different letters; the mystery of letters that made no sound; and the symbols that were used only when they came at the end of a word. As I prepared my bar mitzvah recital and learned to sing those lines of the Torah, I was fascinated by the tiny diacriticals that encoded the changes of pitch and length, making words into something between incantation and song. Hebrew seemed to be full of magic.

As a teenager, I discovered the paintings of Ben Shahn, many of which celebrate letters as images. Here were the shapes of Hebrew, with obvious literal significance, yet with the power of ... what? When I first looked at those paintings, I could no more have asked than answered this kind of question about the impact of text. But now, thinking back, I suspect that the magic of Hebrew, in both my lessons and Shahn's paintings, had to do with an origin of meaning:

which is to say no more than is said by many Orthodox Hebrew theologians and scholars. They have asserted—indeed, have depended upon the belief—that the *aleph-beis,* the alphabet of Hebrew, is "the protoplasm of the universe," the very origin of divine authority and human spiri-tual experience. The theological view insists that the importance of Hebrew is something other and deeper than what its letters and words might mean as items of mere communication. This is a puzzling idea, yet it does reflect, if not explain, how I first experienced Hebrew. And the Jewish sages were no doubt conscious of what they were doing when they decided that the mystery of Hebrew should be sustained by retaining the complexity of its written form. As one modern rabbi has said: "Manufacturers can juggle model names based on market research and they can shift standard and optional equipment for the sake of commercial viability, but the *Aleph-Beis* was not manufactured by man, so it may not be manipulated by man."

Hebrew made a claim, for the children who sat in the *cheder,* to be the language of our original selves. Its shapes and sounds linked us to a remote past and faraway places; they were also an echo of the Holocaust, a reminder of death and survival, grief and triumph. With the learning of this language, we were given the story of creation. And we were defined, a little group of children with this undeniable, perplexing and even embarrassing thing in common. We were Jews in a non-Jewish world. I do not recall having any sense that we were a minority. In some strange way, we were the mainstream, an origin, perhaps, of the world as a whole. Here was the value of being chosen by God. The Children of Israel, with their long and fantastic story, could not be anything but a majority—not of people, not as a matter of mathematics or demography, but as a moral centre, as the launch pad of religion. This was not a matter of evolution or genetics (to use the metaphors of the scholarship we undertook much later in our lives), but of creation itself. In learning the story of ourselves, we were told the story of the origins of humanity.

The starting point is set out in the first part of the Bible, in the opening lines of Genesis. *Berashit bara et ha-shamayayim v'et ha-aretz.* These words ring out not as pieces of history but as the music of the beginning of a world that spoke of my place in it.

בְּרֵאשִׁית בָּרָא אֱלֹהִים אֵת הַשָּׁמַיִם וְאֵת הָאָרֶץ
In the beginning God created the heaven and the earth
הָאָרֶץ הָיְתָה תֹהוּ וָבֹהוּ
And the earth was without form, and void;
וְחֹשֶׁךְ עַל-פְּנֵי תְהוֹם
and darkness was upon the face of the deep.

Talmudic and Kabbalistic ideas insist the world was created so that the Torah, the Jewish law, could be studied and its laws obeyed. The first line of the first chapter of Genesis, God's first announcement, identifies the creation of the word, "et," "and." This word contains the letters ת and א, "aleph" and "tav," signifying the law and the Talmud. Letters with which to create language; language with which to announce the creation of the rest of the world. In this way, language and writing are deemed to be at the heart and origin of everything. The entire world comes from the words of God; these words, caught in the sacred texts of Judaism, are the sum of all knowledge; the world is only that which we say and think and know.

With its place at the centre of both knowledge and morality, as the wellspring of Judaeo-Christian heritage, and with its place at the source of humanity, surely Genesis was a universal story. Not *a* myth but *the* myth. In European culture—be it theology, art or literature—Genesis is the text that stands, somehow without question, beyond challenge, as the myth that carries. It does not tell us about the starting point of human evolution; it is not an essay in biology or any other natural science. Nor does it explain the formation of the earth and the stars, the waters and the firmament; it is a poem, not a scientific treatise.

The creation of the first man and woman is redolent with

metaphorical suggestion. The flow of the story from Eden to the appearance of Abraham, from the creation of the first people to the lives of their descendants, the first Jewish patriarchs, is a succession of episodes that has excited countless imaginations. Adam and Eve, the Garden of Eden, the snake and the apple, the trees of knowledge and life, exile from Eden, the curses on humanity, the killing of Abel by Cain, the flood, Noah's Ark, the first rainbow, the Tower of Babel: how many works of art, music and dance have those first eleven chapters of Genesis inspired?

The universality of Genesis is assumed and implicit. If a figure is Eve, then she represents all women; if a man is Cain, he evokes all murderers; if a boat is an ark, it stands for the salvation of human and animal life on earth; if a tower is Babel, it shows the essence of human conceit and the puzzle of language itself. The images, ideas and ideals of western civilisation again and again take their inspiration and metaphors from the creation story of the Jews. The great geniuses of European civilisation have drawn on the first eleven chapters of Genesis and given them a cumulative, central power and importance.

No one can learn ancient Hebrew as I learned Inuktitut, by being cast into the events and places of which the language speaks. Scholars have spent many lifetimes puzzling over the possible meanings of the Bible's every word. But simple questions can be asked about Genesis. Whose way of life does it reflect and endorse? Whose point of view is being sanctified? What version of the human condition is at issue? This may seem an imposition of secular sociology on the expression of human spirit. Yet if Genesis reveals the essence of human life, then it should be able to accommodate even these prosaic and rather materialist questions.

2

In the beginning there was nothing. Nothing and futility, no form and no light—these are the meanings of the Hebrew words that speak, in the first verses of Genesis, of that which exists before God makes the world. The first and fundamental existence is that of the Creator and his language, as reflected in the famous line of the Gospels, often attributed to Genesis: "In the beginning was the Word, and the Word was with God, and the Word was God."

Words are the beginning of the end of nothingness. In a series of divine pronouncements, the world as we know it comes into being. Let there be light (first day); the arch of the heavens above the infinite water, and hence a division of above and below (second day). Then let there be land amid the water, and the water gathers into seas, and grasses and trees grow from seeds on the earth (third day). Next the light is to take the form of stars, moon and sun, in the heaven, making day and night on earth (fourth day). Then come the creatures of the sea and all the birds in the sky; God tells them to breed and multiply (fifth day). Next God proclaims all the creatures of the land, from cattle to beetles to wild beasts; and, at last, a man and woman. God tells the humans also to breed and multiply, but they are instructed to go out into this new world, to use, subdue and rule over every living thing. They are to conquer and control the things of Creation (sixth day). And then, on the seventh day, God rests.

By the end of the first chapter of Genesis, there are both words and the prospect of restless conquerors who can understand those words. Language is that which defines humans, giving them dominance over all other parts of creation. Yet even by the end of the first chapter of Genesis, the humans and the world they must master do not speak to the universal. These human beings have cattle; but not all the peoples of the world have been herders.

In the second chapter of Genesis, God begins the Creation all over again. This time he makes things rather than announcing them; the genius of manufacture takes over from the power of the word. God now shapes a human being from soil (the words *Adam,* man, and *Adamah,* earth, have the same root) and creates the human soul by blowing divine breath into the man's nostrils. Then God plants a garden, the Garden of Eden, containing all that a man needs. But God also creates two sources of dangerous power: the tree of life and the tree of knowledge. And he issues his first instruction to the man in Eden: eat anything you like except the fruit from the tree of the knowledge of good and evil.

In this second chapter of Genesis, the man is in paradise, a vegetarian and without his herds. The requirement established in the first chapter, that humans go forth and conquer, is set aside. The words of the second Creation establish technical achievements, abundance and self-denial. But the two trees introduce a problem: language has been used to state rules, and rules exist because there is a likelihood of their being broken. The power of words to create has yielded to their potential to cause trouble and conflict.

Then God creates once again, this time by working with the earth, the animals and the birds. He gets the man to name them all—both the cattle, marked once more for special mention, and the other creatures. At the same time, God makes a woman from the man's rib. As Genesis points out at the close of its second chapter, this unity of man and woman—the one being made from the bone of the other—means that the couple will be united, and this will cause them to leave the homes of their parents. They will go forth to make their own way in the world. The power of love is the first origin of exile.

God is the master builder, converting one natural material into another: soil into flesh, breath into life, bone into woman. He also establishes both the terms and the challenge of morality: the tree whose fruit must not be touched. The two humans are given a utopia of plenty, but their life in the garden depends on doing right. The task of the man is to till the ground and watch over all that grows. Add to this the play on Adam and *Adamah,* and we may have the first clues that the human condition taking shape is that of farmer as well as herder. God has also created the man and then the woman, a couple whose lives are inseparable.

Chapter 3 begins with the serpent, the first creature to use words to mislead and manipulate. The woman is persuaded by the serpent's arguments and by the appeal of the fruit: "It was lust to the eyes and the tree was lovely to look at." The important Hebrew word in this verse is *ta'awah,* which connotes intense desire, not the mere "good for food" of the King James version. And the woman persuades the man to eat some too. The power of language has done its worst.

God announces the punishments of Adam and his female partner. These are the first of the divine curses that will shape the human condition.

- The snake will crawl on its belly and be hated by human beings.
- The woman will endure intense pain when giving birth to her children.
- The woman will yearn for the man, but he will be dominant ("And he shall rule over thee"). This is the double curse of love and subordination.
- The soil of the world will be unproductive and full of weeds ("thorn and thistle").
- Human beings must eat "the plants of the field."
- In order to get enough plants ("bread") to eat, the man will endure lifelong hardship.
- All will end in death: "For dust thou art and unto dust shalt thou return."

God prepares the man and the woman, to whom Adam gives the name Eve, for their exile from Eden. Angels armed with burning and whirling swords are posted at the entrance to Eden to ensure that the exiles will never return to harvest the fruit of the tree of life.

In exile, Eve gives birth to Cain and Abel. Cain is the farmer tilling the earth; Abel is the shepherd herding domestic animals. But there is strife between the two brothers. Both want to make an offering to God. Cain brings fruit or vegetables of some kind—nothing special, not the finest he can find—but Abel offers the fattest lambs from his flock. God is delighted by Abel's offering and fails to notice Cain's. Filled with despair and fury, Cain invites Abel to go into the fields, where he kills him.

This first murder results in another curse. God declares that the earth where Cain killed his brother will not yield enough to feed him. Cain and his family must leave for some other land. So Cain goes to Nod, a Hebrew word that denotes wandering, where he sets up his own home. Moreover, he builds the first city, and he has many descendants. These include the first "tent dwellers with livestock," the first musicians, and the first copper and iron smiths. Chapter 4 thus ends with Cain's successful and creative lineage.

What do these events add to our glimpses of the original human society? The one who supplies meat loves God; the one who grows food does no more than his duty, and is also a murderer who, being cursed, is made to wander off to find new pieces of land. But Cain's lineage triumphs. He does find new lands, and his descendants become the city dwellers and the most numerous, creative and successful of people. The one who is forced to roam, in exile, into unknown lands prospers. Genesis tells us nothing of the herder's achievements—he is overshadowed and overwhelmed by his murderous brother, the farmer.

Thus we begin to see the human being as settled *and* unsettled—a person displaced from his home, roaming the harsh earth looking for land to till, for somewhere to live. He can settle in a restless way, building, inventing, shaping, and then, as need be, roam farther afield—repeating the pattern that is the farmer's destiny. Perhaps the curse, or web of curses, lies in God's insistence on both farming and roaming. The power of God's words, in this regard, is the achievement of apparent contradiction.

Chapter 5 of Genesis confirms the success of this contradictory condition, for it is "the book of the generations of Adam." An oratorical formula is at work: the generations beget one another with repetitious incantation. But this formula speaks to a particular system of inheritance and property. Family is the core of all activity and security. For people who are settlers, cursed to leave home and yet working all their lives to make homes, ancestry is that which defines. Not land, not an Eden, not a community, but a family line, a lineage. This lineage or ancestry is patrilineal, recorded through the male line. The men have sons and daughters; the sons have *their* sons and daughters. Men go out and secure the land, on which their women then can live, providing the children—the inheriting sons—of the next generation. Genesis has established a social system. It is striking that chapter 5 ends with a reminder of "the curse upon the ground," which causes life to entail unremitting toil. The social system is that of the farmer.

Adam's lineage brings the Genesis story to Noah and his sons. Noah is given special importance at his first mention: he will "comfort" us in some way. Some biblical scholars have suggested that this anticipates Noah being the first person to make wine—another point where Genesis's endorsement of farming systems is revealed.

In fact, God's human creation has not turned out well. Dismayed by the evil ways of the people of the world, God decides to drown them. Only Noah and his immediate family—uniquely good—are to be spared. It is Noah's job, therefore, to save the other living creatures of the world. God tells Noah he must build a boat, an ark, large enough for seven pairs of all "clean" animals and seven pairs of all birds, but only one pair of all "unclean" animals. Why does God divide the animal world into these categories? It is a reflection of the systems of agriculture, in which there are animals that are domesticated and eaten, and animals that are wild, part of an untamed world. This concern with the clean and the unclean points to the human condition as one in which people, as herders and farmers, depend on the things they directly control.

God now ensures that the water subsides and that Noah and his family reach safety. Once they land on *Adamah,* soil, God says that life should "swarm through the earth," multiplying and being fruitful. Noah sacrifices some "clean" animals. In Robert Alter's translation:

> And the Lord smelled the fragrant odor and the Lord said in his heart, "I will not again damn the soil on humankind's score. For the devisings of the human heart are evil from youth. And I will not again strike down all living things as I did. As long as all the days of the earth—
>
> seedtime and harvest
> and cold and heat
> and summer and winter
> and day and night
> shall not cease."

This promise to the agriculturalists continues with God's instruction that Noah and his sons should themselves "Be fruitful, and multiply, and replenish the earth." Later in the chapter God makes this even more emphatic: "Swarm through the earth and hold sway over it."

Noah and his family are now beset by new difficulties. Noah's three sons—Shem, Ham and Japheth—are identified as the fathers of all the peoples of the earth; they are, after all, the only male survivors. Noah, the first grower of vines and maker of wine, gets drunk. As a result, Ham sees his father naked. He tells his brothers, who proceed to cover their father with a cloak, walking backwards in order not to see him. When Noah wakes from his drunken stupor, he realizes what has happened and proceeds to curse Ham, his youngest son. Ham's own son, Canaan, is also cursed: he and his descendants, the Canaanites, are to be "servants" of the other brothers' lineages. The subsequent subjugation in Jewish history of Canaan and the Canaanites is here given an original justification.

What has happened to cause such a curse? All we know is that Ham has "seen the nakedness of" his father. Scholars have pointed out that in Hebrew literary usage, "to see the nakedness of" can mean "to have sex with." This is linked, perhaps, with the way in which Canaan is associated with sexual excess. Another interpretation of the dispute between father and son asserts that Ham has castrated his father to prevent him having more sons, thus ensuring he gets as much land as possible for himself. Ham is caught in the conflict that the curses have created: on the one hand, the land is hard to make productive; on the other hand, fathers must have many children.

The exile from Eden, the harshness of the earth, the hardness of life, the subjugation of women, the displacement of Cain, the domination over Canaan. The curses of Genesis do not stand alone; their cumulative power and meaning are augmented by God's repeated instruction that human beings must go forth, increase in number, spread out over all parts of the world, and dominate all other forms of life. The curse of the pain of childbirth is given its full poignancy by the divine insistence that humans have as many children as possible. The curse of barren earth is made dreadful by God's emphasis on agriculture. Cain is cursed, but it is his story that prevails: he, the farmer and murderer, establishes a lineage that can take glory from its creative achievements. Noah, the farmer and vintner, is the survivor, but his sons are pitted against one another—two the masters, one the slave.

By the end of the ninth chapter of Genesis, humans are exiles bound to move over the earth, struggling to survive on harsh land, aided by dominance over all other creatures. In their hearts, from first creation, humans rebel against God's laws, are ready to kill one another, brother against brother, harbouring evil. They are farmers, gardeners, city dwellers, metalworkers, music makers, wine drinkers. But they are cursed in these things: they have no sure home and are forever conquerors, making gardens and tending domestic animals in places that are harsh and foreign. Their society is one of restlessness, of male supremacy, of a quest for dominance in which each family knows little trust and seeks always to *father* as many children as possible.

Next come the lineages of Noah, Shem and Japheth—the generations, father to son, that fill the world. They are both one people and many peoples; the descendants of a single patriarch, Noah, and the first diaspora of distinct and rival nations. The scene is set for the last great episode in the poem of Creation: the story of the Tower of Babel.

We learn that all peoples, despite being spread over the world, speak one language. In order to ensure that this continues to be the case, the people decide to build a tower that will reach as far as heaven. They are resolved to make an effort to rise above the human condition, "that we may make us a name lest we be scattered abroad upon the face of the earth." A name, a monument, some permanence: human beings wish for another version of Eden. They seek to escape the results of all those curses. They want to be in a single place, not nomads of some kind; they aspire to a single civilisation rather than restless conquest. To realise these hopes, to save themselves from the human condition laid down in the cumulative stories of Genesis, they bake bricks and use mortar to elevate themselves towards the heavens, to proximity with the Creator.

When God sees Babel, he issues another curse, recounted in verses 6 and 7 of chapter 11:

> Behold, the people is one, and they have all one language; and this they begin to do: and now nothing will be restrained from them. / Go to, let us go down, and there confound their language, that they may not understand one another's speech.

And "from thence did the Lord scatter abroad thence upon the face of all the earth." God is determined that human beings will have neither respite nor permanence; to aspire to these things is to seek divinity. The final, decisive curse is that peoples are henceforth unable to understand one another. They must desist from building the city of Babel, with its tower of human unity. Instead, they are condemned to be farmers and gardeners on a harsh land, to roam and disperse, move and conquer, speaking a babble of languages.

The story of Babel ends with an account of the lineage of Shem. The human diaspora continues, with its restless increase and movement of population. At the close of chapter 11, the succession of generations reaches Abraham and Sarah, in a place called Ur of the Chaldees. Here is the beginning of the story of the patriarchies of Judaism, where the poem of the Creation gives way to the family history that is the remainder of Genesis. At the opening of chapter 12, God speaks for the first time to Abraham, telling him he must leave his father's house and go to another land.

The stories of Genesis bear the signs of having been handed down from narrator to narrator, as oral culture, with all the consequent variety and rhythm that entails. Late in their lives, these stories did, of course, become a book. A sacred text. And the words of Genesis come to us as many translations. The famous English version, the King James Bible, was created in the seventeenth century, and its language carries the literary and intellectual ambition of its day. Twentieth-century evangelists have been determined to render every phrase into the language of everyday suburban America, yielding a series of modern and simplified forms. But most people hear or see depictions of Genesis in literature, in painting, in music or in sermons, rather than actually reading the stories themselves. Despite its status as sacred text, there remains an intriguing fluidity to the Bible. Repeated retellings and re-creations by scholars and preachers, translators and cantors, artists and poets, movie-makers and missionaries keep the stories and text of Genesis alive.

This vast array of interpretations and performances seems to insist that in Genesis humans can discover some of the essence of themselves. Yet the peoples who look to Genesis as the universal creation story do not inquire as to the specificity of this self. They take for granted the loss of Eden, the inevitability of wanderings and conquests, and regard them, without quite knowing that they do, as God-given.

3

The truth of Genesis lies in the profound and disturbing insights it offers into the heart of the society and economy that come with—and descend from—agriculture. Farming has shaped much of the world—its heritage, nations and cultures. Even in places where people do not farm, it echoes in the meaning to many homes of their gardens or, for those without gardens, their window boxes.

Imagine a farm family, busy in the countryside. Mother is making bread, churning butter, attending to hens and ducks that live in the yard and in pens beside the house, preparing food for everyone. Father is in the fields, ploughing the soil, cutting wood, fixing stone walls, providing sustenance. Children explore and play and help and sit at the family table. Grandma and Grandpa, when they are not also hard at work on the land, sit in chairs by the fire. Every day is long and filled with activities. And these activities are contained, given purpose and comfort, by a piece of land, at the centre of which is home.

This family is intensely private, somehow separate from the rest of the world, sufficient unto itself, knowing and meeting each other's needs. Loyalties are as deep and sure as the ground beneath them. There is loyalty, also, to the tasks and expertise and duties that each member of the family undertakes. Children have the job of being happy to help, and then being happier yet to mature into their adulthood and take their places in either this or another farm family.

So many children's books, and much adult literature, celebrate this farm family. Many people around the world have grown up in the vicarious glow of its warmth; have shared, at varying distances, in its peace and comfort; have tasted, if only in their imaginations, the fresh-baked bread and home-cured ham, apple pie and wholesome cheese laid out on the family farmhouse table. These delicious images work a complicated magic on our adult selves: does my family or your family achieve this stability, plenitude, warmth and happiness? The contrast between the lives of the readers of these books and those of this half-known, almost-remembered family makes a poignant contribution to many people's sense of failure. These images of the farm family come close to defining what family is, and what so many families are not; they give a powerful sense of the past, a bittersweet ideal against which it is all too easy to pass sharp judgements upon ourselves.

The farm is the beautiful nest within which the perfect family thrives. Each has its own pattern of fields, with their walls or fences and hedges and gates and gateposts; its own mixture of crops and copses, vegetables and wildflowers, domestic animals and creatures of the wild wood, pigs and cows and sheep and horses as well as a few foxes and badgers and stoats and weasels: the gentlest aspects of material culture and the prettiest, most controllable of nature. The farm may be grand, an estate perhaps, even a park; but its ideal is modest: enough land to provide an abundance of simple foods. A few fields and an array of buildings combine to create the image of a perfect and eternal home. An Eden.

The people of the countryside, from the poorest farmworkers to the richest of landholders, share a profound conservatism: they tend to believe that change would not be of help to them. They insist, as do those who write about them and celebrate their ways, that innovations are for outsiders, rival claimants to the soil, advocates of an antagonistic mode of life.

Yet the family farm is not without its fierce energy and restlessness. Farmers have moulded the landscape and must continue to do so. There are fields to be reshaped, walls to be remade, hedges to be laid, woodlands to be coppiced, lines of saplings to be planted, barns to be added, sheds to be rebuilt. Also, there is buying and selling, of everything from a field to a new farm. This persistent but gentle process shapes and keeps reshaping the landscape. The cumulative actions of the family farm have created the countryside, by which European nature is defined.

By comparison with the farm family, the town family enjoys very uncertain peace and prosperity. Vulnerable to the cacophonous intrusions of other people, living in a place of no natural merit and hence no intrinsic worth, the urban dweller has both a freedom and a compulsion to change and relocate. The very conditions of family life exemplified by farmers—peaceful enjoyment of the unquestioned division and unending experience of labour—are challenged by every kind of urban reality. Images of the harmonious farm family set the conditions to which all aspire.

This description of town/country differences may seem anachronistic and fanciful, a product of old ideas and spurious idealisations. It may appear to be no more than a sketch of quaint notions found at holiday teatime in England for Enid Blyton's Famous Five, or the domestic background to Arthur Ransome's *Swallows and Amazons,* or the norm inspiring the Swiss Family Robinson or, in America, television series such as *Little House on the Prairie.* But what of the farm and family that so draw the hero of D. H. Lawrence's *Sons and Lovers,* or the life that George Eliot's Dorothea longs for in her depths, or the rather grand version of country in much of Jane Austen? And what does Defoe's Robinson Crusoe emulate? The vast literature that works within this framework includes some of the most powerful of English writing. It is a tradition of rural idealisation also found in stories read or told to children in all the other languages of farming cultures—in Asia and Africa as in Europe and the Americas.

The gentle settlements of well-established countryside also have existed in fiercer and more archetypal form. Farmland in God's sole care is forest and savannah. The family farm is a determined, persistent struggle to make sure that God does not get the place to himself: the trees are felled, their roots are hauled from the ground, stones are picked from the earth, invading wild plants and shrubs are rooted out again and again. There is no end to this labour. The soil will grow grass and vegetables and grains only if a great deal else is "kept under control," which means excluded or destroyed. Not only rival plant life, but also wild creatures that harm seeds, seedlings, buds or fruits, or eat the domestic animals that are also part of the farm family. Weeds and vermin. These are the agents of wild nature that have to be walled out, scared off or killed. Otherwise the soil will not yield—more, it will not even exist.

The conditions of this archetypal farm are harsh. This is not Eden but the curse of exile: only by the sweat of his brow does the man provide food for the family. Not only the man, of course; the woman, too, must work all and every day. The children are the labourers who will ease the burden of the cursed land.

The farm family, island of work and production, is also a world-historical centre for reproduction, that other kind of labour. Agriculturalists want large families. The origins of this desire may seem to be instinctive, a matter of basic biology. But the use and purpose of the large family is clear to all: the children will be the workers, and as they grow and marry they will take over the running of the farm, securing for their parents the peace of body and mind that depends upon the continuity of home and holdings. This inheritance may, in the long run, be the privilege and duty of one son; but of the others some can stay and work, some can make alliances among neighbours and relatives, and those who neither inherit nor marry well locally may go far away to take advantage of whatever opportunities come from not being tied to this one corner of the world. Labour, consolidation, inheritance and migration comprise a set of filial duties. Between them, the children must keep the family farm in its full and rightful place.

Agriculturalists are nothing if not fecund: many nations' farmers and pastoralists have succeeded in raising very large families. The average number of children for European rural families has varied from four to eight. Many households have had ten and more children live to adulthood. Even high levels of mortality and periodic catastrophes have not prevented the farm families of the world from contributing to astonishing population growth. If subdivision of land makes it possible to keep many children at home or in the neighbourhood as adults, as was the case in Ireland in the fifty years before the 1846 famine, this process will result in eventual catastrophic causes of emigration (as in Ireland between 1846 and 1900) or a delay followed by a flood of human movement. When all available land is in use or further subdivision of holdings is impossible, the total number of family farms becomes constant. This means that only one child can expect to marry into a farm family. Others may become farm labourers; and many must go elsewhere. In this way the settled countryside—in which the total number of farm families is more or less stable, and available marginal lands are in the ownership of large estates or in use by pastoralists or both—exports people.

The history of much of the world over the past few hundred years—a history that is well known from records and documents—has included great displacements of country people by the concerted efforts and interests of other agriculturalists. The enclosures and Highland clearances are two notorious examples from British history. The movement of Polish and Russian peasants off their lands are examples from Eastern Europe. These drastic processes gave rise to some of the most anguished laments about movement from country to town or from one nation to another, and they have made vivid contributions to the myth of the farm family. These processes have different sets of direct causes, including internal colonialism and ruthless national political measures. But the flow of emigrants I speak of here is intrinsic to agricultural life, one of its continuous and inevitable long-term consequences.

Emigrants go to towns as labourers or to new colonial lands as settlers, part of the great farm-family diaspora. Some of those who move to towns retain or develop a longing for a life on the land, as an idealised alternative to the hardships and tyrannies of urban or industrial labour. The Australian, American, Canadian and southern African frontiers have given the farm opportunity to many millions of European men and women. The settlement of Europeans in the

Canadian West during the late 1800s, for example, was among the largest and most rapid movements of human beings in history. Since the beginnings of agriculture, all over the world, on new-found lands, in the *terra nullus* of colonial frontiers, migrants have made their family farms. There they have many children, who also have many children. These children, in their turn, move on, pushing the frontier outwards until it reaches its limits. Then follows another wave of movement from the frontiers to new towns or to ever wilder regions.

Being willing to go to unknown and harsh places, in defiance of aboriginal resentment; taking part in colonial wars of conquest and "pacification"; accepting the relentless need to remake, with Herculean efforts, a land of forest or marsh or rocks or sand into a patchwork of pasture and fields; knowing little comfort and no respite from hard physical work; setting pleasure at the far end, the distant terminus, of a journey of hardship; making the endurance of this hardship a religious achievement—here are characteristics and abilities that have secured the family farm its place in almost every kind of climate and landscape. These are the qualities that define what Europeans (and other expansionist agricultural cultures) see as the signs and successes of civilisation.

This success is built on opposites. On the one hand, a passion to settle, on the other, a fierce restlessness; a need to find and have and hold an Eden, alongside a preparedness to go out and roam the world; an attachment to all that is meant by home, and an overriding commitment to a socioeconomic system, to some form of profit rather than to a place. The agricultural system is a form of settlement that depends upon, and gives rise to, the most pervasive form of nomadism. The urge to settle and a readiness to move on are not antagonists in the sociology of our era; they are, rather, the two characteristics that combine to give the era its geographical and cultural character.

Many people are aware, in their own psyches, of these conditions of agriculture. When viewed from this angle, on this macro-historical level, the town and the country are not such different systems. In both settings a love of place is secondary to the importance of prosperity. In much of the agricultural world, loyalty to home or lands is romantic and incomplete. Given the right price, everything is for sale. In both country and town, if a new place can be trusted to be better than the old one, then the urge to move will prevail. The measure of success in this system of settlement and migration, conservatism and restlessness, is economic power—the ability to control a pool of resources in the interests of the family.

Counter-cultural movements—from sannyasin to hippies to new-age shamanism—reject the conventions of the farm family and embrace the freedom that comes with reduced attachment to farms or houses, with their bourgeois economic imperatives. They celebrate the sensation of being footloose and on the move—on the streets, in their vans and trucks, as squatters who come and go. Yet these same groups are full of enthusiasm for subsistence vegetable-growing and self-reliant communes, revealing a deep longing for their particular version of settlement in some new kind of place. With the same tension, which is creative as well as contradictory, emigrants create songs and poetry to celebrate the opportunities of new frontiers and to mourn the separation from "the old country." Nostalgia and a sense of loss sit alongside yearning for the perfect garden and dreams of foreign riches or adventure—companions in the agrarian imagination.

The move of country people to another countryside or to town, like the move of townspeople to every other kind of place, is intrinsic to the overall system. Newcomers to the countryside, be they agricultural entrepreneurs or people eager to get away from cities, are part of a continuing flow of life that includes the coming and going of farm families. The city-born family that has lived in the countryside for twenty years has deep attachments to the landscape, the home, the garden, the land where it has made a life. These attachments are not different from the attachments that "traditional" rural people have to their homes and farms, except insofar as the garden of the city-born family may have little to do with actual prosperity. Townspeople and country people live and make their decisions within the same cultural and economic tradition: they move, they settle, they create a home, and they find—or their children or their children's children find—that they must move on. Exile is the deep condition. The longing to be settled, the defensive holding of our ground, the continuing endemic nomadism—I suspect that we share them all.

Who is this "we"? The argument here pays no attention to class or even to nation. Those who are agriculturalists, humans who live by remodelling the land, are the peoples whose story is some version of Genesis. We live outside any one garden that can meet our needs and growing population, so we must roam the earth looking to create or re-create some place that will provide a more or less adequate source of food and security. We are doomed to defend this place against enemies of all kinds: we know that just as we have conquered, others can displace us. This mixture of agriculture and warfare is the system within which farms and towns and nation-states and colonial expansion have an inner and shared coherence. The world view and daily preoccupations of the peasant farmer and the twenty-first–century executive have much in common. The one is able to dominate, exploit and thrive far more effectively than the other. But their intellectual devices, their categories of thought and their underlying interests may well be the same. They speak one another's language, as it were; for all the inequalities between them, they can do business together.

But Genesis is *not* a universal truth about the human condition. Inuit children do not grow up with the curses of exile. Anaviapik would be astonished to think that his descendants were destined to go forth and occupy distant lands. Hunter-gatherers constitute a profound challenge to the underlying messages that emerge from the stories of Genesis. They do not make any intensive efforts to reshape their environment. They rely, instead, on knowing how to find, use and sustain that which is already there. Hunter-gatherers do not conform to the imprecations of Genesis. They do not hope to have large numbers of children; they will not go forth and multiply. Everything about the hunter-gatherer system is founded on the conviction that home is already Eden, and exile must be avoided.

Farmers appear to be settled, and hunters to be wanderers. Yet a look at how ways of life take shape across many generations reveals that it is the agriculturalists, with their commitment to specific farms and large numbers of children, who are forced to keep moving, resettling, colonising new lands. Hunter-gatherers, with their reliance on a single area, are profoundly settled. As a system, over time, it is farming, not hunting, that generates "nomadism." Agriculture evokes the curses of Genesis.

In the history of European civilisation, as in the history of agricultural cultures, the combination of settlement, large families and movement has resulted in a more or less relentless colonial frontier. An agricultural people can never rest—as farming families, as a lineage—in one place. They love home, but they also love the leaving of it. They celebrate stability and security, and yet they are committed to movement. Thus farmers have two ideals, the one of sweet home, the other of conquest and adventure. Not two Edens, but no Eden at all. The family farm may indeed be at the heart of a human and social condition, yet it is not so much an ideal as the cause of that condition's most dynamic element: a readiness to migrate, a nomadism.

4

In 1974, I gave a public lecture at the Commonwealth Institute in London, England. At the time I still had links with the Canadian Department of Indian and Northern Affairs, whose Northern Science Research Group thus far had funded my work in Canada. My mind moved between questions about the nature of hunting systems and the immediate needs of the Inuit in whose communities and territories I had been travelling. I fear that my lecture was an uneasy mix of intellectual speculation and political indignation—a shifting between my naive discovery of the riches of Inuit culture and my urging that the Canadian administration make new moves towards decolonising the North. I suppose the lecture was fuelled by my passion on both issues. I spoke with the enthusiasm that came from having very recently been in "the field."

One of those in the Commonwealth Institute audience was Brian Moser, producer of the Granada Television *Disappearing World* series. *Disappearing World* was an inspired attempt to unite anthropologists and filmmakers. The series, in 1974, had already resulted in some compelling anthropological films, and had created both the film-making skills and the public awareness that gave anthropological documentaries a new place in British television. Brian's response to my lecture was to suggest that *Disappearing World* do one of its films in the Canadian far north. I immediately thought of Anaviapik. What would he make of this idea? And how could we overcome the word "disappearing" in the series title? Anaviapik and his people were determined *not* to disappear, and they had no notion that any such fate awaited them. They battled for their rights, demanded fuller recognition from the world—and had not resigned themselves to a tragic disappearance from its stage.

I hesitated and equivocated and worried. To allay some of my fears, Brian introduced me to the work of Michael Grigsby, the director with whom he thought I should be partnered for an Inuit film. I watched Grigsby's most recent work, a beautiful exploration of the fishing community of Fleetwood in the north of England. Here was a film that made a journey into a way of life. No narration, no intrusive gimmicks, just the place and the voices of those people strong and clear. *Fleetwood* was an inspiration. Here, indeed, was a way of making a film that I could imagine bringing delight to Anaviapik. So I went to Pond Inlet and spoke with him and others about the idea. I met with a measured enthusiasm. The Pond Inlet community, through its newly formed Hamlet Council, laid out some conditions.

The film idea would be welcome, they said, so long as the people of Pond Inlet had a large say in what was filmed, so long as a Pond Inlet elder could check the film before it was finished to make sure that it did not contain errors, so long as the government and the people of Canada got a chance to see it, and so long as the Pond Inlet community was the place where the film got its first public screening. Anaviapik and others told me that they wanted to use the film as a way of getting more Canadians to understand who the people of the Arctic really were, what they had experienced and what they now needed. The film should be a celebration of their place in the North. One person, Anaviapik's son Paulussie, who had often taken me hunting, expressed some uneasiness about the film. His father was old, he said. So many people loved him. What would it be like to see Anaviapik in the film, after he was dead? "*Isumanattuq,*" he said. "It makes you think. It's worrying."

5

Until the 1970s, the Canadian Arctic was isolated, with neither roads nor railways nor scheduled air services. Government officials, mail and supplies came in about once a week on chartered planes or once a year by ship. The only Inuit to have spent much time away from the North tended to be victims of tuberculosis who had been evacuated to hospitals and sanatoriums in the south (many remaining there for years, and some not living to come home). Many children had been taken to residential schools run by Catholic priests; but these schools were almost all in other parts of the North. Though they required immersion in English language and southern culture, they did not entail living outside the geography of the Arctic.

A corollary of the Arctic's isolation was the Inuit concern, in the 1970s as indeed thereafter, that those in the south who seemed to have so much power over the fortunes of Inuit life and land should learn more about those lives and lands. Again and again, Inuit elders spoke to me of their wish that the Qallunaat, the southerners, should know the facts. If southerners knew the real truth, they would never again do anything that was against the interests of the Inuit. Injustice was blamed on ignorance. Anaviapik's wish to educate me had originated in this faith in knowledge. He had told me over and over that I must learn well so I could translate his concerns to *angajurqaat tavvani,* "the bosses down there."

For many people in Pond Inlet, therefore, the film made sense as a way of communicating with "down there," the south, the bosses. The task of the filmmakers was to be true to the facts—of what people did, of how they described their history, of their demand that they be left in peace in their lands. Michael Grigsby was as determined as the Inuit that the film be their voice, their story, their facts.

The people of Pond Inlet took the film crew to hunt seals at their breathing holes and to hunt *uttuit,* the seals that basked on the spring ice. They took us fishing through the ice at the mouth of a river thick with migrating char. They made sure we were able to film at the floe edge, where the migrating narwhal waited for the breaking ice to open enough for them to get into the shallow estuaries where they could gorge on small fish. And they took us, with immense difficulty, through moving ice—first by skidoo and sledge, transporting a boat with us, and then by boat, with skidoo and sledge loaded precariously on gunwales—to the edge of Bylot Island, where the most daring hunters climbed high and greasy cliffs to gather guillemot eggs. These were all expeditions undertaken with the immense knowledge and experience that make the High Arctic into a storehouse of resources for the Inuit.

But the journeys and the hunts were not the sum of the film. Anaviapik worked with other elders to ensure that the film also recorded the most important parts of their modern history. They talked to one another at kitchen tables, as if there were no camera, about the way Qallunaat had sought to dominate Inuit life, and about the kind of fear they felt for policemen, missionaries and traders. They spoke, also, about the way that southern so-called experts had come and interfered with northern wildlife. Polar bears, narwhal, geese, even fish—outsiders were claiming authority over everything. The Inuit protested about the way in which their hunters had been blamed for declines in animal populations. No, they said, it was the Qallunaat, not they, who were doing damage to the land. Their land. This was the central statement: this beautiful place that they knew so well, where they had always lived, was theirs; the Qallunaat had no right to say it was not. "*Inuit nunangaumat.*" "Because it's the people's land."

The film took six weeks to shoot. The edit, back in England, lasted much longer. The material was rich; the task of reducing such eloquence of both voice and image to a television hour was painful and daunting. As the edit reached its last stage, the Pond Inlet council chose the elder who would come and check the film's veracity. They chose Anaviapik.

6

So in 1976 Anaviapik came to London. He had never before left the Canadian far north. He had not even travelled to subarctic Canada in northern Quebec or Manitoba, the provinces which by then had airports serving the eastern Arctic. As the visit was planned, I was full of apprehension. Surely he would find the journey overwhelming, terrifying. He would be homesick, miserable, lost. He would get sick. The whole undertaking began to seem foolhardy. Why had we not arranged for the film to be flown to him? Well, we had talked of that possibility, but we could not fly the whole editing room there, and we would not be able to work on any changes he wanted while he was there. His involvement in the film must be more than a dutiful nod in the direction of the Inuit. For it to be real, he had to come to us.

I got to the airport an hour before the plane was due to land. I went to find officials who might make it possible for me to meet Anaviapik straight off the plane, to ensure that he would not have to deal alone with the long journey from plane to immigration, and then to baggage pick-up. A stern woman with the relevant authority showed a mature indifference to my anxious questions. I had hoped to be able to capture her sympathies, if not her imagination, by referring to the Arctic, sketching the vast distance that Anaviapik had had to travel, telling something of the extreme difference between his language and ours, his world and this one. But she knew better, had the real experience, and was unmoved.

"You've no need to worry," she said. "We've had Aborigines from Australia arriving here. No harm ever came to any one of them."

This left me no argument, no room for appeal.

I waited, in an agony of worry, beside the international-arrivals gate, that boundary between individual scrutiny and anonymous mayhem. I saw neither Australian Aborigines nor, as far as I could tell, any other travellers from distant tribal lands. Just a steady stream of Qallunaat, whose vast numbers and confident haste inflamed my fears. At last Anaviapik arrived, escorted by a British Airways stewardess. He was wearing sealskin boots, with brown trousers tucked into their patterned tops, and a dark jacket. Over his arm he carried a parka, complete with fox trim on its hood. This was eccentric clothing for a hot summer's day in London. But his face was wreathed in smiles. Not a flicker of worry or discomfort. We greeted each other with Inuktitut sounds and words of delight. We walked to my car in the multistorey car park, then drove towards London along the M4.

The traffic was dense and moving fast. I had to concentrate hard on the driving. Anaviapik sat beside me, very quiet, watching as we weaved among cars at what must have seemed like crazy speed. As we were coming through Hammersmith, he broke a long silence, saying: "Now I understand why the Qallunaat come to our land to get the oil." These were not the words of an intimidated hunter-gatherer from the savage wilds. Nor was his comment the next day, when we walked for the first time around the streets of London's Bayswater district: "How amazing that the Qallunaat live in cliffs. I would never be able to find my way here without you."

Over the next two weeks, we often played the game of seeing if Anaviapik could guide us from the Bayswater tube stop to my flat—a distance of about half a block, with one left turn, a crossing of the road and a climb of half a dozen steps to the building's front door. He never succeeded. He always expressed mock dismay at being lost and enjoyed reminding me that, since we lived in cliffs, of course our houses were not easy to tell one from the other.

We went to the cutting room and watched the film. Anaviapik had a number of criticisms. A sequence that we had filmed of him showing us the *qarmak,* the house built of sod and whalebones where he and his family had spent many winters until the 1960s, had been left out. He thought it should be in: how otherwise would Qallunaat understand the way Inuit had lived before the era of the settlement? A scene in the Anglican church, with a glimpse of the Pond Inlet missionary preaching to the congregation, should be cut. Everyone knew that the Inuit were Christians; better to use the time for showing other things. An interview that had troubled us, in which three elders say that the Qallunaat care more about polar bears than about Inuit children, was in doubt. We feared that it would be misinterpreted and might express no more than a partial truth. Anaviapik insisted that it stay in the film: the rules against killing bears were of disproportionate concern to the officials who made the rules in the new North; the risks to children counted for less than the risks to bears.

We worked on the film, making changes and then checking translations, preparing subtitles. In the evenings and on free days Anaviapik and I went on walks in London, tried many kinds of restaurant and visited friends. Anaviapik had said he would like to see white people who were not rich, and had also asked to meet the queen. We spent an afternoon in the East End. But the queen's diary was too full for an appointment to be possible, and we said no thanks to an offer from the palace that Anaviapik come to the front door and "make his mark" in the visitors' book.

To get some relief from homogenised urban crowds, we took a trip to the Norfolk countryside. I wanted to give Anaviapik some sense of an England that is not all cliffs and cliff dwellers. We set off, driving northeast, across Cambridgeshire and through Suffolk. I chose a route that was as rural as could be. He looked out onto the green landscapes and said, "It's all built." He did not see the difference between town and country except as a matter of degree: the one had more people and more houses side by side, and the other had more fields and hedgerows. But all of this, hedgerows as much as houses, was made by people; none of it was "nature"—at least, not a form of nature that he would recognise as such. He was always amiable and interested, but he did not like much of what he saw.

Wherever we went, Anaviapik insisted on quizzing people. Using me as interpreter, he stopped strangers in the street whose appearance intrigued him. He would say in loud Inuktitut: "Hey, you look just like an Inuk," or "Do you know where I come from? Have a guess!" This calling out to strangers was sometimes followed by his insisting that I translate what he had said, and then by very curious conversations between the smiling, extrovert Anaviapik and the Japanese or Chinese men and women who expressed, for the most part, as much delight as astonishment. In shops, he would insist on holding long conversations with anybody we bought things from. Everywhere we went, he persuaded me to draw people at adjacent tables into conversation. Where are you from? Where do you live? Do you have children? Where do *they* live? Do they have children? Are your parents alive? Do you have brothers and sisters? Where are they? These were the invariable start-up questions. If the alarmed stranger was able to overcome surprise—and most did—a long exchange of personal and family stories would follow. In this way Anaviapik conducted a thorough, if not systematic, interrogation of English life.

As time went on, he came to a picture of us as scattered and shiftless. Everyone was on the move. Families dispersed. Children lived far away, out of touch. People dreamed of being able to move to some other country. Many said how much they wished they could live in Canada. No, not the Arctic. But Canada. They had relatives or friends there. They would go there too if they could. In London and in the countryside, he met the nomads or the parents of nomads. He did not make judgements. He was full of admiration for everyone and everything he encountered. But he spoke more and more of the movement from place to place that he sensed and heard about.

If any emotion coloured his comments on what he learned, it was apprehension: might this remaking of environment and shiftlessness of people spread to Inuit lands, to his home? Already he knew, and had spoken often to me about, the southerner's readiness to come and live even in the North. This posed a deep threat. Nomads have no real homes, so they cannot be relied upon to stay away from other people's.

Once, when we were having coffee in a crowded café, we found ourselves sharing a table with a man of about thirty who was reading an Arabic newspaper. Anaviapik urged me to ask the man what this language was whose writing he had never seen before. For a while I demurred. No, I said, it is not a good idea to talk to people who are reading by themselves in a café like this. People liked to be left in peace. And anyway, I could tell him that it was Arabic. But Anaviapik was insistent: he wanted to speak with this man, wanted to tell him that he was visiting from the Arctic. We are both visitors, he said, so it must be fine for us to talk.

I capitulated and leaned across to the stranger. He looked up from his paper, surprised but not hostile. I told him Anaviapik, my friend here, would like to say hello. The man answered in French. So I started again, and soon found myself interpreting a fast and enthusiastic French-Inuktitut conversation. The man said he was from Tunisia and that he was thrilled to meet someone from the Arctic. He had always thought there was a special link between his people and "les Esquimaux." He had seen films of the Arctic, he explained, and had seen the blowing snow. The snow was moved everywhere by the wind. Just like the sand in his country. He had always thought, since then, that the people of the Arctic and the people of North Africa had something very profound in common. Anaviapik was thrilled. Yes, yes. Blowing snow. "*Pirqsirqtualuq*." "It blows snow very much." They exchanged words for this kind of snow or sand. Anaviapik glowed with pleasure. The Tunisian was moved almost to tears: this was an omen, an indication that good things were to happen in his life, in both their lives. To be so far from their homes, and to find that they shared so much!

Anaviapik's visit to London came to an end. He had been with us for three weeks. Long enough to do the work, and long enough for him to feel much homesickness. One morning, as we were having our usual late breakfast, sitting in the room where he slept, he said to me: "I have been sleeping here every night now." There was a pause. "I have never before slept in a room by myself."

"Never?" I said, my heart sinking as I began to imagine that he had been lying there night after night in dark loneliness.

"Only when I used to check my fox traps in winter and had no companion. But not often, not often."

"And has it been difficult for you, here?"

"No," he said. "But I have not changed the time."

I did not know what he meant. "You mean that you have left the time on your watch as Canadian time?"

"No," he said. "I have been thinking about the numbers. Look"—he pointed at his watch—"when it is eleven at night here, it is six in the evening at home. Now, I do not sleep here until four in the morning. That is eleven at night at home. And I sleep seven hours." He pointed to the hands on his watch, showing where they stood after those seven hours. "That is eleven in the morning here. And that is six in the morning at home. And those are my sleeping times there. Eleven until six. So I am living here the same as I do there!" And he burst out laughing—perhaps to divert me from any idea that this was a complaint or an expression of unhappiness, and perhaps to point up the comic absurdity of any suggestion that things could be the same in these two opposite kinds of place.

A few days later I took Anaviapik back to the airport, and we said good-bye at the gate where only travellers can pass. An airline official accompanied him thereafter. He was happy to be going, making jokes. He wrote to me later to say that the journey had been fine, everyone at home was well, the family were pleased to see him, and he wanted to send his greetings to all of us in London. He also told me that he was going to keep a diary, saying something about every day, and that he would send it to me in a year, so that I could share a little in what he was doing. He would be going caribou hunting—it was not too late to find animals with prime autumn fur, and his wife was going to make a new set of winter clothes.

I never saw Anaviapik again. We exchanged letters, and he sent me his diary. He left the Arctic on at least one more occasion—in 1977 he was awarded the Order of Canada, honoured for his commitment to his people, for his faith in the salvation of his land. He never ceased to be a hunter. He watched the transformations of his world—the arrival of telephones and television as well as ever more local government and many forms of social and economic development. But he died in the knowledge that Inuit had the immense advantage, in their struggle against the nomads' frontiers, of living beyond the possibility of farms, and in a region where, thus far, the Qallunaat had no cause to build their cities.

7

Genesis is the myth of agriculture and pastoralism, the story that sets the character and consequences of farming and herding. Where agriculture has no place, beyond the farmers' frontier, there is no such thing as countryside. Instead, there is wild, raw nature, a wilderness. Agriculturalists have much difficulty imagining a human socio-economic system, rather than a few inchoate, animal-like wanderers, existing in this wild beyond. Farmers assume a right to enter the wild, tame it, reshape it, farm it. And there they can reproduce and create that surplus of sons and daughters who will move on.

We who have grown up within the world made by agriculture—and that would include the vast majority of present-day peoples—know within our minds and bodies that mobility is the source of energetic colonial power. We experience the lure of opportunity, keeping an eye open for where else we might better be able to do what we do. Each place seems right and wrong at the same time. Staying at home, having roots—these are important sources of comfort and satisfaction and achievement. Moving on, making progress, wondering if we might prosper there rather than here—these are necessary conditions both for many kinds of individual achievement and for the collective achievement of our social order.

More than ever before, this order seems to depend on restlessness. It urges that moves be welcomed more fully, more quickly. The process is speeding up. But its successes—if this is what they are—rely on the same nomadic potential as ever. We are the new nomads, but we are not newly nomadic. As we move back and forth between town and country, or from one country to another, relocating as need and opportunity push and pull, we participate with a modern vigour in the neolithic system. To this extent, some of the surfaces of our social and economic systems are part of a long-established way of life. Thus we experience our exile from Eden.

Genesis is the creation story in which aggressive, restless agriculture is explained, is rendered an inevitability. Its first eleven chapters are the poem of the colonisers and the farmers. They are not the story of Anaviapik and his people.

THREE: **TIME**

1

Imagine a trail through a forest of birch and pines. The trees are tall, and they sway in a wind at the edge of winter. The leaves have gone from the birches, whose branches seem thin and frail. The pines are dark, and somehow comforting. The trail twists and turns among the trees, straightening through open glades. The ground is soft with a thick growth of moss. The light is fading; the undergrowth is black, in the deep shadows of a subarctic evening. We are moving along a ridge in the foothills of the Rocky Mountains, heading home to the Halfway River Reserve at the end of a day hunting moose. This is the Subarctic, a terrain that stretches across most of northern Canada and into Alaska, a land of immense mixed forests between the Prairies to the south and the tundra to the north. The lands of the Dunne-za.

The Dunne-za, often known as Beaver Indians, are Athabaskans. They are hunters, fishers, gatherers and trappers. Their society and economy embody the hunter-gatherer way of life. The people are egalitarian, with a strong faith in individualism. Their knowledge of the land and its resources is awesome; their oral culture expresses both this knowledge and the ways of thinking and being that ensure abundance and success. The Dunne-za have adapted to the edges of the colonial frontier, making their peace, as best they can, with ranchers, miners and missionaries. Some of their best lands have been taken from them, especially along the Peace River Valley, in both Alberta and British Columbia, and in the area now occupied by the town of Fort St. John. They have endured not only losses of vital areas but restrictions to their territories and humiliation and insult by frontier racists; and they have suffered their own internal disputes and violence. Yet their pride in the skills and knowledge at the heart of Dunne-za life is intense. And they are sure that their lands are where they originated and where they must continue to live. They are concerned to avoid any forms of exile.

It is evening in Dunne-za territory. I am following Jimmy Field, a Dunne-za man in his forties whose hunting skills astonish even his own friends and family. We are on horseback, riding at a fast, determined walk—the gait that Dunne-za riders insist is easiest to sustain for many hours through rough country. Jimmy moves along in front of me, his back very straight, his body somehow held in its strength and yet relaxed. When his horse sways or stumbles, Jimmy's movements are those of the horse; when his horse trots or canters, he does not lift from the saddle, but rises and falls as if he were a part of the animal. I watch him, doing my best to sit and move as he does. But when our horses break from their walk into faster, jerkier movements, Jimmy turns to see if I am all right, and he laughs at my bounces and lurches. He calls out to me: "Agayas'sa.""Real white man."

The *Dunne* in Dunne-za means "people" or "human beings." The Dunne-za word for a beaver is *tsa,* and this may well explain why the people were called Beaver Indians by the Europeans who first encountered them. But the affix *za* means real. Like so many hunter-gatherers, the Dunne-za know themselves as the real people and others as something else. When Jimmy says *agayas'sa,* there is a play on the word. He chuckles every time he uses it and gives an exaggerated stress to the *-sa* on the end—the bit that means "real" or beaver—doing to the southerners in Athabaskan what they did to the Dunne-za in English, confusing authenticity with the name of an animal. And teasing me about my attempts to ride properly.

I watch all of Jimmy's movements, even following the path of his eyes as he scans the trail ahead of us and the forest around. I was surprised when he first suggested that we look for moose on horseback. He had told me many times that he much preferred to hunt on foot, "real way." As he walks, he can read signs and tracks and move with the right kind of quiet, without having to worry about "wild" horses. Yes, he had told me, horses, like cows, are wild. Not like moose or deer or bear. Animals living in the forest can be known. A hunter is in touch with them, can visit them, can come to an understanding with them, can dream them. Horses and cows—you never knew what they were going to do next. But he knew I liked to ride, and on horseback we could make a trip to the far end of the reserve. Jimmy wanted to check out where the moose were moving, now that it was almost winter.

We were at the end of a long day. I watched the light fade, and the black of the trees and the darkness of the undergrowth deepen. Around us the forest was very quiet, even though the wind pushed the branches high above us back and forth. Deep among the trees, the trail was sheltered. An hour or two had gone by since we last stopped, since Jimmy had made a fire and we had brewed tea and eaten dried meat. I was wondering if we would take another break before reaching the houses of the reserve. Then Jimmy's horse stopped. For a moment, I thought we were about to rest. But I saw that Jimmy was very still, and his eyes were fixed on a point somewhere in the darkness. I noticed that his right hand had let go of the reins and was moving towards the scabbard that held a rifle tied beside his saddle. My horse had also stopped.

The seconds of this stillness were very long. Then Jimmy began to lift from his saddle; he slid off his horse with movements that were slow and precise. I looked again at the darkness on which his eyes remained fixed. Then I saw what he had seen: a patch of pale, a hint of white, nothing much more than a lack of darkness. As I focussed on this, it began to take shape. Then I saw that there was another, even less distinct paleness to the side of the first. Two hints of white. I peered harder, with greater care; the pale places began to take clearer, recognisable form. They were antlers; we were within fifty yards of a bull moose. The black of its body was part of the darkness of the forest. Its stillness was a way of being camouflaged, at the edge of night, to perfection.

I looked at Jimmy. He had stepped in front of his horse. He had lifted the butt of his rifle to the level of his shoulder; the barrel, in his left hand, pointed at the ground, and his right hand was around the trigger guard, with his right elbow tucked into his side. The rifle did not break his profile. The moose would see a man, perhaps, and a horse, but not a gun. I looked again at the pale hint of antlers. They seemed brighter, outlined against the trees, now that I knew what they were and could conjure their appearance from near invisibility. They did not move.

Jimmy raised his gun, sighted at the moose, but did not fire. Now the antlers did move. The patches of white turned; for a moment one became invisible, then both were suddenly very clear. The moose had stepped out of the darkest place, was checking, nervous, and for a moment it was out of the shadows. Still Jimmy did not shoot. He turned to look at me and nodded with his head, suggesting that I get off and get my gun ready. I dismounted. But for all the lessons I had received from Dunne-za hunters, I did not have the necessary grace and stealth. By the time I reached the ground, the moose was crashing away through the undergrowth. For a moment I saw its huge body and massive antlers. It was a full-grown bull, perhaps a thousand pounds of meat.

I walked to where Jimmy was standing.

"How come you didn't shoot, Agayas'sa?"

"I didn't see it. Not for a while. Why didn't *you* shoot?"

"Maybe I leave it for you." He laughed. "And maybe too big."

I realised that if he had killed the huge animal, we would have had to butcher it by the light of a fire, then leave the meat cached overnight until we could get back and pack it all to the houses of the reserve. Also, Jimmy would have had to do this with the unskilled support of an *agayas'sa*. This was not an animal that had figured in his dreams.

Jimmy Field, lithe and humorous, was a man who had hunted and trapped all his life. If he hunted with others, they waited on his decisions, took careful note of which direction he chose to move, and listened to what he had to say about tracks. He often travelled alone, over great distances, hunting day after day. Everything he killed he treated with care and respect. No kill was careless, nothing was wasted; everything was known, understood, according to Dunne-za custom. The territory he moved in was the size of an English county, an area reaching some fifty miles north to south and covering several hundred square miles. His knowledge, from trails and gravesites to the movements of birds, animals and fish, was both detailed and wide-ranging. It was knowledge that he had gathered from stories, myths and tales of the hunt as well as from his own experience. When he hunted, he carried no more than a rifle, an axe, a box of matches and a bag over his shoulder with twine and some tea. His clothes were few and simple. He wore moccasins on his feet and had a blue denim jacket. This was enough to stay out for a day or a week.

We came to the last stretch of the trail, at the edge of a high ridge, within a mile of the first house of the reserve. The route was steep. Jimmy stopped and waited for me to catch up.

"I guess we walk now," he said as I came alongside.

He swung out of his saddle, gathered the long, separated reins, and began to step down the trail. I did the same, though I was unsure why we gave ourselves the task of finding our way in the dark when the horses, eager and sure-footed, seemed well able to carry us home. But I was also relieved: my knees and ankles ached from the length of our ride. Walking was a form of rest. Perhaps Jimmy felt the same. More likely, he wanted the horses to arrive without being wet with sweat and therefore more vulnerable to the sharp cold of the subarctic night.

As we walked, we talked. Jimmy had a question on his mind. He spoke, as always when talking to me, in good but simple English that he somehow managed to inflect with a slight but continuous irony, as if to signal his awareness that using this language restricted what we could say to a kind of playful set of basics.

"How long the white man been here?"

"I'm not sure exactly," I said. "I've heard that the fur trade guys were looking for Indians to trade with on the Peace River about two hundred years ago."

"Long time, eh?"

"But the first white people to come and live here, the first farmers, haven't been here that long. Maybe a hundred years at most."

"Not so long," he said. "My father live here before any white guys come through this place."

Jimmy paused. We walked. The horses nudged at us, wanting to get home.

"Indians been here all time. That right, Agayas'sa?"

"Right," I said.

I remember that he laughed. "Agayas'sa," he repeated, with that sharp emphasis on the *-sa.* Then he said: "Way back, long time ago, in the beginning, pretty hard thinking." He paused, then went on: "Some white guys are pretty smart, eh?" And he laughed again.

I had learned early in my time in Dunne-za country that the word "smart" was used to mean arrogant and cunning rather than clever. Jimmy was marking an opposition, an implicit dispute, between what he knew of time and the history newcomers to his land believed.

We walked on, chatting about the warmth of the night and what we might get to eat when we arrived. The dark was made more poignant by our first glimpse of oil lights through the window of a cabin. As we walked towards the light, I was thinking about time, the beginning, and about the timelessness of this place, this trail, this return from the hunt. Jimmy had decided not to kill a moose today because he could rely on a moose—or a deer or a porcupine or a rabbit—to be there, on his trails, tomorrow. We travelled light because in this territory, with his Dunne-za mind, Jimmy knew that there was no need today for the meat of a huge bull moose. He carried all the essentials in his brain.

We came at last to the cabin where Jimmy lived. We unsaddled the horses and let them loose, then went into the house. This was Thomas Hunter's place. Thomas was among the most powerful and respected of the Halfway elders. His wife was Jimmy's sister, Rosie. Jimmy's brother Abalie and his nephew Marshall were there, as well as Sandra and Ronnie, Thomas and Rosie's daughter and son. The wood stove burned hot; its glow added to the light of an oil lamp and two candles. This was the family who had welcomed me into their lives and homes. Although the horses and most of the guns we used belonged to Thomas, and it was he who made many decisions about when and where we should hunt as a family, they were all my teachers. They led me day after day into the forests as well as into the history of the Dunne-za. They were generous with those things in which they were rich—time, knowledge and meat—as well as those things of which they had all too little—saddles, good guns and store-bought food.

Rosie had made a stew from some of the deer Jimmy had killed the day before, and had cooked a heap of fresh bannock. I think they had been waiting for us to get back. The food was fresh and hot; they knew we would be hungry. We sat at the little table at one end of the room and ate. Jimmy joked about the moose I had not seen and then had not fired at.

I had learned that each group of Dunne-za made use of a specific area; later, when I worked with a team of colleagues to map Dunne-za territories, I could see the ways in which particular communities knew and occupied their lands. Combining these maps revealed the total extent of the Dunne-za system, a set of territories amounting to about 30,000 square miles. Even on my first journeys in Dunne-za land, I had a sense that this was a place with no beginning and no end. Sometimes the world of the hunter-gatherer seems boundless, reaching seamlessly into the distance and the very distant past. Nonetheless, these hints of infinity could not prevent my questions about who was where when. A part of me was a traitor to Jimmy Field's saying that Indians had been there "all time."

2

In the 1990s, Canada funded an immense Royal Commission on Aboriginal Peoples. Its task was to hear, and seek solutions to, the problems faced by native communities and First Nations. The commission's work involved much historical documentation of injustices suffered by indigenous peoples in Canada and of the ways in which their stories have been obscured by settler and frontier mythology. My friend and colleague Ted Chamberlin acted as an advisor to some of the commissioners. (There were nine, five of whom were themselves members of Canadian First Nations.) As he learned more of the commission's priorities, Ted realised that there was a need for a kind of history that went beyond the specialist reports of academics. He pointed out that the "stories" and "myths" of each nation, their oral heritage, needed to be woven *into* history. He also suggested that this could best be done through creating a history-making process unlike anything being looked at by the Royal Commission.

Ted Chamberlin's vision was of a new kind of historiography and scholarship—one that grew from the voices and knowledge of aboriginal experts and elders, and one whose form, or "text," at the same time captured the fluidity of oral culture itself. He proposed a history that would give aboriginal voices a place in the country's intellectual heritage. A history designed to allow aboriginal elders to build their heritage into written forms, or "fact," and their knowledge into a continuous history of their nations. He hoped to expand and shift ideas of what "history" could include and what aboriginal history would mean for Canada's sense of the past. He argued that there could be a different kind of record, an alternative way of knowing and telling. He was not sure what this might look like, but he wanted the Royal Commission to explore its possibilities. He had a hunch that a new kind of history could be one of the commission's most enduring contributions. Here was a chance to give aboriginal understanding a place in the wider record, and a chance for those who deal in the records to learn a different historical idiom.

The Royal Commission agreed to hold a preliminary workshop on the history project. It would bring together elders from aboriginal communities with a number of Canadian intellectuals, some of whom were also aboriginal. Ted asked me to join him in organising this event.

We met in a conference centre on an Alberta reserve. It was a large log building, owned and managed by members of the Stoney Indian Band. On the second day of the workshop, we began to discuss the prevailing idea, at least in universities, about the earliest hunter-gatherer migration into the Americas: the Bering Strait theory. This theory postulates that at the end of the last ice age about 30,000 years ago, as the massive sheets of impassable glaciers began to draw back and while sea levels were still low, a land bridge connected what are now Russia and Alaska. Archaeologists speculate that this was the route by which the first inhabitants of the Americas reached the New World. This is the starting point of historical accounts of aboriginal peoples in Canada, the Europeans' story about their story.

One of the workshop participants was a woman from a Cree community who was enrolled in a Ph.D. programme at a prestigious American university. She was not happy about the Bering Strait theory. She pointed out that her people, and most "Indian" people, do not believe that archaeologists know anything about the origins of human life in the Americas. The idea that people first came as immigrants from Asia was, she said, absurd. It went against all that her people knew. Oral cultures set out the origins of aboriginal societies: in place after place, the first people arose where they now live. There had been no immigration, but an emergence; not an arrival from elsewhere, but a transformation from an ancient, prehuman time. She would have nothing to do with so-called scholarship that discredited these central tenets of aboriginal oral culture; the Bering Strait theory should have no place in the workshop.

The group talked a bit about whether or not archaeologists' myths were somehow reconcilable with aboriginal cultural myths: two ways of talking about history, perhaps, or points of historical entry at different levels, in different time dimensions. But no such middle ground was found.

In fact, the workshop became stuck in an intellectual quagmire. Some of the elders did not trust historical work that was not rooted or conducted in an aboriginal language, while mourning the virtual disappearance of those languages in their own communities. Some white scholars attempted to find common ground between the written and oral forms of history. We reached a tentative agreement that maps of aboriginal lands, showing how each nation identified both territories and histories, could be a way of creating a new kind of historical document. But this did not meet the original purpose, did not give voice to a different form of history. We all felt confused and stuck.

The workshop did not result in a project. The Royal Commission did not agree to design and fund an aboriginal history of a new kind. Despite protests by two elders who had been at the workshop and had urged that the work go ahead, nothing was done. An opportunity may have been missed; perhaps no one knew what the opportunity was. For me, the Bering Strait disagreement encapsulated the problem. Questions about where people come from, and how they have moved across the world, are real. But they are perplexing. As Jimmy Field had said, way back in the beginning is pretty hard thinking.

3

In a story that speaks of the creation of everything, there are no beings other than those the story brings into existence. Genesis does not reveal some shadowy universe that predates the world God makes; nor does it speak of those whose lands are conquered by the exiled farmer-herders and their many children. (Though Genesis contains a small exception here: the Canaanites are mentioned when their lands are deemed to be available for settlement.) This absence of other peoples bewilders the literal-minded: where are the wives for Cain and Abel, and do the children of Adam and Eve have to marry each other in some original incest? These worries may suggest a failure to grasp the poetry of creation stories, but they are also part of a deep and authentic unease about the claims of Genesis to universality.

Beyond the story of the lineages of Noah are those for whom Genesis is not the creation: the humans who live by hunting rather than by agriculture. The wild and unclean; the shadow populations of the Bible story. Archaeology and anthropology have their own creation stories. According to these, hunting peoples have a claim to the earth that reaches back a hundred times further than that of the farmers whom the biblical God created and cursed.

Twelve thousand years ago, all human beings lived as hunter-gatherers. This is a catch-all term, of course, referring to those people who relied on local resources, be they plants or animals, rather than on tilling the ground and cultivating exotic species. Hunter-gatherer economies shaped human life in Europe as well as in Tasmania; along the Thames Valley as much as by the Orange River or in the Amazon Basin; by the coasts of the Mediterranean as well by the Indian Ocean. Not every part of the earth's surface was occupied. Humans had not found New Zealand. Antarctica was uninhabited then, as it still was in 1900. Many archaeologists say that peoples from Asia had not yet reached the Americas. But wherever humans lived, they did so by knowing their territories as Jimmy Field knew his, and thus being able to hunt and forage. From deserts of sand to deserts of snow and ice, in tropical forests and on dry savannahs, beside warm seas and cold oceans, hunter-gatherers accumulated necessary information and devised ways of living.

Palaeontologists and physical anthropologists are not in agreement about how long the worldwide spread of human beings has been under way. The earliest evidence of creatures who could be our own ancestors is found in fossils dating back 2.5 million years. The most ancient signs of cooked food come from 1.9 million years ago. The oldest remains discovered thus far of humans who show clear signs of having walked upright, and having other anatomical features that link them closely to modern men and women, go back about 1.5 million years. These earliest glimpses of human evolution come from research in East Africa and central Asia. Evidence suggests that people moved from original sites in Africa and Asia approximately one million years ago, beginning their occupation of the rest of the world. This spread appears to have reached Europe about 500,000 years later.

At what point in this immense span of time can we think of these first humans as the hunter-gatherers whose skills, knowledge and languages represent the probable origins and certain development of human culture and society? Some experts date this at 400,000 years ago; others insist that it is no more than 200,000 years since *Homo erectus* became *Homo sapiens*—the Latin terms that distinguish between humans who are not and those who are genetically the same as modern people. These are matters about which there are no real conclusions, at least not at present. Perhaps genetic analysis will provide new and more convincing dates for human origins. Meanwhile, the fact emerges that for something like 90 per cent of our ancestry we have lived as hunter-gatherers, not as farmers. Archaeologists may one day reach some definitive conclusion about how long human beings have been in North America. Whatever dates come to be agreed upon, they will show that, compared to anyone else, the "Indians" have been there "forever."

4

Ten thousand years ago, the world's hunter-gatherer systems would have displayed an immense range of languages and cultural forms. Hints of this diversity come from the hunter-gatherer populations we know. The indigenous peoples of the Americas speak hundreds of different languages, and all are clear about the many features that allow them to identify themselves as distinct nations or societies. In the forests and tundra of the North American Subarctic and Arctic, where the environment is extreme and landscapes are vast and relatively uniform, hunter-gatherer societies speak a large number of mutually unintelligible dialects that are separable into four language families, families as distinct from one another as the Romance languages of western Europe are from Bantu languages of southern Africa. In California alone, linguists estimate that aboriginal populations spoke some eighty different languages. The variety of hunter-gatherer ways of speaking is itself a sign of the vast spans of time during which these social systems have been alive.

Language creates the potential for an immense panoply of social and family arrangements, an apparent infinity of ways in which people can codify and convey knowledge, beliefs and ideals. For all this diversity, however, there are some characteristics that all hunter-gatherers appear to have shared. These shared characteristics are grounded in the kind of relationship hunter-gatherers establish with the world in which they live. Material well-being depends on knowing, rather than changing, the environment. Many forms of both hunting and gathering have relied on management of the land, from the selective burning of forests to the replanting of roots to ensure abundant growth in the following year. Hunters also insist that animals will agree to be killed only if they are shown respect in both life and death. The rituals and habits of respect are therefore important ways in which hunters and gatherers are not passive harvesters but are engaged in the complicated business of maintaining the world around them to ensure that its produce is bountiful. But the central preoccupation of hunter-gatherer economic and spiritual systems is the maintenance of the natural world as it is. The assumption held deep within this point of view is that the place where a people lives is ideal: therefore change is for the worse.

Another feature of the hunter-gatherer way of life is a deep respect for individual decisions. There are experts rather than leaders, men or women whose skills are revered; but decisions about whether to follow their lead or take their advice are matters of individual choice. A hunt leader does not instruct others to follow or to take any particular direction. The expert makes his or her decision known; others then make their decisions, following or not as each prefers. Social and ethical norms are powerful, but they are reinforced by a minimum of instruction or organised retribution. Beliefs about the effects of human actions on the spirit world contain implied threats: failure to show the necessary respect for animals or to observe important taboos can result in hunger and sickness. But rules and the consequences of breaking them are embedded in the stories and advice of elders or in the diagnoses, after things have gone wrong, of shamans. Thus the individual hunter-gatherer's links and routes to the spirit world, through dreams or other private forms of insight and intuition, are paramount. Choice and freedom are centred on each person, unconstrained by social hierarchy.

The egalitarianism of hunter-gatherer life is further shown in how resources and produce are shared. Every member of a clan, band or village has the same rights to hunt and gather in a large territory; insofar as one individual or family is more successful than others, this achievement yields food for everyone. Pride in success is expressed through giving the results of the hunt to others. Since both distribution and helping oneself from others' supplies confer honour on the successful, there is little difference in standard of living between the most and the least effective.

In this world of egalitarian individualists, hunter-gatherer peoples have relatively small and scattered populations. The need for mobility and the availability of resources appear to encourage hunter-gatherers to limit their numbers. Families do not desire more than two small children at any one time, and to this end they do what they can to space births by about three years. Thus the apparent paradox noted by so many observers of hunter-gatherer societies: intense love for children, yet occasional readiness, at times of shortage, to use infanticide if a child just born is perceived as one too many.

Given that success in hunting and gathering depends on a detailed knowledge of a specific territory, and that the resources of this territory are shared among all its harvesters, geographical conservatism is intrinsic to hunter-gatherer systems. They are in their idea of paradise; to go elsewhere is to face extreme risks. Increases in hunter-gatherer populations did lead to the expansion of territory and movement into new regions. The archaeological record, however, shows that this spread of population took place over a long period of time. The first diaspora of *Homo sapiens* the hunter-gatherer required at least 100,000 years, and perhaps as much as 250,000. The comparable diaspora of *Homo sapiens* the farmer over the same distances, and with a commitment to transform the landscape of each new region, required less than one-tenth of this time.

5

My first and most intensive experiences of hunter-gatherer societies were in the Arctic and Subarctic. The Inuit, the Dunne-za and the Innu of Labrador and northern Quebec speak languages as different from one another as English, Japanese and Hungarian. They occupy territories that are spread across an area the size of Europe. They have customs, technologies and beliefs that are their own. And they have different histories, both ancient and modern. Yet they all live in environments with extreme variations of temperature and long winters, and they all rely on the meat and furs of animals that have adapted to cold land and water. It might be said that the peoples I travelled with and the places they took me to all lie at the margins of the inhabitable world. The domination of their lives by extremes of climate and geography might therefore have given them much in common with each other, and not so much in common with the hunter-gatherer systems that have developed in other, less forbidding parts of the world. Perhaps the voices of Anaviapik, Jimmy Field and others who live in harsh conditions speak to a way of life that cannot be relied upon to represent the great panoply of hunter-gatherer societies or any overall mode of society and economy.

The very history of hunters and gatherers accentuates this possible difficulty. Farmers and herders have always looked for the most fertile and abundant environments—river valleys, gentle shorelines, rich and easily worked soils. So the world's hunter-gatherers have been pushed from these places, driven to live in areas that are not of such interest to invading newcomers. As anthropologists have pointed out, a result of this system of displacement is that the hunter-gatherers who survive do so in extreme environments: the cold of the far north, the driest parts of the Australian outback, the most arid and unforgiving regions of the Kalahari. This reality reinforces the tendency on the part of outsiders to depict hunter-gatherer living conditions as harsh and impoverished. The outsiders, of course, come from an agricultural heritage, from which point of view any landscape that is not farmed is wild and therefore of little economic use—is in fact outside culture. Colonists have often supposed that hunter-gatherers' lives were "nasty, brutish, and short" without agriculture and its "civilising" influences.

Yet these hunter-gatherers delight in their lands and have great confidence in what they see as a wealth of resources. They do not look to other lands with wistful eyes or express a desire to live in the softer climates or gentler landscapes that appeal so much to farmers and herders. Of course, the peoples who originally lived in landscapes that farmers agreed were candidates for Eden were evicted or destroyed in the early days of one or another form of colonial history. This happened in Europe and Asia, as well as in parts of the Americas and Africa, long before European colonists embarked on their conquests around the world. But for those who continue to live as hunter-gatherers, or who recall the hunter-gatherer lives of their parents and grandparents, their lands are—or were—as good as they could be.

The question about where hunting peoples have been allowed to live leads to other questions of place and time. The cultures that have endured in the New World, despite its intense colonialism, are often seen as the living indicators of the most ancient of all ways of life. Thus it is that popular accounts of travel to hunting peoples are often billed as a "journey to the Stone Age," a chance to experience "the most primitive" of human beings. Difference between "them" and "us" is, in this kind of account, a potential comparison between different levels of human evolution.

To spend time in a hunter-gatherer society is indeed to make a journey from one way of being in the world to another. Many writers, from explorers to missionaries to ethnographers, have referred to the difficulties or inspiration or awe they experienced as radical differences were revealed between the hunter-gatherer way of life and their own. Missionaries could not understand why people welcomed gifts with shouts of excitement and delight, then left them behind or threw them away. Explorers were amazed that parents never chided or disciplined their children. Administrators puzzled over a profound reluctance to exchange a condition of "unemployment" as an "impoverished" hunter for a job that would bring a regular and secure income. Political organisers exclaimed in despair when men and women whom they had designated or trained as "leaders" remained committed to anti-political individualism. And everyone commented on the generosity and hospitality that prevailed even when supplies were minimal.

Here were people whose indifference to the order, systems, hierarchies, disciplines, materialism and guile of other forms of civilisation revealed, in the eyes of the colonisers, an unevolved, "simple" human condition. There is evidence of agriculturalists' hostility towards and disdain for hunter-gatherers in Amazonia and in the empires of the Inca and the Aztec. Discrimination against the Twa (Pygmies) by their neighbours is well known and apparently endemic. The herders and farmers of sub-Saharan Africa despise and fear the hunter-gatherers whose lands they have taken and continue to take. Much the same story can be told about the peoples who have used the forests of India, Sri Lanka and Indonesia. In the case of early encounters with Bushmen in southern Africa and Aboriginal groups in Australia, Europeans often insisted that the "primitives" were at the very edge of being human. In the Americas of the nineteenth and early twentieth centuries, both the practices of settlers and the application of legal theory to indigenous peoples' rights to land were based on ideas of human evolution, with the hunter-gatherers at the bottom of a developmental ladder and Europeans seated at its peak. These views condoned the dispossession and even the murder of hunter-gatherers. The disappearance of these people became a by-product, if not a precondition, of "progress."

6

The recognition of hunter-gatherer systems as complex and widespread was something of a revolutionary moment in the history of anthropology. Some eighteenth-century travellers idealised people whose lives were dependent on pursuing game and foraging for wild plants. The idea of hunting as innocent or idyllic, as well as backward, was part of the European notion of evolution at least as early as the Enlightenment. It was not until the 1960s that a systematic and revelatory picture of hunter-gatherer economics and society emerged: systematic in that anthopologists connected a range of cultures that shared important defining features; revelatory in that analysis of these features revealed human genius—here was a way of life with immense technical and social sophistication. Far from being simple or primitive, the economic and cultural techniques of hunter-gatherers were hard to see and difficult to assess precisely because they were meeting needs of mobility, decision making and resource harvesting that were both varied and subtle. Here was a triumph of human achievement, a triumph that spoke to how most of the world had lived for most of human history.

The intellectuals who pioneered the work on hunter-gatherers—James Woodburn at the London School of Economics, Richard Lee at the University of Toronto and Marshall Sahlins at the University of Chicago—reached a number of remarkable conclusions on the basis of both detailed field studies and collaborative exchanges of data. They showed that the routines of the hunter-gatherer way of life allowed more leisure time than those of agricultural systems and secured a good supply of highly nutritious food for most people most of the time. Many of these findings were drawn together at a conference in Chicago in 1968, whose proceedings were published as *Man the Hunter,* a book of immense importance in the history of social science. In 1972, Marshall Sahlins published *Stone Age Economics,* a set of essays about economic anthropology. The first of these essays, "The Original Affluent Society," summarized the main discoveries emerging from anthropological field work: with small populations, low levels of need and expertise in a particular landscape, human beings could eat well, enjoy much leisure and evidence great health of body and mind. The central stereotype of human social evolution was more than contested and undermined: it was turned on its head.

Yet these new insights into hunter-gatherers, for all their force and authority, did not transform debates about land rights and claims; nor did they secure new kinds of assessments by the courts. Those who argued in the 1970s, '80s and '90s for the rights of the Inuit, Dene, Hadza, San and others found that the prejudices and stereotypes of dominant societies had not yielded to the new anthropology of hunter-gatherers.

In fact, to go to a hunter-gatherer society in the late twentieth century created a complicated and, in some ways, disorienting state of mind. The achievements of the people, in their knowledge of their lands and their expertise at harvesting its resources, were there to be seen. Oral history added both the peoples' own view of their "original affluence" and their despair about the violation of their rights. But anthropologists and other outsiders who worked in these circumstances put together the facts knowing that they were likely to be misunderstood. The old prejudices against hunter-gatherers were little, if at all, dented by this set of compelling insights. It was rather like being a scientist in a world that did not accept science itself. Again and again, those who opposed recognition of the rights and claims of hunter-gatherers insisted that anyone who spent time with these peoples was, in effect, visiting a remote and doomed corner of human prehistory.

As I flew over the coastline of Hudson Bay for the first time, or drove into the subarctic boreal forest, or walked onto the dunes of the Kalahari, was I embarking on some kind of time travel? Was I leaving my own era and making a journey into the world as it was thousands of years ago? Vistas of sea ice and tundra, or great spreads of conifers and lakes, or a seeming infinity of sand and camelthorn trees and dry watercourses: were they the world as it was, in some timeless and ancient condition, before anything of modernity—from farmlands to towns, from rifles to airplanes, from telephones to computers—had transformed and perhaps shattered the human spirit? As I listened to the voices of people for whom these places were home, would I hear the sounds of ancient human history, a sort of vocalisation of eternity? Even in the late twentieth century, would there perhaps be a chance, on journeys to the homes of the Arctic or the Subarctic or the deserts or the outback, to come into contact with the most enduring qualities of humanity? Or was I making a set of journeys that would undermine any such ideas, and in relation to which the new ideas of anthropology were already anachronistic? On any journey from a familiar to an unfamiliar cultural milieu, there is the sense of making a leap, a rush of ideas, a sense of awe. When this journey holds within it some potential to be an encounter with the very depths of human history and consciousness, the imagination makes a more drastic set of leaps. Yet these leaps include a complicated enigma, a sort of puzzle.

7

When I first got to know Anaviapik in Pond Inlet, he did not present himself as a human being from some other, more remote time. He gave no indication, and made no claim, of belonging to a different era. He had lived much of his life as a hunter and trapper, making use of snowhouses and sod huts—the *illuviga* and *qarmak* of "traditional" life—and travelling by dogteam and in open boats. His language, Inuktitut, was rich with grammatical sophistication. His family relied on caribou skins for winter clothes, and on sealskins for spring and summer boots. A great deal of Inuktitut material and intellectual culture played its part in his everyday life. But Anaviapik was as modern as anyone. He was a contemporary of mine, not an ancestor. Every part of his life had been changing—"developing" may be a misleading word—for as long as anyone else's. Some of these changes were intrinsic to the Inuktitut system and Inuit lands; many others were to do with historical and colonial processes that originated in Europe. The uses of metal, gunpowder, manufactured clothes, rifles, petroleum—all the things that are the story of European material history—were important to his life as well as to my own. My grandparents' lives were changed by such inventions; so were his. Similarly, the development of the internal combustion and then the jet engine had changed his life as they had mine.

The processes of human experimentation have been taking place for the same amount of time among all peoples; discoveries in one place have always had the power to change life everywhere else. Modern inventions are carried with great speed to all parts of the world. So Anaviapik and I made use of the same raw materials, wore very much the same kinds of clothes, lived in houses with the usual kinds of furniture and facilities. The issues with which he was engaged every day were those of the late twentieth century. Some of the circumstances were the outcome of recent changes, the results of Canadian government programmes and services. One set of facts invited me to be in awe of the links between Anaviapik's way of life—his knowledge, experience or language—and a remote, essential human past. Another set of facts revealed to me that Anaviapik was as much in the present, a modern human being, as anyone else.

Colonialism and racism create and emphasise inequalities. Anaviapik often spoke to me about the differences in material life between Inuit and Qallunaat, and about the power relations between the two kinds of society. He had a strong sense of injustice. But the causes and consequences of injustice belonged in the present, were a feature of modernity. Colonial change ensures that everyone's economic and social lives are interdependent. And colonial beliefs assert that the colonised are inferior. Their supposed place in the past, as an example of some earlier stage of evolution, is used to justify extreme inequality. All hunter-gatherer peoples are aware of the European (or Asian or African) colonists' perception of them. They have heard, over and over again, the colonists' notions; they have experienced the colonists' attitudes and behaviour.

Expropriation and exploitation are compounded by the extent to which the despised and dispossessed can come to despise and dispossess themselves. Self-loathing and self-destruction are features of many colonial encounters. "They," the whites or southerners or newcomers, arrive as conquerors, as strangers with new and terrible diseases, as missionaries with guns and medicines as well as written texts that purport to demonstrate the inferiority of aboriginal peoples. "We," the natives, cannot resist for long the demands and ideas, the guns and diseases, and "we" have an intense need to be part of what is taking place—to acquire new kinds of goods, to get antidotes for the diseases, to learn how to deal with the newcomers. So "they" are powerful while "we" are weak; "they" are rich while "we" are poor. "They" understand and "we" do not.

Anaviapik and other Inuit elders often talked about the riches and skills of the southerners, the Qallunaat. We Inuit, they said, know our lands and can hunt and live here; but the Qallunaat are so clever, able to make so many things. Guns and airplanes, phones and radios. They told me that I must surely be able to make these things. No, I said, I was far less able to make such things than they were. I knew many Inuit who would repair and rebuild guns and engines, and even radio sets. Very few of the Qallunaat I knew could dismantle, improvise spare parts for and reassemble complicated equipment in the way that many Inuit could.

Yet there *is* an important divide between hunter-gatherers and those who have, in relatively recent times, displaced and overwhelmed so many of them. From the point of view of the hunter-gatherer, the relatively recent oppositions between town and country, the rural and the urban, the modern and the tribal overlook a deeper dichotomy. Differences between hunting and gathering and agriculture are at the heart of history.

8

That desolate northern edge, those icy waters and perpetual snow: the Arctic of the European imagination is a terrible, dreary, deadly place. Sailors whose ships were trapped in summer ice and held prisoner for an Arctic winter, or crews of ships that spent too long moving among floes and pack ice and along the endless coastlines of the far north, suffered a slow deterioration of mind and body. Monotony of vista and harshness of climate eroded their souls. They sank into lethargy and experienced darkest despair; their bodies ached and their teeth fell out, until, at last, many died in anguish. No wonder that Mary Shelley's Frankenstein is first glimpsed through the icy mists of Hudson Bay.

The disease of Arctic travel turned out to be scurvy. Depression and weariness are the first signs of a vitamin C deficiency. Mary Shelley needed a place of mythic dimensions for the launch of her terrifying tale. She herself had been to northern Canada—along with many English readers who shared in the armchair search for a Northwest Passage and an abundance of bowhead whales—only in her imagination. The dreadful decline and death of northern sailors was a metaphor for the North and the kinds of fantasy to which it gave rise. In this kind of place modern farmers located hunter-gatherers and insisted that they must be as wild, as primitive, as unredeemable as those terrible lands.

The Arctic, like other, hotter deserts of the world, represents everything that agriculture is not: a lack of soil, an unforgiving climate, a place where not even the toiling of Adam can cause the grains to grow or the cows to give milk. Lands beyond the possible frontier of farming seem indeed to be on the other side of Eden—beyond the scope of human use. No wonder that those humans who turned out to be there appeared to live in a state of simplicity that was, for the romantic, an inspiration of nature, and, for the others, unquestionable evidence of savagery.

Yet the familiar stereotype of the Arctic turns out to be false: it is a place of many environments and all kinds of resources. Anaviapik and his people have always been sure that they were in the best possible place, their version of Eden (though they would not have used the word), and that the southerners, the Qallunaat, endured the real hardships. How striking it is that the Inuit have six seasons, not four: theirs is a world with far greater variety of dark and light, hot and cold, mountain and plain, than any to be found farther south.

The Arctic is best defined by reference to the treeline, the point at which summers are too cool for even willow and pine to be more than miniature, twisting plants reaching no more than a foot in height. This line in North America where forest yields to tundra snakes across many degrees of latitude, reaching far above the Arctic Circle in Alaska and northwest Canada, yet looping south to 53 degrees in Hudson Bay—the same level, at its southernmost extent, as north Yorkshire in England or Copenhagen in Denmark. The Russian Arctic stretches along 5,000 miles of coastline and tundra, an immense area with some fifteen aboriginal cultures.

The area that lies to the south of the treeline and yet is more or less beyond the reach of farmers is also immense. The taiga and forests of the Subarctic are the least-known region of the world, yet they are on a scale that confounds the imagination. Subarctic Lapland, Russia, Alaska and Canada stretch about 8,300 miles in a continuous northern belt around most of the world. The width of this band of forest varies, reaching far above the Arctic Circle in some regions, and as far south in others as the latitude of Wordsworth's pastoral Lake District. The total area is around four million square miles.

At the edge of the treeline, at the beginning of the Subarctic, the forest is sparse, tangled and uniform—a sprawl of jack pines and dwarf willows. A little farther to the south, where the summers are short but hot, the soil thick and moist, there is an abundance of birch and aspen among the evergreens. The silver of their bark and the delicate colours of their foliage pattern the light; in summer, whenever a breeze blows across a protected hillside, the leaves murmur and rattle into the distance.

This immense land of mixed forest is occupied by many different peoples. In Lapland and Siberia, reindeer herders and hunters flourished. They include the Saami, Evenki, Tungus and Chukchi, societies that have moved across the subarctic forests and taiga from northwest Norway to eastern Russia and northern China. In the North American Subarctic, Athabaskan and Algonquian peoples, comprising dozens of distinct groups, lived throughout the forests, in every major river valley and beside each lake system.

Farmers have brought grains with short stems and rapid seed maturation to the Subarctic. The agricultural frontier has nudged farther north. In Alaska, Yukon and the northern provinces of Canada there are new stretches of forest cleared each year; new fields are created, and more farm families move in. But the vast majority of the Subarctic is still the land of hunters, gatherers, fishermen and trappers.

9

I first went to the Subarctic in 1969, following the one road leading from northern Alberta to Great Slave Lake. I travelled with two friends. I remember that we had a protracted and fierce argument about the land. One of my friends, who in later years was to become a distinguished Oxford philosopher, remarked that the scenery, the view from our car as we drove north along the Mackenzie Highway, was "boring." He may well have begun his critique as a reasonable reaction to my continuous expressions of excitement and delight. And I rose to the bait. No, I argued, the landscape was wonderful, inspiring. It was compelling for what it represented, for what we could not see but were bound to imagine. Just think of the extent of this region: the forest on either side of the car stretched for at least a thousand miles. Think also, I said, of the peoples who have lived here. All this, he said, was irrelevant; imagine what I would, the simple fact was that the view from the windows of the car was, hour after hour, unchanging. The same kinds of trees; the same slow undulation to the ground. Boring is what describes this kind of sameness. We argued this back and forth, back and forth. In the end, he said, the point is linguistic: the meaning of the word "boring" makes the word's application to the experience of this drive into the North a clear example of what the word is used to mean. So "boring" it must be. I did not concede the experience of the heart and imagination to this piece of linguistic logic.

It was several years after this argument took place that I first went deep into the Subarctic. I was part of a team of researchers put in place by the Union of British Columbia Indian Chiefs, an organisation set up by that great activist George Manuel. The union represented many of the peoples of the Northwest who shared Manuel's vision, which was a blend of resistance to and negotiation with modern colonialism. The team worked intensively on the history and conditions of the Dunne-za and Cree bands in the eastern foothills of the Rocky Mountains, the peoples who were given a place in a New World frontier and some uncertain guarantees to their lands in 1901 by the eighth treaty Canada signed with indigenous societies. My work in Treaty 8 country included living for about eighteen months at the Halfway River Reserve, the Dunne-za community at the very edge of the mountains where I met Jimmy Field and his family.

In the early 1940s, the United States military, apprehensive about a possible war zone in the North Pacific, proceeded to build, in wide straight lines, a highway along which to carry military supplies from northern Alberta to the Alaska coast. In the late 1960s, most of the Alaska Highway was still a gravel-and-mud slash through the Subarctic. Athabaskan lands lie to both sides of the highway, from Dawson Creek in the south to beyond Whitehorse in the north. The length of these territories is almost 1,000 miles, the width between 550 and 800 miles; the total area of Athabaskan lands in this northwest corner of North America amounts to more than 350,000 square miles. This huge area was known and travelled and cared for by Athabaskan families and communities.

The Halfway reserve lies about twenty miles west of the Alaska Highway, at the southern periphery of the Athabaskan societies of the region. To reach it, I drove sixty miles north from Fort St. John, and then turned off the highway onto a much smaller dirt road that twisted its way along creeks and through the cleared lands of new ranches—the northern edge of the farmers' frontier. Although the reserve sat amongst these ranches, I was soon to learn that for the Halfway hunters and gatherers saw both the area around them, the lands and forests at risk to settlement, and a much larger territory reaching far beyond these encroachments as theirs; and they saw an immense area of forest and hills, reaching a hundred miles or so beyond any settlers, as the lands of relatives.

Archaeological evidence suggests that hunters have been in this region for at least 8,000 years, or since it became habitable again after the last ice age. Groups of hunters could have moved through the region during the ice age, and some artifacts point to human occupancy as long as 25,000 years ago. The Athabaskans of northwestern North America might well represent a continuity of use and occupation of these lands with the first humans to hunt the savannahs left exposed by retreating icefields. The Halfway reserve, sited on level ground beside a river, may have been a summer gathering place and a centre of Athabaskan economic life for as long as people have been able to live in the area.

I went to the Halfway reserve to map hunting and gathering areas. This was part of a land use and occupancy study (to use the terms of what became a distinctively Canadian art) in the course of which a team of researchers had the job of recording the places and resources that have been important to the Dunne-za in the entire region. The results are set out in my book *Maps and Dreams,* which owes its existence to the politics of the northern frontier and the joint efforts of all who worked on the Union of British Columbia Indian Chiefs project. The best of the work, if only because it was the result of so much collaboration, is to be found in its maps. The least developed parts of the book, and the issue that I found most compelling and most difficult to understand, concern hunters' dreams.

The Dunne-za, like many other hunter-gatherer peoples, use dreams to locate both the animals they will kill and the routes along which they must travel to find these animals. The hunt is something other than mere hunting; rather, it is part of a vitally important relationship. The people depend on the animals, and the animals allow themselves to be killed. An animal's agreement to become food is secured through the respect that hunters and their families show to the land in general and to the animals in particular. If animals and hunters are on good terms, then the hunters are successful. If they are on bad terms, the hunt fails, and the animals withdraw into secret and unhappy places.

In the immense landscape of their territory, and with such a deep history of its occupation, the hunters and gatherers of the Halfway reserve carry knowledge and techniques that they have accumulated over many, many generations. Dreamers process this into decisions about the land and the hunt. To find the animals that are willing to be killed, Dunne-za hunters travel along trails that reveal themselves in dreams. How else can they travel such distances; how else process such huge quantities of data; how otherwise make decisions that rely on data that is both voluminous and elusive? Dreaming is the mind's way of combining and using more information than the conscious mind can hold. It allows memory and intuition and facts to intermingle. There is, however, another dimension to Athabaskan dream travel: movement through time as well as over the face of the earth.

10

Hunter-gatherers must store an immense number of facts about their territories and the creatures in those territories. They also accumulate information about hunting and gathering techniques and about events that throw light on the applications of these techniques. They use songs and stories to make the memorising of all this knowledge possible. Elders are of special importance in these societies, if only because it takes almost a lifetime to learn all that needs to be known; elders have the task of passing on their knowledge and experience with the necessary attention to detail, variety and overall coherence.

Yet there is always information that lies outside even the lifetime listening and learning of the most sagacious elder. As in all human systems, there is the occasional need to divine that which lies beyond even the largest storehouse of knowledge. Shamans are the people whose special skills and techniques allow them to move from the practical realm to the spiritual, from the everyday to the metaphysical. Powerful dreamers are the Dunne-za equivalent of shamans. They can receive new information. An elder once told me that those dreamers function rather like tape recorders: they record a dream-song as they sleep, and find that they can play it back in the daytime. The origin of the song is mysterious; its meaning may be obscure; but its existence is not in doubt.

The most powerful of Dunne-za dreamers, when they learned the Christian story of the afterlife, reported that they had found routes to heaven. This was a shamanic response to missionary ideas: if there was a trail to be discovered, the dreamer must find it. And all those who rely on shamans believe it is possible for especially gifted men and women to visit places beyond the reach of ordinary travel; they also believe that shamans can make journeys to other times, be they in the past or in the future.

Follow me, for a moment, away from Dunne-za dreams to the signing of Treaty Number 8. When the Dunne-za and Cree of north-east British Columbia (as well as people in adjacent areas of Alberta and Yukon Territory) say that they are "treaty Indians," they speak with pride. The term is a claim against the rest of the world, based on the deal their grandparents' chiefs or elders signed almost a hundred years ago. The treaty, people say, makes sure that they can live in their own ways on their own lands. The Dunne-za and Cree are two of the many indigenous peoples of the Canadian North who use a treaty to argue for their rights.

This reliance on treaties may seem perplexing or naive to outsiders, who have long come to regard the treaties as nothing other than devices for tricking and, in due course, dispossessing people. Indeed, some of the more notorious treaties did no more than confirm displacement and impoverishment, ensuring that particular tribal groups had no more claims to their territories. But in the Canadian North, treaties seemed to provide more. According to the oral traditions of some of these peoples, and as set out in the treaties, some groups were given rights to extensive areas within which they could pursue their own ways of life and administer their own affairs. This was the case with Treaty 8.

The Crown, in the form of lawyers who specialise in showing that aboriginal peoples have as few rights as possible, has argued that although Treaty 8 guarantees that the people within its area can hunt, fish and trap as they have always done, there is also the treaty's small print: this hunting, fishing and trapping can continue only until such time as the Crown wants these lands for some other purpose. The Indians, in the form of lawyers who specialise in aboriginal-rights issues, have said that the small print was never explained to them, and that their elders' understanding of the treaty has always been that it guaranteed their economy and society forever. The case law that has accumulated in Canada as a result of legal disputes over Indian rights and treaties establishes that the interpretation of a treaty must be made with reference to the time and place where it was signed. Context contributes to meaning, so that to some extent the terms of a treaty must be what they were deemed to be when signed. The small print, existing in written English for people who neither read nor knew English, may be legally irrelevant. Those who signed for the Indians may well have understood nothing about the clauses in small print that undermined the agreement they thought they were entering into. Therefore, argue the lawyers who rely on the treaty as an important part of the guarantee of aboriginal rights and title, the Athabaskans of the entire Treaty 8 area can pursue their way of life when and where they see fit.

How can we know who said what at the occasion of the signing of Treaty 8? Who was there in 1901? What took place? To answer these questions, to get the facts, one of the Dunne-za elders travelled, in a dream, along the trails of time, to be at the treaty signing. He received the story as a song, in his sleep, and sang it to others. In this way the people received an account of what their ancestors had heard and understood when they made the agreement.

I wrote about the trails to heaven in *Maps and Dreams* and spoke there of the ways in which those whom I knew best, and who made it their business to teach me about dreams and trails, would punctuate their explanations with self-deprecating anticipation of my skepticism: "You white men, you say all this is bullshit" and other statements of this kind. I listened with great care, and I struggled to understand the stories and theories that my teachers put before me. I did not disbelieve; on the contrary, I was full of admiration and respect. Yet these were men and women who had, all their lives, experienced the sneers and dismissive rationalism of whites. Many of the older people at Halfway had worked as guides to hunters in the mountains and as cowboys on the newcomers' ranches, and they had spent many hours in the crowded, redneck bars of Fort St. John. They were all too familiar with what white people thought about the land, about animals, about them. They also had been visited by missionaries and schoolteachers. They knew, therefore, that their travels along dream trails, as well as the entire complex of stories and experience in which those trails have their place and meanings, was judged by whites as false or ridiculous.

Athabaskans do indeed travel these trails: on the ground, and in the sky, and through time. But so, to some extent, does everyone else. Stories that are believed and dreams that act as guides—these take us to other places and times. This may be a matter of imagination or, in the case of hunter-gatherers, it can be at the core of what people do in order to live. All cultural systems rely in some measure on oral heritage, and all oral culture takes us on spirit journeys. Protestations of rationalism obscure this truth and deny the experience and power of stories. Rationalism locks human beings into one time, even though they travel to others. Athabaskans do in full what others do in small part.

11

One morning in summer in Dunne-za territory, Jimmy Field's brother Abalie came to tell me that some horses had gone missing, perhaps on the other side of the Halfway River. He suggested we take a ride to go look for them. Jimmy and Abalie's son Marshall would come with us.

We saddled up some of Thomas Hunter's horses and set off. We followed a trail along the river, looking for a place to cross. The water was low, but wide and fast moving. Abalie selected a place where he thought we would have the least difficulty getting across. Jimmy led the way; Abalie rode behind him, then Marshall and me. The horses stepped with surprising lack of concern into the water, but soon they were feeling their way on the round pebbles of the stream bed. The water was deeper than I had expected. The others moved with confidence, somehow keeping their horses at the angle they had chosen to the current. I was slower, and soon found myself taking a route about ten yards farther downstream.

As my horse reached the edge of the fastest part of the stream, it hesitated, stumbled and made a sudden movement, stepping into much deeper water. As the river rose to the level of the stirrups, I pulled my knees up. To my dismay I realised that my horse was heading into a channel that threatened to be at least as deep as the bottom of the saddle. I looked up towards the others and saw that Jimmy had stopped to watch me. He called out that I should ride towards him. Then he shouted: "Better climb on top!" I knelt on the saddle and let the horse half stumble, half swim its way to the shallows on the other side. Jimmy stood there, laughing. "Maybe you want to go swimming," he said.

We caught up to the others and began to look for signs of the missing horses. Instead, we found tracks of mule deer. The horses were soon forgotten, and we spent some time hunting in an area of pines and willows close to the riverbank. Jimmy got off and walked some way up the hillside beyond the river, while the rest of us rode among the willows. After a while we met up with Jimmy. But the deer had moved on, and we returned to the search for horses. None were found, and we soon crossed the river again—this time at a place without deep channels—and rode back towards the reserve.

We followed a trail through the trees, then rode through a wide area of open grass, and there, moving in front of us, was a large porcupine. Jimmy jumped off his horse, pulled out his rifle, followed the porcupine for a few yards and killed it. The hunters were delighted—here was a favourite meat. They told me that to eat porcupine would give me all kinds of good fortune. There were powers to porcupine, things I must know about skinning and eating it. I would taste the strength of that meat, when we were back, but maybe it would be too much for me. I was being teased as well as tantalised.

As we reached the last part of the trail, we saw a pickup truck parked outside one of the houses at the centre of the reserve. A young man was walking from the door of the house towards the truck.

"Who's that?" I asked.

"Booze guy," said Abalie.

"Who?"

"Bootlegger. Comes here some time. Lots of booze, lots of money."

I had heard about bootlegging in the area. Young white men loaded their trucks with the cheapest bottles they could find at the Fort St. John liquor store—sherry and local beer—then drove the fifty or sixty miles to the reserve, selling the stuff at double and triple store prices. If the Dunne-za could not find the cash, the bootleggers offered alcohol in return for saddles, horses or rifles. A Dunne-za man I knew once sold a horse for two cases of beer.

This particular bootlegger had been busy for some time. Several households were already drinking hard. Two of the women in Thomas's family had consumed enough beer to be very unsteady on their feet. They had no interest in our porcupine. One of them asked me if I could come up with a few dollars to help buy more beer. I was filled with rage.

While the others were unsaddling the horses outside Thomas's, I walked off and found the bootlegger. I wanted him to know what kind of destruction his trade could cause. The beatings and the destitution, the deaths. He may have been surprised that another outsider was there; he was, in any case, canny enough to affect a certain bonhomie that was typical of conversation between white men who met for the first time in those wild places. He seemed not to notice my indignation and paid scant attention to my angry protest about his being there. "Hey, they fucking like it! It's the way they're going, isn't it?"

He went on to explain in crude and simple terms the destiny of Indians in that country. They had nothing left, so they might as well have a good time. A party is what you need if you've got nothing else. Anyway, he said, it's hard on these guys, stuck out here on their reserve—nowhere to buy booze, sixty miles to hitch to town, and then how do they carry the stuff back home?

"So I'm doing them a favour, eh?"

And he told me how much he liked Indians—easy to get along with, whatever anyone says. Some guys were a bit rough when drunk, but easy enough to handle. Didn't I find the same thing?

I could no more argue with these notions than I could knock him down. He was louder, clearer and much tougher than I was. I walked back to Jimmy and Abalie. They had got hold of a few beers and were full of jokes and enthusiasm; why didn't I join them, Rosie and Sandra for a party? I talked to Thomas. He did not drink. He was sitting at the table, his face tense and troubled. He did not like parties, he said. I asked him if he thought I could get the bootlegger to leave. No, he said, no good. One day everyone would stop, like he had stopped. One day, not now.

Later that evening, Abalie skinned and cooked the porcupine. The meat was pale, the taste delicate. But most people were lost in a party that threatened to have no end.

12

Anthropologists have pointed out that hunter-gatherers focus on the present. People make decisions based on what they can find, kill or gather now, not at some later time or as a result of long-term strategic planning. This characteristic, which James Woodburn has identified as the difference between peoples for whom gratification is instant and those for whom it is postponed, is inseparable from hunter-gatherers' relationship to time. If the focus is now, then the past and the future circle events rather than define them; hunter-gatherers may allow one time to flow into another, and also allow themselves to move about in time. This helps to explain, perhaps, aspects of both dreaming and drinking. Hunter-gatherer ways of understanding time may also be evident in their view of reincarnation.

I have never been able to believe in reincarnation. Many peoples do, of course, including the hunters of the Subarctic. Parents and grandparents look at every detail of a newborn to see who has returned: there will be a birthmark, some sign to tell them which loved one they are welcoming back. My difficulty with literal belief is not matched, however, by any doubts as to the *importance* of reincarnation. Here is the immortality of body that corresponds to the essential immortality of knowledge. Grief over the death of an elder would be unbearable if the dead took with them to an eternal grave the experience of their long lives. What they did and knew must live on, in stories, in memories, in every elder's contribution to knowledge; these are all the more compelling as the living tissue of a culture if the tissue of those who spoke the oral culture can come back to the world. Cycles of life and death can thus be reconciled, in much more complete fashion, with the expanding cycles of knowledge.

I shall leave the Subarctic for a moment and return to the land beyond the treeline. Inuit do not believe in a complete or material form of reincarnation. Yet the cycle of life is secured for them by a parallel belief.

In spring 1973, while I was working in Pond Inlet, Inugu and his family suggested that I go with them to the floe edge to hunt seals and narwhal. They wanted to take a skidoo as well as a dogteam, and perhaps I would be able to share the job of this double mode of transport. I am sure the offer came more from understanding how much I liked to travel with them than from any real need for what little help I could provide. I was delighted to be given the chance to feel useful, though, the more so when I was assigned to the dogteam rather than the skidoo.

The surface of ice was crystalline under the sun, and seal breathing holes had opened wide into dark patches of sea water. In one area, a lead had crumbling ice at its edges. There were shallow pools of meltwater, and water that had refrozen in sharp fragments created a maze of surfaces and textures. We travelled with care, making many jokes about the nerve-wracking appearance of the ice. In fact there were no real dangers, and we made detours around cracks and leads. Skidoo and dogteam kept more or less together, so help was always on hand if any problems did arise—though the more likely difficulty, in these conditions, was the mechanical failure of the new way of travel.

At the floe edge, we found a group of young men camped there already. Hunting in a dense stretch of pack ice that tides or wind had pushed up to the fast ice, they had come close to groups of narwhal using open areas in the pack to surface and breathe, and they had managed to make a kill. As we drew up to the edge, there was much excitement. Together, back on the fast ice, we ate delicious *muttuk,* the skin of the narwhal, first raw, then stewed in boiling water.

The next day Inugu was not well. He stayed in his tent, too weak to hunt. That night he was no better, and the following morning we headed back to Pond Inlet, Inugu riding on the sledge pulled behind the skidoo. I took the dogteam for the first part of the journey. Later in the day, Inugu seemed much better and took over the dogs. As we made our way along the shoreline of north Baffin, the sun beating down, Inugu reminded me of the time we had gone fishing together. He recalled how we had depended on a fishing rod and reel, and laughed at the memory of how he had thought that such new-fangled devices would not be any use. Then he said, as a matter of fact, without any burden of emotion, that he would not be going to fish there again. I protested. There would surely be a chance even this year; if not this year, then next. He smiled at me and said he was old, not well.

Inugu continued to weaken. When I visited him that summer, he told me that his stomach was "filled by a sickness." I spoke to a nurse at the Pond Inlet nursing station, who told me she thought that Inugu had a stomach upset, some kind of constipation that tended to afflict old people. She had given him a laxative. I went back to Inugu's house and talked to him again about his illness. He said he had understood what the nurse had told him, but he knew something more serious was happening.

Inugu died later that year, at a time when I was working in another part of the Arctic. I learned of his death from one of Anaviapik's letters. I was dismayed. It seemed impossible, somehow unacceptable, that Inugu could die. He had been so vigorous; he was not that old. He had been the person who first took me far onto the sea ice of the North, the person who had shown me the astonishing pleasure of travelling by dogteam. He had made sure after that first trip to Pond Inlet that I always had a warm welcome in his home and a chance of going again with him onto the land.

When I returned to Pond, I went, as usual, to Anaviapik's house. Pitsiolak, his oldest son, was visiting. Pitsiolak hoped to train as an Anglican missionary. He was very committed to Christian teaching, but he had not lost his sense of humour or his very considerable skills as a hunter. As we chatted in his father's house, he told me that I should come and visit Inugu. Inugu? Yes, they had given his *atiq* to their new baby, who had been born just after Inugu's death.

Later that day, I went to Pitsiolak's house and saw the baby Inugu. He was asleep, in a basketlike crib, bundled up in a white blanket. I could see the tiny features of his face, crumpled in sleep. And I felt a surge of relief, a kind of peace, the end of a grief. Here was Inugu. The old man was back in Pond Inlet; at least his *atiq,* some essence of him, was there, growing, making sure that death had not brought a full stop to an important elder.

So did I believe in reincarnation? The Inuit *atiq* system does not force the issue. Nobody said that this was actually Inugu, just his name, and all that names can carry. I did not have to stretch belief too far. Yet my emotions believed, if that makes sense of some kind. My heart was lifted, my sadness gone. The life of the system, perhaps, as much as the life of that wonderful old man, had been sustained. If the essence of elders can be reborn, then the way of life they embodied cannot die.

13

Fatalism and impatience are recurring attitudes when intellectuals, politicians and journalists speak about the destiny of hunter-gatherers or, indeed, of any other indigenous societies. These people are bound to change, run the discouraging arguments, sure to disappear. Impatient fatalists say that whether or not aboriginal life was ideal, the present reality is that hunter-gatherers and other indigenous peoples are on a remorseless journey into Euro-American modernity—be this in the form of alcoholic, lumpen poverty or of farming and wage labour. Moreover, the fatalists continue, the forests or tundra or desert where these peoples once lived their simple lives are now part of a global economy, with demands for timber, pulp, pasture, minerals and space that originate in the developing, industrial centres. Resistance to these inevitabilities is therefore an irritating irrelevance. Instead of attempting to hold back the inevitable, those who care about the well-being of ancient and doomed societies should advocate the full participation of indigenous peoples in the modernisation process. This bundle of arguments endorses the "realist" position; the rival view is duly stigmatised as romanticism.

But what is real? And does the real have nothing to do with what is right? The transformation and destruction of ways of life involve many kinds of process. Some of these are internal: people are attracted to, negotiate with, and become dependent upon those who seek to change or even dispossess them. The relations between the colonists and the colonised are intimate and intricate; material relations, centred as they are on drastic inequality and elaborate ideologies of racial difference, yield connections that are also psychological and erotic. At the same time, despair and anger turn inwards. Just as the colonists can feel intense compassion and desire for those whom the colonial system seeks to overwhelm, so the colonised often turn their despair and rage against one another. These are features of colonial frontiers that Octave Mannoni and Frantz Fanon attempted to explain. Their work centres on Madagascar and North Africa, respectively, neither of whose frontiers involved hunter-gatherers. But all who know the history of Australia or the American West, or are familiar with the Boer displacement of the Bushman in southern Africa, are aware of the tangled interdependency that complicates judgements about who is responsible for what.

Yet the underlying story is clear enough. One kind of economy and culture overwhelms another. The realities of this, the pain and dismay to which it gives rise, and the attempts to find accommodations and alternatives—these are what anthropologists hear about in immense and painful detail. To report this, and to work with or for those in despair, is not to be romantic so much as to be in touch with the real.

Men and women in most hunter-gatherer communities relate histories, often within their own lifetimes, of extreme loss—of life as well as lands—of genocide and of environmental destruction. This is not the stuff of whimsical nostalgia, the implied image of Grandma and Grandpa in their rocking chairs, sitting on a verandah and yarning their regrets about the way the world is not what it was. To say that accounts of loss are nothing more than the drifting pseudo-memories of the elderly is tantamount to denial of the Holocaust. Personal experience on frontiers and the findings of many scholars endorse the voices of the elders. This growing body of data supports those who insist that in "the old days" and "long time ago" their people ate well, lived longer and took better care of one another. Hunter-gatherers throughout the world have indeed suffered terrible losses.

Where, then, is romanticism? In pretending that a particular kind of history has not taken place? Or in giving voice to those whose heritage and experience are part of this history? Perhaps the accusation of romanticism is levelled against the implication that the hunter-gatherer way of life was anything other than miserable. The agricultural, urban or industrial forms of modernity may be flawed, and their history may be strewn with victims, say the would-be realists, but these new ways offer comfort, long life, security and relative health—the very things to which all humans aspire. To suggest that earlier stages along the developmental path were somehow preferable is to disregard reality in favour of fantasy and dreams.

This line of reasoning draws on a faith in progress and posits more or less a priori that human history is composed of changes that are improvements. Yet to suggest that all change is for the better is to pretend that frontiers do not exist, or that they proceed in some benign and innocent way. This is a belief, not a discovery made by social science. To insist that all changes are some form of "development" does not oppose romance with realism. Faith in progress is itself a kind of religion.

Many anthropologists set out as young men and women, in the aftermath of long, confining educations, to live in societies very different from their own. They must make strong connections with one or several families. They find teachers and mentors, whom they later refer to in academic circles as "informants," suggesting espionage rather than social science. They learn a new language and many aspects of an exotic way of life. This is the complicated anthropological double-act of participating and observing. At the end of long days of walking or working, often in great heat or cold, all too often with an admixture of ineptitude and resulting danger, the exhausted anthropologist struggles to write notes, keep a journal, go over vocabulary lists and genealogical charts. Both the unfamiliarity of the setting and the project itself lead towards an intense effort to understand, and the route to understanding leads again and again to listening to what people have to say. The speakers are often elderly men and women whose lives testify to cultural riches, as well as to endeavours either to protect their way of life or to find acceptable accommodations with colonial invasion. As the anthropologist learns about the economy, beliefs and social structure of a people, she or he will also be hearing of terrible epidemics, the loss of land, the collapse of animal populations, the failure of fisheries, assaults upon women, violence, murder. Of course, people will also talk about changes that have not caused them anguish—new materials and technology, medical services, even the coming of Christianity.

The anthropologist lives with people who are, very often, generous and considerate, with one another as well as with the newcomer. Not always, not in all ways. But anthropologists who have worked in hunter-gatherer societies repeatedly celebrate the humour, gentleness and everyday equality they find there. To celebrate the qualities of a system, and to identify the many ways in which that system secures a successful relationship between people and their lands, as well as among the people themselves—this is to identify the real, not to perpetuate the romantic. Nor is it romanticism to express concern about a system's decline, to convey people's dismay about being dispossessed, to affirm their rights to keep their lands, languages and customs. On the contrary: to avoid these concerns, or to write about a people without expressing their achievements, priorities and fears, is misrepresentation.

At the same time, the apparent virtues of hunter-gatherers do not exist because of an impetus towards virtue itself. The qualities of interpersonal and social life that have persisted are those found to be efficient and successful in long-term material ways. This is a point well made by Colin Turnbull in *The Mountain People,* his account of the Ik, former hunter-gatherers of Uganda. Speaking of hunter-gatherer societies generally, Turnbull notes that they "frequently display those characteristics that we find so admirable in man: kindness, generosity, consideration, affection, honesty, hospitality, compassion, charity and others." He goes on to observe: "This sounds like a formidable list of virtues, and so it would be if they *were* virtues, but for the hunter they are not." They are, rather, "necessities for survival; without them [hunter-gatherer] society would collapse." Turnbull, of course, proceeds to describe a society that appears to have become antihuman. His account has become famous, touching in some way the deep sense (or fear) that inside each of us is an unashamed, antisocial, self-preoccupied monster. In fact, *The Mountain People* is a hard book to assess, if only because it strangely lacks the voice of the Ik themselves—there are no interviews with them, no statements from them about either their society's impending collapse or their lost hunter-gatherer history.

Nonetheless, the Ik may remind us that hunter-gatherers, like all other human beings, are capable of violence. Theirs are not pacifist societies. When explorers travelled into the Arctic, they used guides from subarctic communities. There are accounts of these guides falling on camps of other peoples, killing "ancient enemies." When the Gitxsan of northern British Columbia recount the oral histories that establish family rights to particular territories, they describe murderous encounters with neighbouring societies. When the Inuit were not able to get away from individuals causing dangerous levels of trouble, they were not slow to kill them, as in the case of two Catholic missionaries in 1914 and at least one Canadian trader in the 1920s.

The qualities of hunter-gatherers are functional as much as moral: their ways of living are essential both for the economic success of the system and for harmonious interpersonal life. Egalitarianism, respect for the elderly, loving regard for children, diligent respect for the land, plants and animals on which people depend—these are the "virtues," too often missing in the "developed" world, that cause visitors to hunter-gatherer societies to experience deep admiration. To describe these things, and seek to understand them, is not romanticism but the most relevant kind of realism.

14

To contest faith in development, and to oppose unexamined ideas of progress, is to raise central questions about history in general and about hunter-gatherers in particular. Elders in many indigenous societies are clear about the benefits of their ways of life. But these elders also belong to the present—they are not, as I have pointed out, men and women of some other era. Their argument is that the "traditional" system secured important benefits and could continue to do so. Change, they say, is for the most part a result of pressure and invasion rather than an expression of preference. Of course they want to be modern—but on their own terms.

What about changes that took place long before the modern colonial period? If hunter-gatherer populations are so well adjusted to resources, and the home territory is the ideal place to live, then how did hunter-gatherers spread so far and wide? This may appear contradictory. But even with a deep territorial conservatism, a way of life that relies on knowledge of a single territory, change is inevitable. Populations, resources and climates are not constant; each of these can alter to induce families or groups to move to new territories. But this move can only be successful if there is time to accumulate the knowledge that makes any territory productive. So movements of this kind are both reluctant and slow.

All territories have seasonal centres of hunting and gathering. But they also have remoter regions to which individuals or families will travel when things at the centres are not going well, or in search of visions and adventure. These outer areas can, of course, be settled, and become centres in turn. Those who move to them will then create remote areas even farther away from an original centre. If moves of this kind are to areas that turn out to be productive, then the people spread. If they turn out to be unproductive, people die out or return to the centre. This pulse of seasonal rounds, with changing edges and centres, may be very slow, taking hundreds or even thousands of years. But over a very long period of time, and given an underlying success of the hunter-gatherer way of life, the system leads to the populating of the world.

The hunter-gatherers whose circumstances and systems are best known are those at the geographical margins of the world. Bushmen of the Kalahari, Pygmies in equatorial Africa, Aborigines of the arid regions of Australia, Athabaskans of the subarctic boreal forest, and Inuit on tundra and Arctic shorelines: the success of these populations speaks to the genius of hunter-gatherers. But most of the world is not so harsh, and hunter-gatherer populations in other, more temperate places often must have increased with relative ease. Even with a widespread tendency to keep family size small and to stay within a particular territory, the very success of the system meant that it occupied much of the world.

Then came a revolution that transformed almost the entire surface of the earth: agriculture. Unlike the hesitant and conservative spread of hunter-gatherers, the change to agricultural life, from the palaeolithic or mesolithic to the neolithic (these terms all refer to the stones, the *lithos,* from which tools were made, and are as archaic in their way as the worlds they misrepresent), was rapid. Evidence of animal bones and seeds suggests that the first farming appeared in southern Greece in about 6000 B.C. Within 3,000 years, the remotest areas of northern Britain, from the Welsh hills to the Outer Hebrides, were being reshaped by the mixture of tillage and pasture.

This change is perhaps the most profound of all changes in the human story. Farming is based on the growing of crops and the domesticating of animals. As far as botanists can tell, cereals and pulses rich in protein became available in the environment towards the end of the last ice age—10,000 to 15,000 years ago. Within a short time of their botanical availability, human beings in the Near East seem to have begun to cultivate them, making fields and selecting the seeds that would secure ever more reliable harvests.

In effect, agriculture was a narrowing of the resource base, exchanging variety in diet and extent of territory for a concentrated and increased food supply. It must have involved intermediate forms of economy, with mixtures of farming and gathering, herding and hunting. Perhaps a number of societies developed different elements of this new system, making separate explorations of both genetic selection and techniques for working the land, step by step creating species of grain and animal that became the core of a new way of life. Over time, and in region after region, the new farmers transformed both the environment and themselves.

If farming was done well, its yield could supply all of a family's food for the year. But there was a price: the tilling of land is hard work, requiring many hours of arduous labour by as many hands as possible. "By the sweat of their brow . . ." Concentration of the work in one small area, and deep reliance on its yield, made farmers vulnerable—to weather, rival plants, animals that could destroy crops, theft of the produce by other human beings. Farmers had to be aggressive in defense of their way of life. We can see in this the curses of the exile from Eden.

Hunter-gatherers, with their high-protein diets, strong bodies, long lives and egalitarian systems, were overwhelmed by agriculture. If we discount any ideas that attribute to farming a set of intrinsic benefits, a sort of irresistible appeal, then there is a puzzle at the heart of this momentous and irreversible change in human fortunes. Some hunters chose to become farmers, of course. But for very many peoples of the world, perhaps the change was not so much a matter of seizing new opportunities as of having the land itself seized. The spread of agriculture may owe more to domination by immigrant farmers than to a quiet discovery of the benefits of farming.

15

The relationship between hunter-gatherers and farmers is suggested by a surprising line of inquiry: that of the history of languages.

In the 1780s, Sir William Jones, lawyer and oriental scholar, became a judge in the colonial court in India. The demands of the bench left him with both time and intellectual curiosity; he began a study of Sanskrit, the ancient and traditional language of Hindu culture and theology. Sanskrit was known to be the ancestor of many of the principal languages of the Indian subcontinent, including Hindi, Urdu, Sindhi, Nepali and Sinhalese. In 1786, as a result of his scholarly research, Sir William Jones realised that the grammar and vocabulary of Sanskrit bore a striking resemblance to that of both Greek and Latin. He also saw that Sanskrit had many similarities with

Persian. Jones's insights reached still further: he found strong links between Sanskrit and Celtic, Germanic and Slavonic languages.

The nature and importance of Sir William Jones's discovery is described by the Cambridge scholar Colin Renfrew in his remarkable book *Archaeology and Language*. Renfrew also relates how other linguists, through the course of the nineteenth century, explored links between Sanskrit and other seemingly unconnected languages. The combined list of languages showing strong connections and common ancestry came to include all Celtic languages (including Irish, Welsh, Gaelic, Breton and Cornish); all Germanic languages (comprising Danish, Icelandic, Swedish and Norwegian as well as Dutch, German and English); the languages of the Baltic states (Lithuanian, Latvian and Prussian); and the Slavonic languages (including Russian, Polish, Czech and Serbo-Croat).

Then, at the very end of the nineteenth century, a set of languages was discovered to have been spoken by a civilisation that had once flourished in remote but extensive parts of central Asia. This was given the name of Tocharian and was found to be a member of another large group of connected languages. All these different ways of speaking, spread across an immense area of Europe, Scandinavia, the Middle East, Asia and the Indian subcontinent, evidently had a common ancestor. At one time, long ago, they must have come from one people and one place. The family of languages came to be called Indo-European. But who were these people? And where was their land?

These questions have been given many rival answers. Colin Renfrew explains how the evidence of shared vocabulary has been brought to bear: if all Indo-European languages have the same words for some things, then those things might tell us where the speakers of the original language came from. Names of plants or animals could establish an environment where these species were to be found. New names for unfamiliar species might indicate movement away from the home by separated peoples. But no such list has yielded a convincing answer.

Australian archaeologist V. Gordon Childe was among the first and the most thorough scholars to explore the evidence from shared vocabulary. His work spans some forty years of study. But it is one of Childe's earlier works, first published in 1926, that puts together a list of sixty-six nouns common to eight widely scattered Indo-European languages, from Celtic to Sanskrit to Tocharian. These include god, dog, ox, sheep, goat, horse, pig, steer, cow, gelding, cattle, cheese, fat, butter, grain, bread, furrow and plough. Fifteen of Childe's list of nouns are suggestive of agriculture. Three others are metals (copper, gold and silver). And one, mead, is an alcoholic drink. Archaeologists agree that smelting of metal occurred well into the history of agriculture, and anthropological evidence suggests that agriculture or pastoralism is a precondition for the making of alcohol.

A difficulty for archaeologists has come with the attempt to rely on this linguistic evidence to establish a place of origin. Evidence of shared words for species of trees, for example, is inconclusive. And Colin Renfrew, whose review of these arguments is lucid and compelling, and who scrutinizes both the archaeological and the linguistic evidence, concludes that no location of the Indo-European homeland can, as yet, be deduced from the evidence of language.

But if the question is changed, the linguistic evidence may be more telling. Instead of asking where the starting point of the Indo-European story might be, we can speculate instead about the nature of the story. We can set aside the puzzle of where the people came from and consider, instead, how the original speakers of a single Indo-European language lived. What was their economic system? Childe's vocabulary list suggests the answer. They were farmers.

The archaeological record—built from the remains of plants, animals and implements—does suggest a place where farming began. The original crops appear to have been specific varieties of wheat and barley, in association with sheep, goat, cattle and pigs. Archaeologists can identify the pollen and seeds of these plants and trace their dispersal. On this evidence, the earliest agricultural sites, dating from about 10,000 years ago, have been found between the Mediterranean and the Caspian Sea in northern Iran and Iraq, northeast Syria and southern Turkey. These findings suggest that agriculture originated in these regions, perhaps moving from the first development of the crucial grains and techniques in Iran and Iraq. The archaeology shows that this movement was in all directions, with the strongest flows to the northwest, across all of Europe; southeast towards the Indian subcontinent; and to the east, into central Asia.

The story of this spread of agriculture correlates with the spread of Indo-European languages. So can the story of one also be that of the other? Colin Renfrew is worth quoting at some length:

> It is perfectly reasonable to view the coming of farming to Europe as a single process, albeit one with many phases, with pauses and sudden advances. For if we were to take the wheat sown in Orkney in the neolithic around 3000 B.C., and ask where each year's seed corn had itself been reaped, we would trace a line across the map of Europe that would inevitably lead us back to the Greek early neolithic, and from there back across to western Anatolia.

Renfrew gives a picture not only of the grains moving in these lines, across Europe and elsewhere, but also of the people who plant them. The dynamic of farming societies, with much greater population sizes than their hunter-gatherer contemporaries, is the key to this movement of economy and people. Renfrew argues that

> the new economy of farming allowed the population in each area to rise, over just a few centuries from perhaps 0.1 person per square kilometer to something like 5 or 10 per square kilometer ... [Thus] with only small, local movements of twenty or thirty kilometers, this would gradually result in the peopling of the whole of Europe by a farming population, the descendants of the first European farmers.

And he continues:

> If that was the case, we would expect that the language of those first farmers in Greece around 6500 B.C. would be carried across the whole of Europe. Of course it would change in the process ... divergences would emerge and dialects would form. Over a period of millennia, these would separate into distinct although cognate languages.

Renfrew's focus is the spread of Indo-European languages and farming in Europe. Greece, with the earliest neolithic sites in Europe, is therefore his point of departure. But the communities in Greece were emigrants or descendants from people farther east, who were the hypothetical first farmers and the speakers of an original Indo-European language.

This combination of economic and linguistic history is compelling. It makes sense of many kinds of data from archaeology, linguistics and anthropology. The model of a growing population with a recurrent demand for new farmland and pasture may reveal the process by which hunter-gatherers, despite the strengths and achievements of their forms of economy and social life, disappeared. The implication is that they did not adopt agriculture so much as lose to it. This loss may have been in the form of wars of conquest and resistance, with farmers driven by the need for land enforcing their demands with large armies and much aggression. Or hunter-gatherers may have sought to adjust to the newcomers, blending their discoveries with some elements of their own societies. But the domination of one language family does not encourage a picture of hunter-gatherers who adopt a new economy, learning from others and gradually changing their way of life. People do not give up their way of speaking unless they are overwhelmed, by numbers or some related imperial processes.

16

Some of the ancient languages of Europe did not disappear. Basque, Estonian, Caucasian and Finno-Ugrian languages, including Hungarian and the Saami languages of northern Scandinavia and Finland—none of these have Indo-European origins. They belong to several linguistic families, and they are as different from one another as each is from any Indo-European language. But the speakers of these non–Indo-European languages are all farmers or, in the case of Saami or Lapps, pastoralists. These may well be the descendants of communities of hunter-gatherers the agriculturalist newcomers did not overwhelm. They are a relatively small minority of the peoples of Europe, but their story may well have a different and less terrible quality than that of the majority of peoples who, 10,000 years ago, were speaking what must have been a great array of hunter-gatherer languages.

The fate of many hunter-gatherers may well have been "terrible" if only because farming makes little, if any, allowance for rival ways of using the land. There are no records that yield any insights into what took place. But there are many reasons for supposing that it was ruthless, and few for thinking it might have been benign. Farmers displace hunter-gatherers bit by bit, region by region; given the numbers of the farmers and their intense dependence on very specific pieces of land, resistance was most likely to have been overwhelmed, perhaps with great violence. There are striking links between agriculture and warfare: farmers compete with one another as well as with hunter-gatherers, and their skills were brought to bear on the development of weapons as well as farm implements. From the very beginning, they must have learned how to beat their ploughshares into swords.

All this suggests that the conquest of hunter-gatherers by farmers in the Old World may not have been all that different from the advance of European agricultural frontiers in the New World. The European discovery of Australia, North America and southern Africa was, for the most part, an encounter between agriculturalists and hunter-gatherers. The spread of European settlers across these new lands is the story of a devastating displacement, destruction and transformation of hunter-gatherers by farmers. The majority of settlers have been speakers of Indo-European languages. And in every region where this conquest has taken place, the farming newcomers have urged or forced the indigenous peoples whose lands they have taken to give up their own languages in favour of English or French or Russian or Hindi. Much colonisation, therefore, can be seen as the most recent extension of the triumphant progress of those who speak Indo-European languages into the lands of those who speak, or once spoke, hunter-gatherer languages.

In Australia and North America, this process is well known. An analogous process has also taken place in southern Africa, where the spread of Indo-European–speaking peoples was preceded by the movement of farmers who spoke another agricultural and pastoralist set of languages. The neolithic revolution was not the exclusive achievement of the ancestors of Indo-Europeans. Rather, farming may well have had several different, though more or less contemporary, starting points. The spread of Polynesian and Chinese languages could well be linked to a very early neolithic people whose agricultural skills and population increases gave them the momentum and ability to occupy and transform the Far East and Southeast Asia. Another such group, originating somewhere in what is now Cameroon, appears to have made the same kind of colonising journey into southern Africa. In each case, the impact of these peoples on hunter-gatherers was probably much the same—dispossession, displacement, destruction or absorption. Systems were overwhelmed; languages lost.

17

All human groups incline to the view that they lead the best possible way of life. From this follows a more or less casual assumption that those pursuing any other way of life would, if they could, change from their inferior to our superior society, religion, morality, economy and forms of knowledge. Ethnocentrism of this kind appears to be the norm in judgements people make about one another. But these judgements have received a rationale and purpose from within the needs of agriculture: insofar as farmers systematically require additional lands, and may also gain immense benefits from cheap labour or a supply of slaves, the supposed superiority of their ideas and practices has been a justification for dispossessing and subduing others. Farmers have no difficulty answering questions about history: hunter-gatherers become farmers because farming is better in every way. And what is better? There is more food, greater security, longer life, less brutality in both everyday dealings among people and between societies; there is more order, with systems of management and law. These kinds of claims, when listed in this way, show themselves to be a mixture of truth and fiction, an amalgam of the conceit and realities of the "modern" when opposed to the "ancient." Yet sermons, schoolbooks, evolutionist tracts, World Bank directives and the conditions of International Monetary Fund loans all contain this amalgam. They answer questions about the ascendancy of "our" way of life by explicit insistence that those without the benefits of agricultural civilisation are at immense material and social disadvantage, and by implicit suggestion that there is a civilised condition to which all should aspire. Faith in agriculture is somehow beyond question, a faith about faith itself.

18

The spread of farming through the world has meant the occupation and eventual destruction of most of the world's hunter-gatherer societies. Yet there is one characteristic of farming peoples that has caused more death to others than even their genius for warfare. They have created and spread disease.

Agriculturalists stored large quantities of grain, and lived in concentrated and permanent communities. The vectors of infectious diseases, rats and mice and fleas and bugs, therefore reached populations of unprecedented density. Farm families lived all year round on one piece of land. They had large numbers of children. And they shared crowded houses with domesticated animals, in which human and animal waste was sure to accumulate. In these conditions, typical of almost all farming, microbes flourished as never before. Moreover, they crossed from the animals to the people.

In *Guns, Germs and Steel,* Jared Diamond gives this list of illnesses and their probable species of origin:

Measles:	cattle
Tuberculosis:	cattle
Smallpox:	cattle or other livestock
Flu:	pigs and ducks
Pertussis:	pigs and dogs
Falciparium malaria:	birds (chickens and ducks?)

Infectious microbes of these diseases, finding their way into their agriculturalist human hosts, became endemic. But enough humans developed antibodies to ensure that these diseases were not so deadly as to eliminate all the humans in whom they thrived. Thus there was an adaptive balance between the causes of disease and the development of resistance to the diseases themselves.

Hunter-gatherers had never developed these illnesses. Nor, therefore, did they develop any resistance to them. When farmers met hunters, there was a likelihood that the hunters would be infected with diseases that could devastate them. Thus the exiles from Eden, the cursed roamers across the world, the people who went forth and multiplied, were also the vehicles for another curse. The microbes that came with the conquerors and settlers to the Americas killed far more indigenous peoples than did European weapons. No doubt much the same thing had happened in the course of other, earlier encounters, as farmers displaced or absorbed hunters in almost every other part of the world. Farmers brought this new curse, invisible to all, in their bloodstreams and breath.

In many known cases, and probably throughout the neolithic triumph over the lands of hunters, infectious disease achieved what aggression and force of arms might otherwise have found more difficult. Contact meant catching flu, measles, chickenpox and many other diseases—and dying from them. In some parts of North America, 75 per cent of the indigenous population was wiped out by newcomers' illnesses. In all areas, weakness and fear remained for those who survived.

19

Supporters of the colonial process have cited the apparent "nomadism" of native populations to justify advances of the settlement frontier. They have made much of the fact that hunter-gatherers lack year-round permanent settlements. They insist that these are peoples without the institutional life of the village; they equate a relative indifference to possessions and an absence of manmade monuments with a low level of human evolution. Colonial occupation of tribal lands has also relied on a broad theory of manifest destiny which says that the taking of "nomadic" peoples' lands by civilised farmers is ordained by fate or God. This doctrine is at the heart of American history. It was most clearly articulated in the 1840s by a politician and journalist called John L. O'Sullivan, who declared that settlers had the right "to spread over and to possess the whole of the continent which Providence has given us.... It is a right such as that of the tree to the space of air and the earth suitable for the full expansions of its principles and destiny of growth." This theory underpinned American wars against Mexico, but it also justified the dispossession of all indigenous populations. A corollary of the idea of manifest destiny was the view that Indians existed "as beasts of the field," roaming around without any apparent form of society.

Both British and American courts have at times made the existence of organised society a test of aboriginal claims to ownership of their lands. Hunter-gatherers have duly been failed, on the grounds that they lack the minimally necessary social institutions. The accusation of nomadism has been a repeated, if rather general, way in which colonists express this failure of others to achieve their self-styled levels of society. Setting aside for the time being the grotesque charge implicit in the organised-society test—that these are peoples who lack what it means to be human—the proposition that nomadism is evidence of lack of rights to land is a great and terrible paradox of the agriculturalists' judgement upon others. Great because it speaks to the entire range of frontiers that displace peoples whose territories are wanted for new farms; terrible because this process gives rise to, and then relies upon, a racism that is relentless and purposeful.

The profound dichotomy that has shaped the agricultural era may indeed therefore lie in an opposition between nomads and settlers, between people for whom home is a place of timeless constancy, a centre in which humanity itself arose, and those who are on the move and, if at rest, rest only while preparing for further movement. The paradox, of course, is that this is a divide between the settled hunters and the nomadic farmers.

20

There was both a modernity and a timelessness to Jimmy Field. Of course he relied on his gun, but he lived by tracks and trails and snares. He moved in his lands, rather than waiting for animals to come to him. He did not intercept the creatures of his world but followed them, sharing their landscapes, knowing where they would be. Trusting his dreams. He liked to go to town and drink a few beers. He wore a cowboy hat and spoke everyday English. But he was an Indian and lived in an Athabaskan language. He was the toughest, gentlest, most supportive and humorous of human beings. He shared everything and had no claims to any status. He went his own way and cared for others. The more I knew him, the more I felt his quiet authority and his genius as a hunter.

White people of the new frontiers often gossip about men and women they have known from the aboriginal world. There is a certain pride, a sort of special claim to be conversant with "the natives" that comes from saying with absurd confidence that this or that man is "the best hunter of these parts." This idiom, which sets the hunters so far away by claiming an implausible intimacy and making impossible comparisons, had always filled me with unease. Yet I did think, as I came to know Jimmy Field, that here indeed was "the best hunter." Better to say, perhaps, that he was the most complete hunter it is possible to know; for he was and claimed to be nothing else. He was part of a family and a community. Yet of all that family and community, his skills were somehow consummate, and his quiet excellence was both inspiring and moving.

In what time did Jimmy Field's life take place? He was my contemporary; there was no evolutionary divide between us. We lived at the same point in history; yet our antecedents, our sources, reached back into different kinds of time and different ways of understanding time itself. He was, in many ways, my elder. He carried in his mind the skills and understandings of a way of life; I did my best to learn from him. He belonged to that land, those territories, as those territories belonged to him. I was a visitor, an anthropologist, a child needing to learn, neither belonging to nor owning any definite place. He was settled there, and I was the nomad. The places he took me into were places where Athabaskans, or people very like them, had lived for at least 8,000 years. Jimmy would say that they had lived there through all of human time. Yet we travelled together in a modern world.

Jimmy Field laughed at the old and the new, with different kinds of humour. The old Indian ways—"long time ago," as he put it—were real to him and full of truth. He laughed about them, with bewilderment and awe as much as to tease. But they were indeed old ways. And the new frontier, the real cowboys with their ranches and quarter horses, the logging trucks, the cutlines of the oil and gas industry, the pipelines—these also made him laugh, but with unease and a little bitterness. He knew what was at stake. He had no illusions, because he was in both kinds of time: that of which his people were made, as hunters, as gatherers, as Dunne-za, and that in which they now endured, in a small and uncertain part of the frontier, as "Indians."

After I left the northeast corner of British Columbia, I lost touch with Jimmy. Occasional pieces of news reached me, and I managed to send a few letters to his brother Abalie. I wrote care of Arlene Laboucane, a friend who had worked for the Union of British Columbia Indian Chiefs and had coordinated the organization's work in the region. Then, in the spring of 1998, Arlene wrote to tell me there was bad news.

Jimmy had gone into Fort St. John for a medical appointment. He had also done some grocery shopping. But he had never arrived home. His family had gone around Fort St. John asking about him. Had he returned to town for some reason? Had he gone somewhere else? But the little they heard confirmed their fears: he had gone to the highway and not come back.

More time went by. More searches were made, more questions asked. Messages were put out on the local radio station. Nothing was heard, and Jimmy seemed to have disappeared into nowhere.

Then a young man walked into the Fort St. John police station. He said that he and two friends had come out of a house in Fort St. John, where they had been partying, and had had difficulty starting their car. Jimmy was out there. One of the boys walked across the street to talk to him and offered him a ride to a local bar. Jimmy got into the back of the pickup, and the three boys got into the front. But they didn't drive to the bar; instead, they drove across town and to a field off the road. They drove round and round, trying to bounce Jimmy out of the truck. He clung on. They stopped the pickup, got out and battered Jimmy into unconsciousness. Then they drove to a bridge over a creek and threw him into the water.

The police went to the river and searched. At the side of the road, by the bridge, they found a faded blue denim jacket. It was Jimmy's. But his body was gone, downstream, lost somewhere under the winter ice. In spring, when the snow melted and the rivers flooded, Jimmy would no doubt be swept away, towards the Peace River, on currents that carried the logs and rocks and earth of the interior towards Great Slave Lake and, beyond, towards the Arctic Ocean. No one expected to find him.

The police questioned the other boys, who confessed. Two young men were charged with second-degree murder. Both pleaded guilty to manslaughter. The one deemed "most culpable" received a sentence of five years in jail; the others got three years and eighteen months. The crime was familiar, so much a part of the frontier, the long story of white men disposing of Indians. The leniency of the sentence is of a piece with this story.

Let us put to one side the quality of laws and the mind-set of those who somehow shrug at this murder. Instead, imagine, albeit without giving those boys a chance to have their say, the events at the side of the road. And think of those four minds. The one moving in a country that was his, of which he knew every corner, that he could explain and explore and use and delight in. The other three looking out at "wilderness," with a tough indifference or a resolve, somewhere and somehow, to get hold of some part of this place and transform it from nothing into something, from mere land into money, from frontier into a ranch or an oil well or something, just something. Think of the one mind filled with knowledge of place and the others empty of any such knowledge, filled instead with the anger and determination and craziness of youth, reinforced by restlessness and nomadism and technologies from elsewhere.

This meeting of minds, this beating and murder, could have taken place—has taken place—on all frontiers between settlers and hunter-gatherers. In this sense it is an event of any time, from the beginning of farming to its continuation today. This murder belongs in any of a hundred centuries, on any of the world's continents, in any of thousands of hunter-gatherer cultures.

This time, in 1998, in this place, at the edge of Fort St. John on the Alaska Highway, it was Jimmy Field, Dunne-za. He was a man who travelled time in dreams and moved on his lands with awe-inspiring skills. He was a great man among his people. His death was the more shocking, the more frightening, for that. But the pain of his death contained all such deaths, in all places, in all times. The bodies of hunter-gatherers continue to be swept under the ice, frozen, dead, towards the oceans.

FOUR: **WORDS**

1

In the summer of 1987, I drove from Hazelton, in the Skeena Valley of British Columbia, to Aiyansh, headquarters of the modern Nisga'a tribal government, in the Nass Valley, a distance of about 300 miles through pine, fir and cedar forests. The Skeena and the Nass are large rivers, each of which drains a large area in the northern part of the province. Their sources are close together on the western slopes of the Rockies; they then separate, shaping two immense landscapes on their way to the North Pacific Coast.

Many of the tributaries of the Skeena and the Nass are themselves powerful enough to have cut wild valleys in the mountains. The trees, where they have not been felled—shaved might be a better word—into vast open scrubland and plantations of replacement saplings, spread up the mountainsides in a continuous mantle of dark greens. The road passes for all of its length through this country, crossing rivers, following the shores of long, deep lakes, and, for the last fifty miles or so, becoming a mud and gravel surface that goes through a mixture of industrial forestry and the remaining stands of old trees.

The scale and drama of the land adds to the reputation of the Nisga'a. However much it has been savaged by logging, this terrain is vast, unnerving. I stopped for a rest near the Tseax River and walked a short distance off the road. Fir and cedars towered a hundred feet above a floor of roots and mosses. The forest sprawled across the valley, then climbed steep, high slopes. The great straight trees yielded to short, slim evergreens that had rooted in crevices of sheer rock. Above these, where the heights of the mountains level, I could see clumps of gnarled pines and the strange beauty of alpine meadows. Patches of snow lay on high ridges and covered the peaks. It was late summer in the valleys between the Rocky Mountains and the Pacific Ocean. I wondered what it would be like to fish in these rivers, to hunt for elk or black bears in this forest, to climb those peaks to search for mountain goat.

I was going to Aiyansh to talk to tribal leaders about making a documentary film for the television series *As Long as the Rivers Flow.* My producers, James Cullingham and Peter Raymont, had been in touch with Nisga'a leaders; an appointment had been made for me to explain the project. I was scheduled to meet with Alvin McKay and Rod Robinson, president and vice-president of the Nisga'a Tribal Council, both also hereditary chiefs. The nearer I came to Aiyansh, the more apprehensive I became.

Like other societies of the North Pacific Coast, the Nisga'a have celebrated and institutionalised the power of the word. The young are trained in public speaking. In the potlatch—the famous Northwest Coast event at which inheritance, territory, names and disputes are adjudicated—to be able to talk with authority is part of *having* authority. To speak with power in public is a central part of being a Nisga'a chief.

I anticipated a difficult negotiation. Why would the Nisga'a want this film? What benefit would it bring to the people? What control would they have over what was in the film and how it was distributed? How many Nisga'a would we employ?

As I waited in the tribal council offices, I was overwhelmed by shyness, an uneasy sense that far too many white strangers had no doubt already made their appearance in the village. At last Alvin and Rod were free to talk with me, and I was shown into Alvin's office. They were imposing men, relaxed but a little formal, conscious of the positions they held as well as their importance as statesmen within Nisga'a society. With the skill of statesmen, they left it to me to explain myself. I talked a little about the series, explaining that a film about Nisga'a history and culture would speak to the longest struggle for recognition of aboriginal rights within Canada. Alvin and Rod listened. They did not interrupt, and they did not ask the difficult questions I had anticipated. I finished what I had to say. There was a pause. Then Rod said: "When do you want to begin?" I hesitated, wondering if I should first answer questions that had not been asked. For a moment I was at a loss. Was this absence of interrogation a quiet withdrawal, a way of saying no? Or did it signal a form of go-ahead? My mind rushed around the possibilities. But Alvin and Rod's graciousness was unfailing. They smiled and sat back comfortably in their chairs, making me feel at ease enough to ask about eulachon.

Eulachon are sardine-like creatures of no more than six inches in length, so rich in fat that they were known to Europeans as candlefish—apparently they could be dried, stood on end and lit like candles. The Nisga'a caught them in the tens of thousands, piled them in large open containers, and allowed them to soften and begin to rot. Then they carried the matured eulachon to large clay-lined vats set over wood fires, where the fish were boiled until the grease in their bodies rose to the surface in a thick, oleaginous layer. The rendered oil was scooped off and set out to cool. It was used as an all-purpose sauce, as a medicine and even, I was to learn in due course, as a mosquito repellent. Nothing in the economy of the region has had the exchange value of grease—as the Nisga's also call it—and it has always been among the most important of all gifts to be handed over at feasts or potlatches. In acts of grand extravagance, it was also spilled on the fire in the feast hall to create an explosive burst of flame that represented the great and generous wealth of the hosts.

Almost all this grease came from the Nisga'a eulachon harvest at Fishery Bay, close to the mouth of the Nass River. I knew that its production continued to be a source of immense pride to the Nisga'a, and that the grease was still of trade value between them and neighbouring groups. And I knew Fishery Bay to be a place of extraordinary beauty. So I suggested that the harvest there could be a good starting point for the film. They listened, did not demur, and left me wondering again what they thought. With some hesitancy I asked: "If we did begin with the eulachon fishing, when would we have to be here to do the filming?"

Rod replied that in the old days we would have filmed at Fishery Bay in early summer, since the eulachon and the spring salmon came up the river at the same time, in May and June. He went on to say, in the same tone of voice, with the same directness, speaking of matters of fact with due matter-of-factness, that things had changed. I did not write down his answer, but I summarised it in my notebook after leaving the office. It went like this:

"The eulachon were waiting out in the bay, at the mouth of the Nass River, waiting for the right time to go up the river. They did this every year, gathering together there, lots and lots of them, in the sea water. Now the people along the river were hungry. It was the end of winter, and they were short of fresh food. There were not fish in the river yet, the ice was still there, and it was a hard time for them. So Xemsen, who was like a creator for the Nisga'a people, and who had made the Nass River, made the sun send bright sunbeams down the river, towards the sea where the eulachon were waiting.

"Seeing this bright light, the eulachon began to wonder if they had got the time wrong. They began to wonder if it was getting late to go up the river. They began talking to one another, saying, 'Hey, maybe we should be going up there.' Then Xemsen got a seagull to pick up some herring and fly up from the river carrying it in his mouth, as if it was eulachon. So the eulachon waiting in the sea got nervous. Some of them started a rumour: we've been left behind; many eulachon have already gone up the river to feed the people. And the rumour went all over the bay.

"The eulachon got more and more worried. They began to dash around and splash on the surface, thinking that they had made a mistake, and were too late, and had been left behind. After a while, the worry and the rumours were too much for them, so they all rushed up the river. The Nisga'a along the river saw them come, and so the eulachon were able to feed the people. So the Nisga'a were okay, and their hard times were over for that year. Then, when they had finished all they had to do in the river, and had fed all the people, the eulachon turned round and headed back to the sea.

"When they came to the estuary, back to the salt water, they met the spring salmon. These salmon had now come in from the ocean and were gathering together before beginning their journey up the Nass. When they saw the eulachon heading towards them, coming from the river, they asked them what they had been doing. Where were they coming from? Why were they not waiting out in the bay, along with the salmon? Those eulachon were pretty pleased with themselves, and proud of the way they had been able to feed the people.

"But this annoyed the spring salmon. They laughed at the eulachon. They told the eulachon that they weren't much in the way of food, that they were only little things. 'Look at us,' they said. 'We spring salmon, one of our heads gives more food to the Nisga'a than a whole lot of you put together.' The eulachon didn't like these jokes and got kind of angry. In the end they said that they would never travel with the spring salmon again—from now on they would always go up the river early, long before the salmon arrived in the bay.

"And that's what they do, arriving here pretty early in the year. So you had better begin filming in March."

2

Imagine a river that flows hard through the Nass Valley, under a canopy of cedar and pine. Its bed and banks are black and green, granite and moss, with beaches of dark gravel. The water turns and breaks around rocks and spills over stone. Floods have carried dead and broken trees from far inland. There is lightness and darkness, a viscous quality that now reveals and now obscures whatever is beneath the surface. The late-September air is dank, heavy with mist and the scent of the forest. A fecund time.

The Nisga'a artist George Gosnell had brought me to this riverbank to watch coho, the species of salmon that abounds—or used to abound—in every creek and river along the Pacific Coast. Alaskans call them "silvers," for when they first appear in the river mouths in late summer and early autumn they have an extraordinary brightness. Nisga'a and Gitxsan call them *u'ux*. The Haida call them *taay.* The Nisga'a and Gitxsan are part of what anthropologists have identified as the Tsimshian culture area, speaking close dialects of the same language and living on a similar mix of salmon fishing, hunting, and gathering of abundant berries and roots. Their cultures are famous, along with those of the Haida, Tlingit and Kwakwalla, for the elaborate designs and immense power of their art; this is the land of totem poles, carved boxes and jewellery with intricate symmetry and remarkable aesthetic power.

George was a gentle, thoughtful man, and a good carver. He had been helping us with the film, guiding us to the best individuals for each part of the work. One of his brothers, James Gosnell, had been a powerful and inspirational Nisga'a political activist in the 1960s. Another brother, Joe Gosnell, was the Nisga'a chief and main spokesperson during the final stages of the landmark 1999 Nisga'a Land Settlement Agreement. But George was more philosophical than political, not in the sense of being indifferent to questions of tribal rights, but by virtue of taking a quiet and introspective approach to things. His art, which is celebratory of Nisga'a culture and themes, was where he spoke to and for his people's rights.

At the moment, we were working on a sequence with some of the boys who liked to fish for coho in the Tseax River, at a point where it flowed not far from George's house. I asked him about the Nisga'a words for different phases in the coho's life. What was a parr?

A smolt? And were these the same as the words for chinook or steelhead parr or smolt? George said he did not know; we would have to ask some of the elders. But even they might have difficulty. So many had spent their lives using English.

We had been hearing, from almost every Nisga'a we talked to, about loss of language, and about the schools that many said had caused it. I found myself thinking about the grief that comes with loss of words. What must it mean not to be able to put names in your own language to the things that you care most about?

I had been meeting with George every day to talk about what we might film, or how we should be covering aspects of Nisga'a history. He had a clear and often inspiring sense of what the film might say and look like. He made sure that we went to the places and people most expressive of Nisga'a culture.

One morning I asked George to suggest whom we might interview about being forced to leave the Nass Valley to get an education. George thought for a moment. He had been to one of the residential schools, he said. It had been a terrible experience. He had been so small when he went there ...

George began to tell his own story in a quiet voice, and I realised that his was the interview we needed. Fine, George said. He paused while the crew got the equipment in place, then he began again.

"I was born in Canyon City in 1944, and my parents were in the fishing business. My dad was a fisherman and my mother was a net mender—a net woman it was called in them days; it's still called that today.

"In 1952 I had word, they had word, that they were going to send me to Lytton Indian Residential School. In 1952. And I only heard then what the residential school was; but I never knew exactly what to expect. It was nice, the train ride that accompanied the getting from Sunnyside Cannery to the town of Lytton, which is in the Fraser Canyon. I got on the train sometime in September. The train ride, at that time, was three days long. It was very, very ... how would I say it now? Sort of an adventurous type of trip on the train that time. I really enjoyed the trip.

"And we were out in Lytton, that was the town of Lytton. The residential school was about three and a half, four miles from the town of Lytton. This was the St. George's Residential School.

"When we got to Lytton, this is the ... one of the frightening parts of the trip that are coming in, settling in on me. I already was missing my parents. See, that's three days away. And the stay at the residential school in the olden days was one year. You're away from your parents for one year. Your sisters, your brothers, uncles, grandfathers. Very difficult time.

"We got off the train at Lytton, and I didn't know how they were going to transport us from the train station to the residential school. At that time I hardly knew anything about vehicles or a bus. So we all stood there. It must have been about thirty-five, forty students at the train station at that time. So a few minutes went by. Then there was this truck, there was this dump truck that came around the corner. Everybody looked at each other. It was a dump truck that they used, it's an ordinary dump truck where they put us all. We all stood on the back of the dump truck. And remember, this is about 1:30, maybe 2:00 in the morning. It was a very chilly morning, that morning, and we weren't dressed to stand up on the back of a dump truck. That was our transportation from the town of Lytton to the Indian residential school known as St. George's."

George paused. He had told all this in a flat tone, putting together each bit of information, giving the bare bones of the story. He did not have to remind us that Lytton is in the interior of British Columbia, in an environment very different from the coastal forests of his home. And we could imagine that group of children, standing in the dark, cut by the cold. He waited for me to ask a question. I asked George what he saw when he first arrived at the school buildings.

"We were all standing in the back of this dump truck, for maybe about ten, fifteen minutes—I guess it took about that on the trip to the actual school. We arrived there and there was only about three or four lights that were in front of the school at that time. I don't know how to, how to ... I could explain that it sort of felt like, I guess you could say that you were going to go inside like a prison, like, that's what the feeling of it, the frightening feeling, come into you.

"We were there and we were introduced to the principal at that time. We all walked into the school, the waiting area they called it at that time. You could, I could feel the sadness coming in. I could feel the sadness coming into me, 'cause I knew I was going to be confined in this place for one year. The smell of the school was nothing like I ever smelled before. It was the sort of heavy kind of medicine type of smell in there. And the time now is about 2:00, maybe 2:30. This is the time. And we were advised to follow this one lady up to number one dormitory.

"That's where I was put, in number one dormitory. There was about three other, three other kids with me that time. The other students went in number two or number three or number four dormitories. Our dormitory was right on top.

"I walked into the dormitory and there was about forty, forty children laying on nice neat little rows, beds there, maybe forty, maybe fifty students in the dormitory. Dormitory number one. And nobody got up that time, but we were told where our beds were, and I didn't know what to do. I didn't take my clothes off, 'cause there were so many children about there."

George paused again. He looked through the window, to the Aiyansh street, to the trees of the North Pacific forest. I asked him if he was okay to go on talking. He nodded. I asked him what it was like the next morning.

"The next morning, it was about, oh, I'd say about 7:00, 7:30 maybe, or 6:30. I forgot the exact time. Maybe 6:30. I hardly slept that night, maybe not even an hour's sleep. I was too frightened. And everybody started getting up.

"A lady comes through the door and hollered at everybody. Told everybody to get up, wash up and get ready for breakfast. I was in there; I looked around; like I said, there must be forty, fifty children in that dormitory, dormitory number one.

"The first thing I noticed right away was how they looked. They had a very strange look in their faces. Their hair was completely cut right off. They had very short haircuts and their eyes didn't look, they didn't look very happy at all—very sad-looking eyes. Very frightening thing to see. First time I've ever seen anything like that in my life.

"And then we went down to what they call the dining room after we, after we washed up. Again, this is a very, very—I would use—a strange place. I've never been in a dining room before where there was so many children. It had big long tables. It must be twenty to thirty students there in one table. And it had one guy that's on the end that passes the food. He is the table—oh, the guy that looks after the table, anyway.

"How the food came in is, it came in big, big buckets, maybe big pots, that was passed on to that older guy that sat on the end. He was the one who served the breakfast at that time. It was mostly, what do you call it? Oats, or mush, or whatever you call it. Small little bowl, two slices of bread and one glass of milk, and that was our breakfast. It didn't take me long, that morning, I got really hungry. I'd got used to eating all the time. You can't do that in the residential school. You only eat at a certain time.

"So after that, after the breakfast, we went out. The main school was here, and where we had our classroom was a little bit outside, about fifty feet away from the main school. We were introduced. We were known as the northern group of native children that was there. The people that came from the southern part were called the southern group. So this is how it went all the way, all the way through the year, we were known as the northerners.

"And I didn't know how to speak English at that time. I knew very little English language. And I tried to use our own Nisga'a language. That's when I found out the harsh realities of being confined in a residential school. I didn't expect to get strapped that time, but I did. I went to the principal's office and I got strapped for using our own language. Strapped once on each hand. The second time they catch you speaking your language, you are strapped twice on each hand.

"All through the year I was caught using my language four times. So the strap was a very big—oh, something like a leather belt, very wide and heavy. And when the principal strapped you he didn't do it in a gentle way. He did it in a very harsh way. He talked to you very sternly. Push you around. 'You are not supposed to use your language in this school,' was his words. 'You did not come here to learn your own language, the native language. You are now here, inside these school walls, to learn education as we see it,' is how he said it.

"And my heart was beating very fast at that time. I didn't understand what the meaning of his words were. But a few minutes passed after I got strapped. I didn't want to go back to the classroom. I sat in the hallway until after dinner. I sat there for about two hours. After I got strapped I didn't want to go back to the classroom. I didn't want anybody to see me cry. So they—sometimes they'd send you into this little room where they, I guess, where you could go and cry, I guess. Very quiet little room, small little room where they sent me, and I don't know if they took the other students there. I would imagine they did.

"That's when you get strapped by a strange person in strange surroundings. It's very difficult to try and grasp what the, what they really wanted you to do. I remember my parents. I wasn't the only one that got the strap. There was quite a few children, both northern group and southern groups of children got the strap for using their own language.

"I don't know why the residential school. I don't know why they had such far distant places for education. To get torn apart from, from your parents and your brothers and your sisters to educate us.

"They educated us all right. They made us forget our own

language. God gave us our own language. It was taken away; they were trying to take our language away."

George turned again to look out of the window. I asked him why they wanted to take the language away.

"I don't know; maybe to try to forget our past. We have a very colourful past. But the language is very important to express yourself, to be able to talk about your grandparents, the chiefs, the important things that happened in the past. The Nisga'a language is very important to talk about—you cannot really, cannot use the English language 'cause meanings disappear in English language. Not like when we use our own Nisga'a language; meanings come out crystal clear when you are speaking."

I asked him what it was like when he came back from the school and lived here again.

"Christmas was a very hard time for the residential-school students. Especially the northern students, 'cause the southern students were allowed to go home. But we had to stay in the school during the Christmas holiday. Very sad time, Christmastime, in the residential school."

George stopped. He was clenching his jaws, close to tears. He looked out the window, then turned back towards us.

"And then in the summer, after the school was finally finished with our one year, spending our time in the residential school was over, we came back on the train. Again the trip took three days. I got off the train. I looked in my mother's face. And I used English ..."

He stopped again, struggling with tears, trying to keep the story going.

"She asked me why I used the English. I told her that's what we went away for.

"I forgot the Nisga'a language that time. It took me many years to use it. Through the help of my father, and my mother, I relearned the Nisga'a language.

"See, we were brought up in the residential school to use the

English language, and they said you cannot use your own Nisga'a language. And when we finally got back home, my mom in the Nass Valley, where we were using the English language, said, 'That's not your own language.' She said, 'Use your own language.' "

They had taught George English—though what this English was, in the mouth of an eight-year-old who had been in that school for just one year, is not easy to imagine. He had suffered separation from his mother, the loss of his father. This was about the loss of language as a loss of family, of home, of culture.

It was upsetting, terrible to hear, and yet, in that brutal way of filmmaking, I was delighted. We had, as they say, got it. And George managed to end all that he recalled on a joyful note. He spoke of his children learning to speak the language. They came home from the new Nisga'a school in Aiyansh able to speak more fluently than their father. "I could only count in Nisga'a to twenty," he said, "but my daughter, she comes home, and she speaks to me in Nisga'a, and she can go way past my little twenty." He glowed with pride.

During the interview, the sound recordist had signalled to me that there was a problem; I had noticed her looking around the room in a desperate attempt to see the source of some noise. But I wasn't going to let a technical difficulty interrupt what George was saying. As soon as the interview was over, the sound recordist handed me her headphones and played back the last part. I could hear a faint thudding noise in the background. As he had talked about the end of that first year at school, his return home, and his mother's dismay at his speaking English, George's heart had begun to pound.

3

It is possible to travel through the vast forests of the Pacific Northwest, look up at the wild beauty of the coastal mountains, stare into the clear, fast waters of all those rivers, and hear a kind of silence. This is not the silence of wild and empty wilderness, of remote mountains—these are extremes of geography, nature without culture.

There is, rather, a silence that marks the loss of the words that give this place—like many such places—its fullest and richest expression. The loss of the people's own names for their hills, rivers, lakes, bays, peaks, slopes, islands, trails; and of their ways of evoking the origins, significance, humour and poignancy of the landscape. The loss of those meanings, as George Gosnell put it, that "come out crystal clear when you are speaking."

Grief about what took place in residential schools is familiar to all who spend time in Indian communities in modern Canada. There is also incredulity: how could "the government" have done this? And often a simple statement: they took our language away. The silence of wordlessness is a silencing of who people have been, of where they live, and therefore of who they are. It is a silence between parents and children, making it impossible for the experience and knowledge of the past to be carried into the future. The ability to name is hard to separate from the right to have, use and enjoy that which is your own.

4

In the mid–nineteenth century, English was the language of the world's dominant empire, the very symbol and apotheosis of British power and superiority. The sun never set on those who spoke this language of the Empire. To become an English speaker was seen as an unqualified good, an attainment of civilisation in and of itself, giving you a place in a larger world, access to the most important kinds of words.

It is not surprising, therefore, to find that the Canadian residential-school policy had, at its centre, an insistence that pupils be educated exclusively in English. But the policy went further: children would not be permitted to use their own languages at all, even with one another.

This determination to eliminate indigenous languages in North America may have origins in conflicts over land on the Plains. In the

American West, outright war was waged against the many peoples settlers lumped together as "Indians." In the 1850s and '60s, settlement on the vast and rich prairies west of the Mississippi River was fraught with real danger. Farmers attempted to enter, occupy and plough lands where buffalo hunters mounted fierce resistance. The balance of power between the U.S. Cavalry and the societies who developed hunting on horseback was at times strongly in favour of the hunters. But as tribes succumbed to war, smallpox and the systematic slaughter of the animals on which their hunting systems depended, the colonial administration looked to education as a means to achieve a final solution. This was the political and military climate in which the Indian day school in the United States was entrusted to the churches. Aboriginal children would be remade as farmers and Christians, in the image of the story of Genesis.

In Canada, fear of the "Indian problem" was less acute. Settlement of the Canadian West was less a matter of warfare than a case of gradual incursion, new infectious diseases and negotiation. A period of fur trading with white newcomers effected something of a transition between life as hunter-gatherers and participation in an imperial economy. But the administration of Canadian Indians was much influenced by the experience and government processes of the United States. American and Canadian organisers of agricultural frontiers shared a point of view; their ideological and practical preoccupations constituted a single project. The "Indians" had to be subdued or prevented from causing trouble, and they needed to receive the benefits of farming and Christianity. Or, to express things the other way round, farmers and Christians needed the land, and therefore opposition had to be eliminated. In Canada, as in America, the mid–nineteenth century was a time when Indians had "education" thrust upon them. Their own education, at home, from their own elders, was stigmatised as savagery—something from which Indian children should be cut off.

So concerned were the educators about the subversive influence of home on Indian children, so fearful were they that a few hours each day with hunter-gatherer parents or elders would undermine colonial education and its Christian message, that they proposed an education system built on a residential model. Some advocates of this system argued that children should not go home at all, even during the usual school holiday times. In the end, the proponents of a short summer holiday prevailed over those who said there should be no interruption to the task of remaking and reforming Indian children.

A section of the 1867 British North America Act, which established a Canadian nation, gave the new federal government the power to legislate for Indians and their property. John A. Macdonald, Canada's first prime minister, made it clear what this meant: his administration had "the onerous duty of ... [the Indians'] guardianship as of persons underage, incapable of the management of their own affairs." The legislation and theory that established Canadian residential schools were in place by 1870. After two decades of educational experiments on Indian children, ranging from church-run day schools to schooling for children billeted with white families, the long-term plan began to take shape. In 1896, Canada passed additional laws to create the overall framework for the management of Indians. Residential schools, funded by government and run by churches, were given pride of place. As churchmen of the day put it, the time had come "to get the savage out of the Indian." The "Indian problem" had been defined; ten months of each year in a boarding school was the "solution."

This form of education for Indians became a system of abuse. In his book *A National Crime,* the Canadian historian John Milloy has documented the stages and range of this abuse in scrupulous detail. Children as young as five or six were forced to go to the schools, staying there until they were teenagers. Siblings were either kept apart within a school or sent to different schools. Disregard for health, clothing, nutrition and sanitation yielded appalling rates of disease and mortality. At one point early in the twentieth century, administrators of the schools discovered that 25 per cent of their pupils died at school or soon after returning home. Throughout the life of the schools, some of which lasted until the 1970s, the litany of complaints by both parents and the inspectors who had the dismaying task of assessing the schools continued: children were in rags, miserable with cold; they lived in squalor and were underfed to the point of widespread and chronic malnutrition. The regime in many of the schools also relied on fierce physical punishment: children were strapped, whipped, locked in dark rooms, chained in corners.

Many survivors of the residential schools are now willing to describe what happened to them there. Their accounts, like the work of John Milloy and other historians, reveal unbearable details of deprivation and suffering. They also reveal the repeated failure of the government to respond to complaints. In recent years, revelations of widespread sexual abuse, and the legal prosecution of some of the abusers, have drawn more attention to the schools. Indian children—vulnerable and lonely—endured the erotic as well as the sadistic attentions of their "educators."

5

Residential schools like those in North America have been used in Australia for Aborigines, in Scandinavia for Saami, and in parts of Africa for Hadza. "Education" for many indigenous people has been a means of enforcing the things that Europeans believed in and getting rid of the things they did not. Educators spoke of the need for "improvement," "development" and "civilised religion." But the real objective was to break rather than to create.

Perhaps all education has as its objective some form of breaking: the perceived wilfulness and even wickedness of children has been as much an issue for many educators as ignorance. Ideas about a child's social development, be it in potty training or language learning or tidiness, evoke wider, evolutionary theories of the stages of necessary development. Immature to mature, simple to complex, primitive to civilised, backward to modern, underdeveloped to developed: there are many terms that suggest seemingly natural, desirable or inevitable forms of supposed progress.

The concept of "poisonous pedagogy" described so brilliantly by the Swiss psychoanalyst Alice Miller invites many comparisons between educational and imperialist projects. She points out that pedagogy can all too easily be wielded to control rather than to nurture, out of hostility rather than respect. She argues that pedagogy of this kind will cause a loss of human potential and many forms of personality disorder. Alice Miller's insights into the nature of much education suggest that indigenous peoples have been subjected to an extreme version of much that has been endured (and no doubt still is) by children who are deemed to be "backward" by virtue of social class, heritage or intelligence. If there has been a twist of the knife of education into the heart of being Indian, it may be as an extreme case of a pedagogical norm, a greater poison than usual.

In residential schools, children experienced extremes of social as well as pedagogical abuse that are inseparable from racism. The idea of the "savage" or the "primitive" places millions of people at the lowest level of a human hierarchy. Everything about them—their use of resources, their diet, their dress, their religions and their languages—is said to be "wrong." The word is ambiguous: wrongness is associated with both moral and practical failure, with evil as well as incompetence. Yet European views of indigenous wrongness hardly needed to disentangle the ambiguity. Colonial discourse deemed the savages' ways of being in the world to be bad for them in both senses of the word—they are damned *and* destitute. So of course they must be changed.

Colonial theorists and administrations have often insisted that savage societies are dangerous because they are wrong; lacking the necessary intellectual skills or moral principles, flouting the essential rules of civilised society, they could not be trusted. In the case of hunter-gatherers, moving from one camp to another, seemingly nomadic and anarchic, European colonists and theorists declared that their moral and social codes lay outside society itself. With regard to the Maya, Aztec, Iroquois, Apache and other tribal peoples whose social systems and military organisation gave them real potential to resist European invasion, the colonial view was that aboriginal customs were savage and warlike.

It is easy to see how government and churches often shared an overlapping, if not rivalrous, desire to subjugate the heathen and create the Christian. It is also easy to see the links between education and agriculture. If North America's Indians (or South Africa's San or Amazonia's Amerindians or Australia's Aborigines) could be persuaded to become farmers, especially if they owned and worked their land as individuals or became labourers on such farms, they would take vital steps up the ladder of social evolution. Government and church tended to agree that this was a good thing, in and of itself. Family farmers on family farms would be part of the system, taking a settled, cooperative place on agricultural frontiers that the government endorsed and defended. They would not resist other incursions onto their former land. And family farms would be Christian homes. Further cause for satisfaction, more evidence of "progress," more complete integration into the newcomers' society—and even less likelihood of resistance.

Agricultural settlement precipitates conflict; colonial governments have the job of dealing with resistance to invasion. They have relied on two ways of meeting this resistance: overwhelming by force or converting to the agricultural ideal. Military and missionary zeal are two ways of achieving the same result. As the exiles from Eden invade and transform other people's lands and lives, they rely on both soldiers and divine authority. They are cursed; and they require that the curses be shared, their destiny adopted, by all others. To educate the hunters is to achieve ideological and military victory.

6

The abuses in residential schools can be listed. Protests can be made against the abusers. The worst of the abusers can be taken to court, maybe sent to jail. The system can be exposed for its failures to deal with parents' complaints and children's sickness and death. One can imagine a residential-school programme that was not blighted by the sadistic and perverse activities of its teachers and staff, but it would still be a source of grief and rage. For the fundamental, irreducible commitment of the undertaking was to eliminate ways of life—and to do so through getting rid of peoples' languages. This was managed by brutal and corrupt institutions; but it is the intention, not the corruption or brutality, that is the deepest wrong.

Why was a campaign against languages at the centre of residential-school policy? If real education was the objective, then a policy of multilingualism would have sufficed. Administrators in other parts of the British Empire—in India, for example, or in much of Africa—did not seek to eliminate languages they encountered. On the contrary, many British administrators took pride in their ability to speak Swahili or Hindustani. Yet in North America, Australia and some parts of southern Africa, no such use or endorsement of indigenous languages is to be found.

Perhaps the colonial use of Swahili or Hindi solved the problem of a lingua franca. In imperial territories that are not so much nations as continents, where people speak hundreds of different languages, the urgent need is for a single language that can unify and facilitate administrative control. Thus the use of English in North America and Australia might be seen as an analogue for the use of Hindi and Swahili in India and Africa. The difficulty with this analogy, however, lies in the colonial attitude to people speaking more than one language. Hindi and Swahili, along with English, were indeed shared languages in regions with immense linguistic complexities. But in most of Africa and India there was no great insistence that all other, more "primitive" languages be eliminated. Hindi and English were not intended to displace the web of tribal languages that are a feature of Indian intellectual and cultural life; nor were Swahili and English supposed to overwhelm all other African languages. Much of the British Empire was polyglot—as were French, Portuguese and Dutch colonies. The drive to achieve monolingual English in tribal communities is an aspect of one particular kind of imperial frontier rather than another.

What is the difference between a frontier where English is taught as an addition to existing languages and one where English is established and enforced as the only possible language? I suspect that the clearest answer points again towards the nature of the frontier between hunter-gatherers and agriculturalists. In most of India and much of Africa, English-speaking imperialists were not the first conquerors. The spread of farmers through much of Asia and Africa occurred several millennia before the British Empire reached these places. The hunter-gatherers of much of the Old World had long been overwhelmed by farmers and herders.

In many regions, therefore, the British Empire dominated agriculturalists—large, sedentary populations whose mode of life was a version of their own, albeit one that the British were ready to stigmatise as "native," "savage" and "backward." In a profound sense, the imperialists and those they conquered spoke the same language. They shared ideas about the exclusive ownership of small parcels of land; they shaped and managed such land to grow domestic plants and support domestic animals; they built relatively durable and permanent villages; their societies were hierarchical. They could locate one another, fight coherent wars and negotiate deals that everyone understood.

In much of Africa, India and Southeast Asia, European conquerors were dealing with farmers like themselves. They captured and exported those people they wanted to sell as slaves, and they dispossessed those whose lands they coveted. But the overall project was the extraction of profit from existing populations and societies.

In this scenario, pre-existing languages did not pose much of a threat. Indeed, a proliferation of languages meant a complex of societies, each to some extent pitted against the others. The principle of divide and rule invited colonial theorists to welcome diversity of language, rather than to seek its obliteration. In South Africa, for example, the apartheid system used linguistic divisions as a way of reducing its opponents' ability to unify in opposition. The new South Africa's eleven official languages are an outcome of this administrative acceptance, even encouragement, of diversity. India's linguistic complexity may be explained in a similar way. Many colonial administrations did everything they could to control rather than annihilate native populations, on whose labour, rents and resources colonial wealth depended.

In much of the New World, on the other hand, the newcomers encountered hunter-gatherers. They met human beings and social systems quite alien to them. They heard languages that were as remote from their own as human languages can be. The economic and material counterparts of these differences are central to the approach colonists in North America, Australia and parts of southern Africa took to the peoples they sought to conquer.

In these places, hunter-gatherers occupied large territories over which they moved with great freedom and ease. New settlers wanted this land, which, through their European eyes, appeared to be empty or, at best, randomly and minimally occupied. They found the hunters' flexible use of land both bewildering and threatening. The Europeans were looking for a place to convert into home—at least for a while, for a generation or two. The hunters seemed to be everywhere and nowhere, making sudden appearances out of the forest or desert or outback or hills, opposing the occupation and transformation of their lands and causing trouble. At first, of course, these strange people were useful. They could show the newcomers where to find water and edible plants, how to travel difficult terrain, and how to hunt the animals of the region. They were even willing to share their understanding of the gods and spirits of the place. But once the newcomers had absorbed such vital knowledge, they could dispense with their teachers. The underlying, enduring issue was land. Rival claims to the earth itself had to be obscured or obliterated.

Inasmuch as this is an explanation of genocidal tendencies on frontiers between farmers and hunter-gatherers, it is also a way of looking at the history of many modern societies, and of the languages these societies do and do not speak. Indigenous populations could raise questions about the rights of newcomers, asserting *their* rights to the land. A small family farm, isolated in wild country, is a vulnerable thing: a group of angry hunters could destroy a decade of hard work in a single quick attack. Any opposition to farming had to be checked, made impossible. The enemies of settlement had to be silenced or removed. This is the story of the United States, Canada, Australia and much of southern Africa.

The residential school was part of a process of ethnocide. The plan that shaped these schools, and the attitudes that informed their daily regimens, emerged from the agriculturalists' need to get rid of hunter-gatherers. These schools represent a dedicated and ruthless attempt to transform the personalities and circumstances of "native people" into ... well, what? Farmworkers and industrial labourers? Domestic servants and housewives? All of these, and yet the project is easier to understand as a negative rather than as a positive undertaking. The intention was to stop people being who they were—to ensure that they could no longer live and think and occupy the land as hunter-gatherers. The new and modern nation-states make no room for hunter-gatherers.

In the history of the "education" provided by these states, there is no acknowledgement that hunter-gatherers had a right to be on their lands, nor a jot of concern for their skills and knowledge. So they were silenced. Broken down. Transformed or killed. The abuse of the schools embodied the abuse of the colonial intention and process. And at the core of this abuse was a determination to destroy language. To secure an uninhabited land, there must be no minds in the way, no rival words that imply enduring presence and deep claims to the place.

7

The hunter-gatherers' tendency to leave landscape open, unchanged, worked against them. They did not make fields and hedges, created few monuments to demonstrate their ownership or use of the land. And though they used fire to manage both grasslands and forest, to settlers this practice seemed destructive, a kind of pyromaniac irrationality. So it was easy for newcomers to speak of open, unfarmed territory as "wilderness"—the *wild déor* place, realm of the wild deer, a symbol of land that is beyond human habitation, without human voices. In this "wilderness," the voices of the hunting peoples were likened to the calls of the wolf or the hooting owls: resonant, beautiful, haunting, susceptible to much sentimental and nostalgic interpretation, but not quite human. If the wild hunter-gatherers could be made to speak a *real* language, they, like their lands, could be turned into something of use and value to the settlers.

Agricultural settlement and religious evangelism, the endeavours and theories of frontier, treat the sounds of wilderness much as they do the trees or grasslands—by cutting them down, uprooting them, ploughing them under, transforming them from "worthless" to "valuable." Making them yield surplus. Thus is the tabula rasa of empty, silent wilderness given substance. Hunter-gatherers who survive the attacks against them and their territories are given words for agriculture and words for God, words for local government and advisory committees. In recent times, at the outer edges of the administrative frontier, words for migrant labour and heavy-equipment operating. Words in a new kind of language. The old language must be discarded—or, like parklands within frontier development, it may become an island of folk culture, somewhere to be visited and enjoyed that must never be too noisy.

8

Hunter-gatherers read and write. They did not have the alphabetical or pictorial scripts that agricultural societies have developed in relatively recent times. They did not use letters to represent sounds. But all hunters read tracks; everyone who lives by hunting or gathering must notice, read, interpret and share the meanings of signs in the natural world. And where carvings establish family histories, people read images on totem poles and house posts. These are also forms of literacy.

Similarly, all sections of all societies rely on the spoken word. I wonder, for example, where the decision makers of a modern corporation fit on the oral/literate spectrum. They may well rely more on listening and speaking than on reading and writing to convey ideas and make crucial plans. And what of politicians: do their ways of expressing "truth" and making decisions rely on the spoken or the written? Many people whose lives appear to be controlled by endless flows of paper in fact depend for their authority, judgements and success on speaking.

Hunter-gatherers, however, depend on the spoken word for almost all forms of history, spirituality and practical knowledge. For this reason, their storytelling is of great importance; all adults are expected to be able to speak well. Among the peoples of the North Pacific Coast, this is a skill that is inseparable from public status: chiefs have to be orators at potlatches in order to affirm prestige as well as land rights. They are relied upon to set out in public—and, if necessary, in full—the histories of both territories and families. In these narratives, as in so much hunter-gatherer speaking, the distinction between the mythic and the practical is blurred.

In many of the hunter-gatherer communities I have known I have been aware of shyness and a reluctance to speak, a wariness of too many words, a scorn for light-headed chatter. The spoken word is taken seriously; when someone decides to speak, he or she does so with care and often at considerable length. It is also striking how little speakers tailor what they say to accommodate proprieties of the kind with which we agricultural and urban types are all too familiar. Stories are told in full. Anyone can listen. Those who are ready to understand do understand; others are free to stop listening. Children are not "protected" from things that may frighten them or be sexually "shocking." People tend to say what they really think about one another; and things are said in public that sound quite startling to those of us whose lives are shaped by dissembling and concealment.

Anaviapik's readiness to say which of his children he liked best was an example of Inuit forthrightness. X was always his favourite, the one he loved most. Y was the best of the daughters—yes, he loved her better than Z. And W was a child he had never much loved, certainly his least favourite of all the children. Anaviapik went through these preferences with three of his children in the room. One of them was W, the son given the position of least liked. Anaviapik spoke with ease. He was telling me what everyone else already knew. The least-loved son laughed at the familiar truth of his father's words and commented himself: "Yes, I never was much loved by him. My grandmother loved me." For Anaviapik, to speak was to deal in facts and truth, not in protective or manipulative evasions.

Incentives for deception are few in hunter-gatherer communities. But it is important to avoid oversimplification of the social organisation of hunter-gatherers and their ways of sharing facts. Many groups live in small bands, often of no more than one or two extended families, with seasonal gatherings bringing together larger numbers of people. Groups living on concentrated and predictable resources (salmon runs, for example) have larger villages and more elaborate systems of social stratification and control. As I have suggested, this may have been a much more frequent pattern of hunter-gatherer society at a time when hunter-gatherers occupied those parts of the world that agriculturalists most wanted to settle and farm.

In the hierarchical hunter-gatherer systems that are well documented, institutions exist to ensure that produce is distributed widely—as in the potlatch. A striking feature of the potlatch, however, is that it functions as a forum in which facts also must be shared. Holding a name of importance is tied to having responsibility for a territory. The definition and management of this territory are made public by the obligation on the holder of the name to speak fully at potlatches. This institutionalised openness corresponds to the open sharing of information in other, less complex hunter-gatherer societies.

The underlying reason for a sharing of facts applies to all hunter-gatherer economies. Those who rely on wild plants and animals need to know all they can at all times about as many aspects of their lands as possible. To conceal information is to create serious risks. People have to know when plants are ready for harvesting in certain areas; when fish are running and where; what animal populations are doing. There is therefore a central economic rationale for openness and truth that cuts across all other qualities of hunter-gatherer society.

A hunter-gatherer family shares what it has, whether that is information or food. To give to others is to be able to receive from others. Knowledge and food are stored, as it were, by being shared. By contrast, in societies where social ambitions and personal rivalries are systemic, distortion and secrecy are used to manipulate others. There is wariness about allowing others to see what goes on in either the larger economy or the home. People are secretive about how much they earn; they are anxious about how their neighbours might judge the quality and significance of their car, their carpet or even their food. "Scandals" are kept secret for fear of losing influence or position. In many political arenas, there is a need to manipulate what voters believe.

In societies that are competitive and hierarchical, words are dangerous because the truth is to be feared. The curses of Genesis—the combined chemistry, perhaps, of the fruits of the tree of life and tree of knowledge, or the immediate effect of that apple of original sin—

required the telling of lies. Social life in all societies, and the efficacy of language itself, depend on a certain level of honesty: people must be faithful to their words in daily life, and faithful to the rules that govern the uses of those words. But in hierarchical and competitive societies, the reliability of facts and the veracity of statements of personal history tend to be observed in the breach. This is not so among the Inuit or the Dunne-za nor, I suspect, in any other hunter-gatherer community.

Verbal untruth has the power to render us neurotic. Social convention or personal vulnerability can displace truth. A parent says she loves all her children to the same degree, even if this is far from the case. Or a husband insists he loves his wife, even when there is deep animosity between them. These are examples where dishonesty is somehow normal. Yet these lies are often detected as such at a preverbal level. Something deeper than words can understand that words are false. In this way, a tension or contradiction may arise between words and behaviour.

The strain between what is said and what is known to be real is a powerful cause of conflict. Tension between reality and what is said about reality becomes a matter of profound importance, of course, when dealing with children. Those for whom language is a recent acquisition are also the most sensitive to non-verbal or pre-verbal messages. A child reads feelings from the subtlest of an adult's gestures. The child knows much of what is "true." Yet the words of many adults contradict these truths. What can the child trust? Words or behaviour? The words themselves may also confound one another. What does the child do with a multiplicity of conflicting messages? Where are words *or* facts when the words about the facts, or the words themselves, do not fit together?

The disorders that psychology associates with the dissonance between what parents say to children and what children know to be reality—from deep insecurities to chronic anxiety to depression—are not to be found among the hunter-gatherers I have known. This is not to claim that they are people who know nothing of mental illness.

Rather, it is to look at the absence of a particular kind of illness, one that in my own society is somewhere between common and the norm. The apparent sturdiness of the hunter-gatherer personality, the virtual universality of self-confidence and equanimity, the absence of anxiety disorders and most depressive illnesses—these may well be the benefits of using words to tell the truth.

All the more tragic, therefore, are the colonial processes that cause hunter-gatherers to lose confidence in their ways of raising children, their ability to live well on their own resources, and their languages. The cycle of this breakdown is familiar: children who are taken from their homes and raised in residential schools; parents who are themselves enduring dispossession and demoralisation and so do not learn how to care for their children. Hunter-gatherer communities are all too easily and profoundly broken in this way, causing anger and disarray for adults and children alike.

9

Hunter-gatherers have not disappeared altogether. Some of their communities, cultures and languages have survived at the margins of, or beyond, agricultural frontiers. Their cause is allied to that of other indigenous peoples. All around the world there are voices demanding that nation-states recognise the fate of indigenous societies, whose peoples have been invaded, dominated and, in many cases, murdered by colonists. Everywhere there are calls for the honouring of agreements, for the entrenchment of rights, and for the acknowledgement of systems of justice based on the enduring power of old words. Yet there has been a paradox to indigenous peoples' campaigns against colonialism: for the most part, the people who lead them, including the most assertive aboriginal leaders and activists, use the language of the newcomers—be it English, Spanish, Portuguese, Indonesian or Swahili.

The Nisga'a have been at the forefront of these campaigns for justice. Their story is at the core of Canadian aboriginal history. In the 1850s, at the same time the residential-school policy was being designed, the Nisga'a were the first northern people to defy the colonists. They said there had never been a land deal; there were no words on paper, for they had never signed a treaty. Therefore, even under the law of the newcomers, Nisga'a chiefs (along with other aboriginal groups in the Canadian West and North) have always maintained that their people have never ceded any part of their land. Early on, Nisga'a leaders took the argument to the Privy Council in England, created petitions, contested their case in every possible forum. Their protests recurred throughout the twentieth century. In the 1960s they brought the matter to court, with hopes of persuading the colonial powers that even British justice gave the Nisga'a basic rights to their own lands.

In 1973, the Nisga'a won a split decision in the Supreme Court of Canada; half the court acknowledged that Nisga'a aboriginal title was indeed extant. During the 1980s, however, Nisga'a leaders were still having to argue that their title, supported by the Supreme Court, should be recognised in a set of new arrangements. In 1999, the terms were at last agreed upon: the Nisga'a and the Canadian government created a modern treaty, which was ratified in April 2000. Public events in Nisga'a communities, including their potlatches and other ceremonies, included speeches by elders in the Nisga'a language. But the internal Nisga'a negotiations, and then the deal itself, were all done in English.

Nisga'a tribal leaders were able to explain to me why filming of the eulachon harvest should begin in early March by telling a Nisga'a myth. In this way they expressed their heritage, relying on the texture and qualities of oral culture. Yet they spoke in English. They did not use Nisga'a words or grammar. So what is lost by their making a translation, as it were, from one language to another? If Nisga'a or Inuit are unilingual English speakers, is there something at the heart of being Inuit or Nisga'a that is bound to have disappeared? There are tribal villagers in India who speak Indo-European languages, yet live beyond or at the margins of the Hindu and national system, with a heritage reaching back to pre-Hindu societies. There are Bushmen in southern Africa who speak only Nama and Afrikaans, the languages of herders and farmers. And there are trappers in the Mackenzie Delta, in the Canadian western Arctic, who lived on and from their land long before the advent of intensive administration but who have spoken only English for more than fifty years.

Have all these peoples lost their cultural hearts? Are they in some way separated from their own heritage? There are many ways of answering these questions. I want to look at just two. Both lead back to exile and Eden. The first kind of answer is to do with naming things; the other looks at why people change the languages they speak.

The Subarctic is a good place for grouse, plump and delicious birds that live on seeds, buds and shoots. Most change their plumage in winter. They are remarkable for tending to stay in the Subarctic all year round, living in and even under the snow. They are much hunted, therefore, by all northern peoples. In Dunne-za country, ornithology identifies five species: spruce grouse, sharp-tailed grouse, blue grouse, willow ptarmigan and rock ptarmigan. The first three are most common at lower elevations, in the forests and clearings that are the main hunting areas of all Athabaskan peoples. In Dunne-za Athabaskan, the words for spruce, sharp-tailed and blue grouse are *tchi'-djo, tchi'tchah* and *chiskonsteh*.

Each of these words begins with the syllable "tchi" or "chi." This is not a root that means "grouse," and it cannot be separated from the word in which it appears, any more than the syllables "pot" or "poten" can be used apart from their meaning in the words "potential," "potentate" or "potency." When referring to a grouse, therefore, a Dunne-za speaking Athabaskan has to identify which of the species is being referred to. The same principle applies to the words for fish in Inuktitut. There are terms to mean arctic char, arctic char that are running upstream, arctic char that are moving down to the sea and arctic char that remain all year in the lake, as well as words for lake trout, salmon and so on. There is no word that means "fish." Similarly, there are Inuktitut words for ringed seal, one-year-old ringed seal, adult male ringed seal, harp seal, bearded seal and so on. There is no word that means "seal." Speakers of Inuktitut have to have precise information about the seals to which they refer. These are ways, perhaps, in which hunter-gatherer languages express, celebrate and enforce the importance of detailed knowledge of the natural world.

When the Dunne-za I travelled and hunted with used English, they took advantage of a coincidence of sound: since "tchi" is at the beginning of the words for three species of grouse, they referred to all grouse as "chicken." They could therefore speak about grouse without having to say whether they meant spruce, sharp-tailed or blue. In fact, very few of the men and women I knew had learned those English names for individual species of grouse. Thus, in speaking English, they simplified the world, reducing the detail and therefore the knowledge encoded in their own language.

The Dunne-za could have used English, as can anyone with some ornithological knowledge, to identify each species of bird, just as Inuit could use English to speak of any kind of snow. The limitations on knowledge and communication, in these examples, are not intrinsic to language so much as a feature of using languages in a limited way. There are words, of course, that cannot be translated as single words. Just as in the case of some Inuktitut words for snow, there are concepts that do not translate word for word even between closely related Indo-European languages. Hence the examples of *Schadenfreude* and *Weltanschauung* in German, *sympathique* and *simpático* in French and Spanish, *katcha* and *pukka* in Hindi. These are words that do not have concise equivalents in English.

Between hunter-gatherer and agricultural languages there are many such examples, some of which have profound implications for cultural translation. It is difficult to convey the meanings of the English words "vermin," "fence," "advocacy," "hierarchy" or "bequeath" in Inuktitut or, I gather, in Athabaskan, Algonquian or San. There are also problems to do with the language of private property. Inuktitut has a way of speaking of ownership, with a root *(nangminiq)* signifying that a place or thing is for the use of someone. By attaching personal possessive endings to *nangminiq,* anyone can make a word that means "mine," "yours" and so on. But there is no verb form equivalent to the English "I own, you own, he owns." Equally noticeable is the absence of more than basic numerical language from almost all hunter-gatherer vocabularies and grammar. Inuktitut counts to five, and has words for ten and twenty. But there is no continuous sequence of number words, and no linguistic system for multiplying and dividing. In the San languages of the Kalahari, this lack of words for counting is even more striking: some have the numbers one, two and three, whereas others have only one and two.

A switch from one Indo-European language to another can mean some loss of vocabulary and a related change in conceptual possibilities. But such examples are few, and can often be dealt with by taking words from one language into another—as has been the case with the French *tête-à-tête* and *pied-à-terre* and the German *Schadenfreude.* When it comes to a change from a hunter-gatherer to an agricultural language, the problems of vocabulary and conceptual translation are immense. Yet, as the case of Inuktitut words for snow showed, they are not insurmountable. Given sufficient expertise in both languages, a readiness to use circumlocution and inventive explanation, and a strong appreciation of what the grammar of the languages allows—in other words, given highly motivated and sophisticated bilingualism—Inuktitut and English, for example, can be translated back and forth.

When it comes to the relationship between agriculturalists and hunter-gatherers, however, the very thing that does *not* exist is a commitment to bilingualism. Farmers may learn a ragged use of a hunter-gatherer language—to give orders, for example, or to buy and sell things. But the linguistic requirement at the agricultural frontier is that the hunters learn the newcomers' language. This is a corollary of the farmers' conviction that theirs is the superior language. Farmers have tended to see, and to celebrate, language change as progress.

The second way of answering questions about language change and cultural heritage begins with a reflection on the links between languages and status. Many people lose their mother tongue. But individuals do not often lose the ability to speak; virtually no one loses language itself. The loss of mother tongue usually occurs between generations, and takes the form of an exchange between one's own and someone else's language. This exchange may be perplexing to outsiders, and even to those who have experienced it. Why would parents not speak to their children in their own first language? Why would they discourage their children from learning the language of their ancestors? Yet a simple formula explains many, if not most, instances of language replacement: if a society is overwhelmed by a more numerous and/or more powerful population, and the original society is deemed by the majority or the powerful to be inferior, then its people are likely to lose their language. This loss may be achieved by direct measures, from genocidal attack to purposive administration, or by some combination of these, as in the case of Canada's residential schools. Apart from extreme forms of language suppression, loss of original language is often the result of a series of choices that parents and children in the subordinate population make about their short- and long-term interests. These choices can be followed by regret, and even trauma; but they have a certain inevitability. If material well-being, social advantage and overall security come from being adopted into a dominant society, then the impetus to speak that society's language tends to be fast and decisive.

Those who are overwhelmed and despised welcome employment and marriage with the richer and more powerful, or with those who do not seem to be cut off from riches and power by ethnic difference. "Good" marriage and new kinds of employment can be helped by, or result in, knowledge of the new language. This appears to set the scene for some degree of bilingualism. In patriarchal and patrilineal social systems, where men dominate and inheritance of goods or status is from father to sons, marriage is much more common with the women of an "inferior" people than with its men. But in patriarchal families, women nurture small children with little help from husbands and fathers. In cases of intermarriage, the official language of the house tends to be that of the men (that is, the dominant culture), with the language of child care being that of the mother. So sufficient conditions for bilingualism, rather than language loss, may exist. Moreover, the less patriarchal the dominant society, the better the chances for bilingualism to persist in a child's life and thus be passed on to future generations.

Other kinds of social division can also influence bilingualism. A subordinate population can be encouraged to speak a language that marks (and perhaps sustains) their low status. In this case, employers and "masters" may use bilingualism to establish and reinforce "superiority." This happened in aristocratic Russian households, where French was used for social or family life and Russian for the farm or kitchen. Similar kinds of bilingualism are common among many wealthy families in India. A less familiar example of this division of language by social function can be seen in the case of sacred and secular life. For many Jews, Hebrew has been a language of religious observance and biblical studies, whereas the language of the country of residence becomes the first language in all other spheres of activity.

But even when bilingualism exists, the pressures that come from being overwhelmed in numbers, and from being despised, do not end. Children who are born into a mixed heritage and have the apparent advantage of bilingualism may still experience the pressure not to speak their mother tongue. If this pressure is only slight, and is redressed by some degree of respect, status or economic advantage accorded to bilingualism, then bilingualism may last from generation to generation. This was the case in Norman England from 1066 until the time of Shakespeare, in the Jewish diaspora for hundreds of years, and in India for the entire colonial and postcolonial era. But these mitigating circumstances are not the twentieth-century norm; more familiar is a strong tendency on the part of children with mixed linguistic heritage to embrace the language that achieves the highest levels of status and security. Pressure to this effect can come from bilingual mothers determined to ensure that their children do not suffer, as they may have done, from having the "wrong" ancestry.

These explanations have a technical, abstract quality. In reality, of course, the forces that cause language exchange and loss of mother tongue are associated with many forms and degrees of exploitation, brutality and anguish. Slaves brought from Africa to the New World were deprived of their languages in a calculated and cynical belief that this would ensure a more complete enslavement. They were separated from others with whom they could speak their own language and expected to achieve a servile, minimal English (or French or Spanish or Portuguese). In the history of Jewish relocation, many refugees who settled in England and America encouraged their children to be monolingual in English. Both the language of their original countries and the Yiddish that gave an additional linguistic identity to European Jews were often dropped.

Between the enforced language change of slavery and the relatively benign and free shift to English by many Jewish refugees lies all that has happened in the encounter between hunter-gatherer and agricultural languages. We know very little of what occurred in the first five or six thousand years of agriculture; perhaps there were long periods of bilingualism. But agricultural peoples are numerous, aggressive, patriarchal and patrilineal. And they tend to see hunter-gatherers as inferior beings. The combination of farmers' skills, cultural systems and evolutionary beliefs means that any extended encounter between them and hunter-gatherers creates the conditions of language exchange. At every such frontier, the hunters who survive adopt the farmers' forms of speech.

There is a need here for some caution. The organised, relentless administration of change by modern nation-states should not necessarily be taken as a long historical norm. Language change can be brought about at a frontier between economic and cultural systems where there is no plan, no central organiser, no missionaries or schools. The occupation of hunter-gatherer territories in much of the world preceded the era of modern colonialism during which European nations and economies have launched themselves on Africa, Australasia and the Americas. For the most part, the events that determined the spread of Indo-European languages alongside farming occurred several thousand years ago and are a matter of difficult speculation. Through long periods of economic change, including the shift from late palaeolithic to early neolithic life in Europe, there has been a switch to the languages of farmers.

10

When I lived in the eastern foothills of the Rockies, I wanted to learn the Dunne-za language. It is a member of the Athabaskan family of languages spoken by the subarctic peoples of northwest North America as well as by the Navajo in the southwest United States. There are many dialects of Athabaskan, most of which are mutually unintelligible. But all are notorious for being difficult for Indo-European language speakers to learn. They are tonal; words are modified at the beginning and middle as well as at the end, and verb forms vary according to categories unlike anything in European grammar.

Dunne-za men and women encouraged me to learn and were patient with my difficulties. I managed to acquire some basic vocabulary and some useful phrases, but could not make that essential move from words to grammar. I learned sentences, but could not fix their principles in order to make sentences of my own. After a while, I sought out a missionary in the region who spoke the language very well. He was sympathetic, and encouraged me to persist. Towards the end of our conversation I asked him how long it had taken him to be able to live and work in the language. He thought for a moment. "About twelve years," he said.

I gave up on anything more than the most elementary Athabaskan. All the Dunne-za men and women I knew had spent time with settlers, had worked as guides and traded furs or, in the case of younger people, had been to school. They were all able to speak my language far better than I could hope to speak theirs. We could do interviews, collect the data for maps of hunting territories, make jokes, chat, and get through day-to-day life using English. Nonetheless, I began to see how this created limitations—Dunne-za could talk to me about their feeling for the land and could explain how they used it, but much of the elders' knowledge was too different, too sophisticated, to be expressed in their fur trader's English.

One day Abalie Field and I took shelter from a thunderstorm. We stood together and looked out at the weather, listening to the thunder booming in the distance and watching clouds to see how much rain was heading our way. Abalie began to talk about the causes of thunder. Was it true that the whites believed thunder was something to do with electricity? Yes, I said. How crazy, he said. Thunder is from the swan. The swan? Yes, said Abalie, then hesitated. He turned to look at me, and I saw frustration in his face. There was something he wanted to say, but he could not find the words. I encouraged him. What is the story of the swan? Perhaps he thought I was going to be incredulous and skeptical. I asked if he thought I would not be interested and assured him I was. No, this was not the problem. "English no good," he said. "If I talk in our language you can hear many things. Good stories. True things."

"About the swan?" I said.

"About Swan. About everything."

I knew from anthropology, especially the work of the remarkable ethnographer Robin Ridington, that the Dunne-za have a complex and powerful creation story that centres on the transformation of a swan into a culture hero. The bird, flying to and from the sun and moon, becomes a youth who has power to create things, to change life. The story's sophistication is immense. Ridington shows that even in English words, put together with scrupulous care, the story of Swan is full of mystery, elusive and magical. But Abalie and I could not get beyond basic English. He could not tell me what he knew, about Swan, about thunder or about anything else that went beyond the surfaces—the bare facts—of Dunne-za land and heritage.

When indigenous people go to court, in defence of their rights or laying claim to their territories in defiance of colonial incursion, they must speak English or rely on an interpreter. Their authority and knowledge are turned into a simple English text—the court transcript. In many such cases, a great deal of evidence has been led by expert witnesses—anthropologists and historians. These witnesses make some translations and explain meanings that are not self-evident, using sophisticated English. But the sophistication of the hunter-gatherer language is lost. Moreover, the words in which rights and powers are expressed are not always, and perhaps not often, susceptible to translation into literal prose and linear argument. Behind the words, or within the words, there are meanings that are more akin to poetry or music.

Oral culture blends fact and metaphor. The line between myth and information is not easy to draw. Those who come from an oral tradition and listen to its stories can place themselves at this line. In many Northwest Coast societies, the singing of a song and the telling of history are central expressions of a people's rights to, and their management of, territory. If someone must be convinced of a right to a territory, the deed can be presented—the history recounted and the song sung. The ownership of these stories and songs goes along with a territory, in much the same way as only the owner of a property in European legal arrangements possesses its title deed. On the Northwest Coast, storytelling, knowledge and poetry are thus combined in a formal system: that which is implicit in other, more informal hunter-gatherer economies is there made explicit.

But in all hunter-gatherer societies, stories establish and communicate rights.

In the 1980s, two Northwest Coast societies joined forces to bring a legal action against the government of British Columbia. This was the case of *Delgamuukw v. the Queen.* Chiefs of the Gitxsan and Wet'suwet'en, neighbouring and interrelated societies with territories in and around the upper Skeena, Bulkley and Morice River systems, argued that they, not the provincial or federal government, had jurisdiction in their 45,000 square miles of territory. The chiefs and their lawyers were aware that a possible paradox lay in this legal move: they would challenge a jurisdiction by following and accepting the legal process of the very people whose jurisdiction they challenged. This entailed many potential compromises—rules of procedure, rules of evidence, conventions of cross-examination and preconceptions of the judges could all turn out to obscure or mute the chiefs' words.

To minimise this problem, as well as to make sure that the legal action achieved some parallels with Gitxsan and Wet'suwet'en procedures, the people and their lawyers decided that the chiefs would give as much evidence as possible according to the laws of their own society. Each chief would speak only of his or her own territory. The histories that encapsulated and expressed rights to territories would be told one by one. This would take an inordinate length of time, causing immense strain to all involved. But the people chose to take the risk of alienating and confusing the court rather than disregard all the rules of their own society. After expressing some unease, Chief Justice Allan McEachern, the judge who heard the case in the Supreme Court of British Columbia, decided he had no alternative but to hear the Gitxsan and Wet'suwet'en presentation of the evidence. In the end, *Delgamuukw v. the Queen* became the longest case in North American legal history: the evidence of the chiefs, along with cross-examination and procedural arguments, took 243 court days. The transcript for the entire case amounts to 126 volumes.

The timescale of the case, as it turned out, posed less of a difficulty than the application of rules of evidence. Histories that reveal rights to, knowledge of and feeling for territories are not title deeds. They are spoken words. Moreover, they are words that express an authority transferred from one teller of the history, one owner of a story, to another. In the Gitxsan and Wet'suwet'en systems, inheritance of land, and inheritance of the stories that establish rights to the land, are inseparable. For the peoples of the Northwest Coast, as to any hunter-gatherer society or, indeed, any oral culture, words spoken by chiefs are a natural and inevitable basis for truth. The validity of such stories has been tested by detailed knowledge and the assent given to the stories by those who hear them.

In oral culture, the transacting of an agreement, the confirming of a piece of information, occurs between the one who tells and those who listen. A chief is accepted as a chief because his or her accounts are upheld by all who have heard them, and are recognised as consistent with the stories of other and neighbouring chiefs; acceptance by those who listen and are witnesses legitimises the authority of that chief, and the story is "true." Expertise in relating histories is matched by expertise in listening and assenting to them. The tests of validity are in the relationship of narrator to listener, and in the accumulation of these relationships over long periods of time.

It does not take a specialist in jurisprudence to see that these ideas of storytelling, authority and truth are at odds with the rules of evidence. In particular, the hearsay rule asserts that evidence is inadmissible if it is something a witness heard about rather than witnessed for himself. The function of histories in Gitxsan and Wet'suwet'en society runs counter to this rule. A chief knows because he or she has been told by the culture's experts in truth—other chiefs, elders long since dead, from whom the histories and the land are inherited. These are truths because others have said that they agree with the truth of what is told. The dependence on what has been told, and the circularity of the tests, mean that none of this is, prima facie, admissible in a court of law. The heart and substance of Gitxsan and Wet'suwet'en deeds to their territories are thus deemed not to be "evidence."

The trial judge agreed to hear all this hearsay, deferring a ruling on its admissibility until making his judgement. So the chiefs had their say, and their many, many days on the stand. Their words went into the record. The hearsay issue raises profound questions about truth, but not, directly, about the nature of the words themselves. Gitxsan and Wet'suwet'en languages, Tsimshian and Athabaskan, could, in theory, be translated. However, there was a moment of great crisis in the early stages of the trial where the very meaning of meaning was at issue.

Among the most authoritative and important of Gitxsan chiefs was Mary Johnson, one of the most senior members of the Fireweed Clan and a much-respected elder in the Gitxsan village of Kispiox. She was a strong, energetic woman, with a wide face as alert and watchful as it was beautiful. She was almost eighty years old at the time of the trial, but she gave evidence for several days and dealt with intense cross-examination. As a chief who held a large territory, she was both passionate about and proud of her culture.

I had spent time with Mary Johnson when working in Gitxsan country, had sat in her home and listened to her explain the history of her name and some of the history of her territory. She had taken me along the Kispiox Valley to see places where loggers and farmers had occupied and transformed parts of her people's lands. I had felt the authority and power of her experience; I had sensed something of her confidence, her delight in who she was and who the Gitxsan are. She had a large personality, a compelling presence.

I was in the court, in the northern British Columbia town of Smithers, when Mary Johnson gave her evidence. I was dismayed to see how isolated and vulnerable she seemed. She was physically small, and made to seem smaller by all the furniture, literal and metaphorical, of the courtroom. Four teams of lawyers. The judge on a raised platform. A crowd of spectators. And the witness box, where she stood, an elderly woman, very still, somehow expectant. She had to defer to the process, had to wait while arguments broke out about applications of rules of evidence and while details of her life were given in a deadening official language—no trace of the calling of names, the symbolic moments of respect, that she would expect in the feast hall or at any public occasion among her own people. I felt nervous for her. I feared that she would not find the words with which to set out her place in the Gitxsan world and the place of her territories in the total system of Gitxsan territories.

I need not have worried. As soon as Mary Johnson was asked to talk, her voice came steady and sure. She addressed the judge, speaking to him with no sense of intimidation and with a dignified politeness. She did call him "your majesty" a few times, but this was a confusion about the English words of respect in a foreign courtroom, not an attempt to offset uncertainty with humour. And Chief Justice McEachern listened, as judges do, with a mixture of attention and apparent ennui: the slow, measured, discursive idiom of Mary Johnson was not easy for him to listen to. As I looked from the judge to Mary, I wondered if the space between them could be narrowed; might she capture his interest, catch his imagination? Here was a chance for the judge to build a fuller understanding of the Gitxsan view of the world, and to get some better sense of the idea of a Gitxsan jurisdiction. Mary Johnson explained the laws to him, and the way she had inherited and knew her land.

The directness, clarity, dignity and obvious honesty of Mary Johnson were very moving. I felt—as I had before in her presence—the force of who she was, and awe at the way her words had survived the colonial onslaught. She used a mixture of English and Tsimshian. She was one of the elders most confident in Tsimshian and most determined that success in the courts should also mean survival of the language despite its erosion. Watching and listening, I thought: her pride must communicate itself to the judge as well as to her children and grandchildren. But I had the impression that the judge's eyes were often drawn away from Mary. And he seemed to have difficulty with the nature of the facts she was presenting. Perhaps I was too quick to interpret every shift of his eyes and attention. I was nervous but, along with many others in the court, I trusted to the power of Mary Johnson.

She told the history that establishes her ownership of her territory. She came to a point in the story where two sisters and their brother are starving. The brother dies. Then the sisters hear the drumming sound of a grouse—the beating of wings that are part of a mating display. They know a grouse that is displaying will come back to the same spot over and over again. It is an old log, covered with thick moss. So they know where to hide. One of the sisters slips under the moss, alongside the log. She fails to catch the grouse. The other sister then hides. The grouse flies back again to its perch on the log and is caught: at last the sisters have something to eat. Once they have eaten, the two girls remember their dead brother; they compose a song for him, a lament.

Peter Grant, the lawyer who was leading Mary Johnson's evidence, asked her if this was the place in her history where she would sing the sisters' lament. When she replied that it was, Peter Grant said, "Go ahead, you can sing the song."

The transcript of the evidence then reads as follows:

> The Court [i.e., Chief Justice McEachern]: Is the wording of the song necessary?
>
> Mr. Grant: Yes.
>
> The Court: I don't want to be skeptical, but I have some difficulty understanding why the actual wording of the song is necessary.
>
> Mary Johnson: Do you want me to sing the song?
>
> The Court: Are you going to ask the witness to now sing the song?

Mr. Grant: The song is part of the history, and I am asking the witness to sing the song as part of the history, because the song itself invokes the history.

The Court: How long is it?

Mr. Grant: It is not very long. It's very short.

The Court: Could it not be written out and the witness asked if this is the wording? We are on the verge of getting off track here. To have witnesses singing songs in courts is not the proper way to approach this problem ... I just say, I've never heard it happen before, I never thought it necessary, and I don't think it necessary now. It doesn't seem to me she has to sing it.

Mr. Grant: It's a song which itself invokes the history and the depth of the history of what she is telling. It is necessary for you to appreciate ...

The Court: I have a tin ear, Mr. Grant. It's not going to do any good to sing to me.

Mr. Grant: I would ask, Mrs. Johnson, if you could go ahead and sing the song.

Mary Johnson: It's a sad song when they raise the pole, and when the pole is half-way up they told the chiefs that pull the rope to stop for a few minutes, and they sang the song and they cried. If the court wants me to sing it, I'll sing it.

The Court: No I don't, Mrs. Johnson. I don't think that this is the way this part of the trial should be conducted. I just don't think it's necessary. I think it is not the right way to present the case.

Mr. Grant: You can go ahead and sing the song now.

Mary Johnson sang. Her voice was strong, and the sadness of the lament was clear, anguished and startling. My eyes filled with tears. I had heard the story and the song before. It did not belong in this court, against the opposition of the judge, resounding in his tin ear. Yet it was somehow perfect, a complete expression of Gitxsan language, in all its senses, with all its meanings. I looked around me. Everyone had their eyes fixed on Mary Johnson. The sound of her voice was the only sound in the room. The drabness of the place was transformed. The Gitxsan people in the court were mourning a starvation, long long ago. Another time, another kind of place. Mary Johnson transported us. Not all of us, no doubt, and not, I could see, the judge, whose indignation was conveyed in his hunched shoulders and angry lack of expression. He would not and could not listen—a matter of judicial role, no doubt.

We sat in the court, listening to Mary Johnson's song and sensing the tension of the moment. A judge with the task of reaching conclusions that went to the very survival of a society that had already suffered extreme loss at the hands of European settlement. An elderly woman revealing herself, her people, her heart to this man, this process—and that tin ear. The power of the judge and the vulnerability of the woman. The lament held in the air. And so many different feelings. Those of the Gitxsan and Wet'suwet'en who filled the back rows of the court. And those of the judge, the teams of lawyers, the ushers, the clerks. The history of the region and of the nation, of the encounter between colonists and indigenous cultures, the story of whites and Indians, the deaths by smallpox, the losses of life, of land, of hope: the girls who caught the grouse had made a song for all of this, and the tears carved on the totem pole now stood for all that had taken place, in the court as well as in Gitxsan territory.

The song was slow and full of sadness. Mary Johnson reached the end. There was a moment of strange silence, a moment for us all to return to that place, the court. Chief Justice McEachern sat low in his seat, looking away from Mary Johnson. He looked instead at the group of lawyers, sitting a little below him in the front row of the court's seating.

The Court: All right Mr. Grant, would you explain to me, because this may happen again, why you think it was necessary to sing the song? This is a trial, not a performance … It is not necessary in a matter of this kind for that song to have been sung, and I think that I must say now that I ought not to have been exposed to it. I don't think it should happen again. I think I'm being imposed upon and I don't think that should happen in a trial like this … I see no reason whatsoever why it was necessary to ask her to sing that song. Go on with the evidence, please.

The judge was "embarrassed," in both the personal and the technical sense of that word. He carried this embarrassment, and perhaps some anger, into his judgement. He concluded, two years later, that none of this oral cultural evidence was admissible, and that he needed neither the voices of the chiefs nor those of the anthropologists and archaeologists to decide the matter. He did not believe the Gitxsan or Wet'suwet'en could claim aboriginal rights to their territories, concluded that they did not have links to more than the land of their reserves and that they lacked all "the badges of civilization … had no written language, no horses or wheeled vehicles." Among his conclusions he noted "that aboriginal life in the territory was, at best, 'nasty, brutish, and short.' "

I quote the judgement to arouse incredulity as well as anger. It seemed to be an exercise in preconception, a ruling in astonishing disregard for the evidence. In due course, in late 1997, the Supreme Court of Canada overturned it in almost every respect, and ruled, in particular, that the British Columbia judge's failure to pay due attention to the chiefs' oral evidence invalidated all his findings. The Supreme Court duly ordered a retrial. The tin ear was recognised as deaf indeed. But the relief of the reversal was not available, and was anticipated by no one, when the first judgement came down.

I often wonder if the chief justice has ever sat in his study or strolled in his Vancouver garden and thought back to those moments in the court, when the Gitxsan and Wet'suwet'en had faith that their words would be heard. Does he think now, especially in light of the Supreme Court rejection of his findings, that there was something he failed to grasp? Does he ask himself what the words were that he could not hear? Or does he still console himself with the thought that the legal game had run its course? He followed the rules, he might insist, worked his way through the legal conventions, left it to the superior courts to reach a final verdict. He may well continue to believe that life in the past was without "civilization," was "nasty, brutish, and short." The judge looked out on the mountains and valleys of Gitxsan and Wet'suwet'en country and saw emptiness, wilderness. He heard nothing but its apparent silence. To his mind, shaped by the notions of roaming farmers, eyeing a landscape to see where there are opportunities for settlement, the silence was vast, even wonderful. His words, his judgement, confirmed it.

The indigenous peoples of the Americas (some of whom were themselves farmers), many cultures in Africa and Asia, and all the hunter-gatherers of the world have suffered a holocaust. Those who live now, in villages and reserves as well as at remote frontiers of modern nation-states, are survivors. They live in post-ethnocide circumstances: their parents or grandparents or great-grandparents experienced disease, murder and expropriation that reduced populous and strong communities to fragmented and demoralised families. For those who are survivors of this holocaust, some kinds of silence are a condition of life; yet this silence is also unbearable.

The silence of the ancestors; their ghostly absence; memories of memories. Who can speak of what happened? The witnesses themselves are silenced. The few who do speak must relive a past that is a long account of losses, or recollect times when so much was so different. And for many, perhaps most, the very language of the witnesses is dead or dying.

The ignorance of the others, *their* silence about the past, is also unbearable. How can those newcomers to Canada (or Australia, or South Africa, or Brazil, or Irian Jaya) move around on the land, set up their farms, build their homes, enjoy the richness and beauty of these places, and not speak of the peoples they have displaced? Is it possible that they do not know? Is their silence deeper than a reluctance to speak? Ignorance of what has been done, unawareness of the losses: these are another silence. Primo Levi, writing as a survivor of the European Holocaust, described his need for everyone to know, his dismay that they might not know. Thus it is for the survivors of all those cultures for whom annihilation is also their story, the story of their era and their homes.

Where a language survives, and there continue to be families in which it is the language of everyday life, the sense of loss is not so absolute. If names are given, stories are told, and the world is described in people's own, original language, then in one important way those people are able to feel at home. As a nation they may have lost lands and endured the deaths that came with new diseases, and their aboriginal rights are still often contested and denied. But they take pride in who they are and have been.

For those who speak only English, pride is more difficult, more fragile. They must reconcile themselves to a profound loss at the centre of who they are, then move forward to assert their rights to lands and resources in the language of the settlers. But though this is difficult, it is not impossible: the Gitxsan and Wet'suwet'en brought their case about jurisdiction in the settlers' courts, relying mostly on English, and, in the end, won a judgement that is widely held to strengthen the rights of all the First Nations of Canada.

11

Towards the end of making the film with the people of Pond Inlet, Michael Grigsby, the director, wanted to set up some interviews. We had filmed many kinds of hunting, had seen the land, had explored much of the external, visible world. But we needed the recollections and thoughts that would give the external its meanings. The film could only come to life with words. I spoke to Anaviapik about our concern. He agreed that the film had to have people speaking. But could they use Inuktitut? Could they talk about anything at all? Of course, I said, we would use subtitles to translate the Inuktitut into English; and it was for the people to choose what they wished to talk about. I remember that he asked me again: could they talk about the things that were of real importance? Yes, I said, those were the very things that the film would need. "All right," said Anaviapik, "then we will talk, and you can tell them when to film. We will pretend not to know that they are there."

We spent a good deal of time discussing how these interviews would be done. Anaviapik and the elders who came to visit him would often sit at his small kitchen table to exchange plans, reminisce and tell stories. Whoever happened to be in the room would listen, punctuating what was said with questions and shouts of surprise. But those who listened did not interrupt with changes of topic or switch the conversation to themselves. A remarkable feature of Inuktitut, in fact, is the absence of devices for fending off possible interruption. There are no Inuktitut translations for "er" and "um." The person who speaks can rely on speaking for as long as he or she wants. Oral cultures depend on respect, on allowing full space for people's words. Everyone liked the idea of filming Anaviapik and one or two other elders sitting at his kitchen table, in the dilapidated metal chairs they always used.

One of the women who sometimes came to visit Anaviapik's household was Utuva. She had a boisterousness that gave pleasure to everyone. She was much loved, and much admired for her eloquence. I had often visited Utuva at her home, and she was a patient teacher of the language as well as a wonderful teller of stories.

One of the men who used to visit Anaviapik was Qanguk. Perhaps a little older than Utuva or Anaviapik, and more reserved, he had a way of speaking with remarkable stillness, relying on words and not gesture, using the power of the language itself rather than any raising or lowering of his voice. His own feelings came through in the way his eyes gleamed. Like most Inuit of his generation, he had a moving combination of personal grace, sure dignity and a readiness to laugh until the tears rolled down his cheeks. Though I had never been out on the land with Qanguk, I had often had the chance to talk with him, and knew that he too was a wonderful storyteller.

So Anaviapik invited Utuva and Qanguk to sit with him at his kitchen table and talk. The cameraman sat in a chair at one side of the room, holding the 16mm Eclaire in his lap. No intrusive tripod. The sound recordist attached a tiny microphone to the end of a fishing rod and dangled it overhead. Michael Grigsby crouched on the floor beside the cameraman. Anaviapik asked that I sit at the table with them, so that if there was anything the film crew wanted to ask, I could translate for them. Ivan and I calculated where to put my chair, a little to one side, out of the shot.

Thus we arranged ourselves. When all was ready, I said to Anaviapik, Utuva and Qanguk that I could start things off with a question. "No," said Anaviapik, "we don't need a question. We are going to talk first about the Qallunaat, what they were like and what we thought of them in the old days." And they spoke. Each told some story, a recollection, in turn. They did not interrupt one another apart from words of agreement. But when one person had finished, another took up the theme.

In this way, they talked about the kind of fear that Qallunaat had inspired in them. "Ah yes, we were so scared of white people. They seemed to us to be dangerous. They intimidated us. They came up here and did all kinds of things they were not supposed to do, things that were against their own rules. They did not care about our culture. They wrote down lies and sent them south, to their bosses in the south. We Inuit were scared of them. We knew that they wrote lies, but we didn't dare say anything."

They spoke of the way their children had been taken from them to go to schools, and how they had felt unable to resist. "*Aakkaurumalaurpugut, aanguralualluta. Ilirasulaurpugut.*" "We wanted to say no, but we said yes. We were intimidated."

This, they said, had been the cause of many families leaving their homes in camps, out on the land, and coming to live in the settlement. They knew that life was less healthy in this place. "We used to come and visit, to trade, and we would see the Inuit who lived here, and we would see how pale they were, as if they were recovering from an illness, so pale and bloodless. Then when we saw those who lived out on the land, they looked beautiful. Beautiful and full of blood!"

Anaviapik, Utuva and Qanguk spoke of their fear that this land was still being taken from them, their way of life being disregarded. What would happen then? "If we no longer have our land, our beautiful land, we will be poor, orphaned, pitiful." Then there was Inuktitut, the language, the way of being: that could be lost, too. Then there would be nothing of their way of life, no stories.

The first time Qanguk spoke, he told a story, a childhood memory. It was 1923, at the time when some Inuit were going to be tried for the killing of a white fur trader. Police came to Pond Inlet for this event. His parents' generation, Qanguk told us, had used the police as a kind of bogeyman to scare Inuit children into being good. The "Helloraluk" or "Polisialuk," the great Hello or Policeman, would get you if you didn't do this or that: here was a threat that Inuit children could not ignore. ("Hello" is a word used for whites in several parts of the Arctic.) The threat that a *polisialuk* would come to the house to get the child would have been made in a teasing way—Inuit parents rarely, if ever, used an aggressive tone or manner with their children. But it was nonetheless a threat that would have inspired real fear. Qanguk continued:

"When the police arrived for the trial, they put on their red tunics and trousers with stripes. They had long knives swinging at their sides.

We saw those knives. One of the policemen in these clothes came to visit our house. In he came—the *polisialuk*. I saw the sword. I thought for sure that some threat was about to be carried out. I had done something wrong. Something I had done had been found out. My parents' warnings were in my mind. This large man had come for me.

"The *polisialuk* has brought his huge knife. He is going to cut my head off. I thought I was going to be beheaded! I ran behind a chair, crying, trying to hide. But I was to be more frightened. I was terrified: the *polisialuk* reached behind the chair, he wanted to pick me up. This was it! He took me in his hands. I was going to be beheaded. I screamed and cried in terror. The policeman had to hand me to my mother. At first I could not tell her what I was afraid of, I did not have the words in my head. Then I did say, I told her that I thought I was going to have my head cut off. And everyone laughed.

"Ah yes, we were so scared of white people. They seemed to us to be dangerous. They intimidated us."

The symbolic beheading in the story was not quite apt. The Inuit have not been beheaded. Qanguk told the story in Inuktitut, with the words and ways of telling that are at the heart of Inuit oral culture. The Qallunaat war on the language had not been won—it had not even begun in the North when Qanguk was a child, though it was likely implicit in what many representatives of the northern imperial advance conveyed to the Inuit about the destiny of their culture. Soon there would be an educational system in place that set about replacing Inuktitut with English. It is a tribute to the strength of elders, as well as a function of the isolation of their communities, that Inuktitut continues to be spoken in many homes in the eastern and central Arctic.

After Qanguk finished his story, Utuva and Anaviapik each talked about the ways in which southerners misunderstood Inuit, and how Inuit had so often acquiesced in southern plans for them. "That was in the old days," said Anaviapik. "If it were now, we wouldn't care what they said." And they all laughed.

Inuktitut grammar and word formation are governed by clear rules that allow almost no exceptions. Yet Inuktitut is also a language of extraordinary poetic and metaphorical potential. Perhaps this is a feature of all language. The languages of hunter-gatherers, however, may have a special commitment to and reliance on metaphor. The logic and the poetry of words are, for many peoples, inseparable. Something essential to the human mind and human expression, and therefore human well-being, may well be clearer here than elsewhere. Hunter-gatherers accumulate immense bodies of knowledge and use this knowledge to make critical decisions—hence the importance of songs (to aid memory and create new insights) and dreams (to allow decisions that draw on a blend of facts and intuition).

The attack upon, disregard for, and destruction of hunter-gatherer languages is a symbol, an important metaphor of its own. The beheading of Qanguk. The image of decapitation might be too well defined, though, too sharp to describe the assaults that have been mounted. It has been more a case of electroconvulsive therapy than of beheading—more a disruption of thought than the removal of the means to think at all. The clearing of minds is inseparable from the securing of lands. And the loss of the words that hold history, knowledge and heritage devastatingly compounds all other forms of dispossession.

FIVE: **GODS**

1

All societies have myths that describe the creation or emergence of their people. I once heard one of these stories told in answer to a question about power.

Towards the end of filming in Nisga'a territory, we began to feel that our material was strong in history but weak in mystery. The facts of the colonial experience were prevailing over deeper histories. We looked for someone to give the Nisga'a view of the people's origins. We were told that Harold Wright, an uncle of Rod Robinson, would be the man. Harold lived with his wife in a tiny house. George Gosnell took me to visit him there.

Harold sat in an old armchair alongside a wood stove. He was bursting with life and eager to talk. He was quick and sharp and often laughed at his own irreverence. Yes, he would tell us old stories, but did we have enough time to listen? Was there enough film in our camera? We should begin with the flood.

So we filmed Harold telling of the creation of the Nass River. Much of this he did out on the lava beds—the fields of volcanic flow that spread for some hundred square miles to the east of Aiyansh—where he could link the events of his narratives to the mountains around him. One day, driving back towards Aiyansh along the dirt road that crosses the lava beds, we chatted about the story of the volcano. Harold had explained that the eruption occurred because children had been fooling around with humpback salmon, making them into living candles and then watching them swim up the Tseax River. The volcano erupted as a warning that fish must not be shown disrespect. I told Harold that there are old logbooks from European sailing ships that record a huge eruption of some kind being sighted far to the east, more or less where Aiyansh stands today. Perhaps that was the volcano of the Nisga'a story.

No, no, said Harold; it was long ago, long before the time of ships. And it was to do with the power of spirits, not the power of the land. Might they not be the same thing? I ventured. Harold replied that "power" is what some of the people have, what they had "in the old days."

I paused, wondering what to ask next. We drove along for a while in silence. Then I said to him: "What is power?"

Without a blink of hesitation, he answered: "Transformation. Transformation is power."

The next day, we visited Harold at his home. I was eager for him to say, on film, what he had said in the car about transformation and power. In reply, he told the story of people he called "the demised ones" and of the hero child who recovers the light. Along the way, he gave a version of Nisga'a origins. He spoke in English, though in a form that was often hard to follow and must have been shaped by the original Nisga'a. Here is a summary of some of what he said.

Long ago, long time ago, the people were like dead, demised, ghosts. They lived in a village on the shore, and waves of the sea broke onto the edge of the land. Like they do now. Making the same noises, hissing on the gravel. But it was dark, and the people had no substance.

One day a young man came to the village. He came from some other village, far away. He was also one of the demised. There was something different about him. He was a stranger. But he liked to go hunting, out onto the sea. He liked to go hunting for sea lions. The people took him with them. They got on well with him. But at night, he often slipped out of the house where he stayed. And sometimes when the hunters left, he stayed behind. He gave some reason. He would stay behind in the village.

The other hunters began to wonder what he was doing. They were suspicious. One day they asked the young man if he wanted to go sea lion hunting with them. He said he would stay behind, and that they should go without him. The hunters had made a plan; they had decided to trick him and to see what it was he was doing. So they left him behind, as he wished, and set off without him, pretending to go out to sea to hunt. Then, when they were out of sight, they doubled back, returning to the land. They hid their canoe and came back secretly to the village, to see what the young man was doing.

They hid and watched him. They saw that he went to a place where people were buried in burial boxes. In those days, the people put the bodies of the dead in boxes and set them in a tree. The young man went to one of those trees and climbed up to the box. He got into the box. The hunters recognised that it was the box of a woman chief who had died some time ago.

They saw that the young man climbed into the box and had sex with the dead woman.

The hunters ran to the tree and waited for the boy to come down. When he saw them there, he ran away. As he ran away, he called out: "In the tree is our child." Then he disappeared, going up the coast. They never saw him again.

Some of them climbed up the tree to see the body of the dead woman. They looked into the box and saw that she was pregnant. Soon after, she gave birth to a child, who was born from sex between her and the young man who had now run away.

They took the new baby to the village. The women there gave it different foods. Whatever they gave, it refused. They offered it boiled meat, and soup, and plants they prepared. They offered it the breast so it could have milk. The baby refused everything. It got hungrier and hungrier. Then one of the women offered the baby the sap of a pine tree, the resin. Immediately the baby fed and was satisfied. So for every meal it ate pine resin.

This child grew very strong. It had more strength than any other child. It became a young man and a good hunter. But this child seemed to know everything. Its knowledge was more than any other of the people there. It had powers, too. As it grew into a young man, it showed that it could transform itself into other beings.

One day, the young man said: "It is dark here. How come there is not enough light?" People said that their grandfather must have the light; he kept it, in a box, in the sky. The young man said that he would go there and get the light. The others told him that it was not possible. No one could get through to the grandfather's house, where the box was kept in which the light was hidden.

The young man persuaded one of his friends that they should go together to the sky and find a way of going through to visit the grandfather. They would see the box and get the light. To do this, said the young man, they would turn themselves into birds. Then they could fly to the sky. His friend agreed that they would do this together.

So they became birds, and they flew to the sky. It was a long way. When they came to the edge of the sky, they saw the entranceway to the grandfather's house. It was opening and closing, like the beating of a heart. Everything was blood red. Opening and shutting. The young man saw that if they flew in at the wrong moment, they would be caught by the closing of the entrance and would lose their lives. So he said to his friend: "We must count the beats, and as it closes you must fly towards the entrance. Then when you get there, it will be opening again." That way he thought that they would be able to get through.

The young man went first. As the entrance was shutting, he flew at it. Just as he got there, and just as his friend thought that he would fly right into the blood and be killed, the entrance opened. And he went through. The friend then had to do the same thing. But he was too afraid to fly at the closed entrance. He waited a moment, to see if it was opening. Then he flew. But it was the wrong moment. As he reached the entrance it began to close, and he was caught and lost.

The young man was now alone. He went through the entrance and found the grandfather and grandmother. Sure enough, they had a box. And he could see light leaking through the thin gap where its lid closed. Now he transformed himself into a small child. The grandfather and grandmother were happy to see this child. They were lonely for children. They began to play with the child, calling him "grandson."

The child, seeing that the grandfather and grandmother loved him, and knowing that they feared he might go away again, asked for many things to play with. Eventually, he asked for the box to be opened, so that he could play with whatever was in there. At first, the grandparents refused, saying that the box must not be opened. The child kept on asking. The grandparents were afraid that if they kept on saying no the child they now loved would leave them and go back to the earth below. So in the end they agreed, and allowed the child to open the box and to play with whatever was in there.

In a flash, the child turned himself into a bird again and pecked at the light, swallowing it. Then he turned and flew out of the grandfather's house, and through the blood-free entrance, and down to the land, and at last came again to the village by the sea.

Once he was there, he turned himself back into a man. As he did so, the light was released from his body. For the first time, the world where the demised lived was bright with sunlight. Not all the time—the grandfather and grandmother catch the light and make nighttime, so that they are not always thinking they have lost the light from their box.

This light changed the people. They were no longer demised and ghostly. They became solid and real. They became the kind of people who are alive now. And the young man went on to play many more tricks, making things.

2

Christian missionaries have had great difficulty translating the Bible into the languages of hunter-gatherers. Finding equivalent terms for all the features of ancient Palestine, from camels to vineyards to orchards to millstones to lions, is hard enough. Harder still are translations for the moral concepts so integral to Christian teachings, like evil, redemption, sinner and disciple. The Inuktitut version of the Bible, which I have tried to read, is bewilderingly complex, full of both grammatical constructs and terms far removed from the forms of any Inuktitut conversation I have ever heard. Yet whatever the quality of their translations, Christians have had huge success, both in the Canadian North and in many other hunter-gatherer societies. Obscure as the language of the Bible may be, its events, or story, or message have been welcomed and adopted.

Why have peoples whose own lands, customs and beliefs are so removed from those of Christianity been so receptive to God and Jesus? Look again at the way farmers have created God, and at what it means for hunter-gatherers to hear about him for the first time.

In the Judaeo-Christian narrative, God creates the world. Genesis is the beginning of a set of myths in which God is pivotal as an omnipotent and omniscient power whose dictates have absolute authority. Those hunter-gatherers whose societies have been documented had no one creator, no supernatural power that rises above other parts of either the natural or the supernatural world. When missionaries sought hunters, they encountered spirits and forces, but none that was an ultimate being, a creator who made all that exists from the nothing that preceded it. Missionaries did not encounter a rival God. This made their task much easier than it might have been: they could tell a new story without having to displace an old one. Here is the starting point for understanding how the Christian God of farmers has been welcomed in places where farming, and all that it implies, is utterly foreign.

3

Europeans first met hunter-gatherer societies of the continent that was to become North America about a thousand years ago. Vikings travelled from the west coast of Greenland, crossing the neck of sea that later explorers would name Davis Strait to reach the shores of North America. The Vinland Saga, the story of some of these journeys of exploration, gives tiny glimpses of people the Vikings called "Skrælings." There may be stories deep in the oral traditions of the Inuit of Labrador, or the Mi'kmaq of Newfoundland, that record events from the Inuit or Mi'kmaq point of view, perhaps with some revealing name that the hunters of those coasts gave to the Vikings. But no such stories have as yet been found. They may have died out with the peoples who told them, for the fate of the Mi'kmaq, in the aftermath of later encounters with Europeans, was devastation: many died in a series of conflicts with armed settlers and from new diseases.

The hunters of the Atlantic coasts lived well. The waters were rich in cod, harbour seals and porpoises. The rivers had abundant runs of salmon. The landscape is very similar to that of the Scottish Highlands: windswept hillsides, wooded valleys, gannets gleaming white in the sky and puffins nesting in their burrows on offshore islands. Northern Europeans knew the look and the creatures of this kind of land. And they reached it with relative ease. First the Vikings from Greenland, then cod fishermen and whalers from Spain, then Moravian missionaries from Germany, then settlers from southwest England. The Vikings and fishermen went home without creating more than temporary worksites on the new shores. The missionaries and nomads of agriculture, looking for places to pasture their flocks or create their fields, sometimes succeeded and sometimes failed to establish outposts of Europe on this particular edge of the Americas.

This was a familiar terrain, but it was harsh. In 1534, when the explorer Jacques Cartier travelled the coast that would later be called Labrador, he was dismayed by its desolate qualities and lack of soil. He called it "frightful and ill-shaped ... the land God gave to Cain." So the Europeans who did attempt to settle there found themselves living on fish and game, using the land more as hunters than as farmers. To this day, there are the descendants of Europeans living along the North Atlantic coast, in tiny villages and "outports," as Newfoundlanders call them, who wear the clothes and speak the language of the Inuit.

It was on the Atlantic coast that the Christian God first reached hunter-gatherers of the Canadian Subarctic. On the Pacific Coast, Christianity spread later, from Siberian outposts in the north and the outposts of American evangelists to the south. Later still, Christianity made its way into the High Arctic. By 1900, every part of northern North America was within one or another diocese. Harold Wright, like Anaviapik, had grown up at a time when missionaries were working among his people. For the grandparents of these men, Christianity had been something of a rumour, glimpsed at Sunday morning rituals on board the ships of whaling crews or at makeshift services held by roaming explorers. The people of Labrador, however, had been hearing the stories of both Genesis and Jesus for many generations.

4

I first worked on the Labrador coast in 1976. The Labrador Inuit had begun a study of their relationship to their lands, and they were mapping all the places they had used and known. I had the job of talking to elders about the ways in which Inuit life had changed, or stayed the same, since the arrival of Europeans. Many of the changes had come from the Moravians, German missionaries who set up churches, trading posts and brass bands in several villages along the coast. The Moravians were the first Christians to establish themselves among the Inuit, and they had been there since the end of the eighteenth century. By the mid-1970s, every Inuit family was Christian, and many were intently serious about going to church. Shamans had been disgraced; people told me that shamanism had long since been driven from their minds. They believed in God and Jesus, which the missionaries had shaped into the Inuktitut forms "Guti" and "Jisusi."

The Germans had had other impacts on the language. Many Labrador Inuit counted *ainsilik, tsuvailik, turailik, virilik, vunvilik,* from the German *eins, zwei, drei, vier, fünf.* The word for potatoes, which Moravians first brought to the area, was *kartupalik,* from the German *Kartoffeln.*

Even though they adopted these and other German words, the Inuit of the Labrador coast did not give up their language. They continued to live as hunters, with an admixture of trading, even as they adjusted their seasonal movements to spend more time in the Moravian communities. Farming did not displace hunting. But the Inuit did come to believe in the Christian God, and they accepted many parts of the Christian story: Jesus's birth, the crucifixion, the resurrection, and the division of the afterlife into the territories of heaven and hell.

5

The Inuit are not the only hunters of the North Atlantic coast. Another culture, with territories reaching from the Arctic tundra of northern Labrador to the subarctic forests of the St. Lawrence, has hunted and gathered inland alongside the Inuit for several thousands of years. They are the Innu, a people who speak languages of the same Algonquian family as the Cree and Ojibway. The similarity of the names Inuit and Innu is an odd linguistic coincidence. The words both mean "person" or "human being"; but the languages themselves, Inuktitut and Innu'aimun, are mutually unintelligible. Their respective language families, Eskimoan and Algonquian, are as remote from one another as English and Japanese.

Between them, the Innu and Inuit of Newfoundland occupied all of the region that became Labrador, and also travelled into part of what is now Quebec. The Inuit are based on the coast, hunting sea mammals; the Innu centre their lives inland, relying on the animals of the forest, tundra and freshwater lakes. Both peoples depend on caribou for clothing and meat. Inuit and Innu hunt the same herds, but they do so at different times of year or in different places. At the northern edge of the caribou migrations, the two peoples' territories overlap. But they have always avoided one another and, for the most part, have systems of life that are complementary rather than competitive. Archaeological and linguistic evidence suggests the region and its resources have been shared in this way, with a minimum of conflict, for a very long time. Each group knows how to find and harvest the animals on particular territories; neither seeks to move beyond these territories, into what would be, for them, a dangerous unknown.

Missionaries came to Innu territories in the 1700s. They urged the families they met to stay at their summer gathering places, near the coast, where they could attend church and immerse themselves in Christian teaching and practices. But the Innu economy required that people continue to spend much of their time in their territories in the interior, hunting caribou and other species, securing meat for food and skins that they could trade with white newcomers to get the guns, ammunition, tea, tobacco and sugar on which they soon came to depend.

All northern hunter-gatherers have patterns of movement on the land. The seasonal round, in its basic form, provides a period of time during which families gather together in relatively large groups. This is when storytelling, pooling and exchanging of information, spiritual life (including gambling games) and marriages can all take place. When families leave these large communities, they spread out in much smaller groups to a succession of hunting areas and campsites.

The Innu gathered together in summer, when families moved to the coast. In summer, the coast is cooler and windier than the interior, and people could escape the worst of the mosquitoes and blackflies that swarm in untold billions when the air temperature rises above freezing. In late August or September, Innu families moved out onto their lands, spreading in family groups across a huge area for the rest of the year. For many northern groups, including the Innu, the coming of traders and missionaries resulted in a second yearly gathering, in midwinter, again to sell furs and buy European goods as well as to participate in Christian rituals.

In Innu coastal villages, the Church put in place a regime of belief, worship and education that exemplified the colonial wish to transform hunter-gatherers into some version of Europeans or Canadians—men, women and children who dressed, spoke and worked like "everyone else." Christianity urged people to "settle," to leave their territories. Meanwhile, these territories were declared to belong to "the Crown," and then could be occupied and transformed as and when the colonists deemed fit. The Innu no longer were seen as the occupiers, much less the owners, of their own lands. The consequences of this jurisdictional shift were to be revealed in the late twentieth century: a vast dam was built in the heart of their territories, NATO set up a low-flying jet-bomber test zone across the entire area, and tourist developments were located on prime fishing spots. The Innu were expected to be grateful for the chance to live as Christians, with the benefits of education and a permanent address.

So the Innu, like many other hunting peoples, adapted both their seasonal movements and their spiritual lives to accommodate European ideas and economics. It was a balancing act that, as the stories of both residential schools and "development" have shown, can endure only so long as the people continue to spend much of their time in their own lands and own languages, some distance from the newcomers. With their long history of dealings with missionaries, and their ongoing attempts to achieve a balance between rival ways of living in their lands, the Innu might well be expected to have yielded more completely than other hunting peoples to the ideals and commands of the new God.

6

In 1988 I worked on the film *Hunters and Bombers* with the Innu of Sheshashiu, the largest Innu community in Labrador and the one most affected by the NATO low-level flight training centred at Goose Bay, about ten miles away. Sheshashiu is a sprawl of houses connected by gravel roads set on the shore of Lake Melville, near the mouth of the Churchill River. It is a "settlement" like so many others, a place where most houses were built at the lowest possible cost and where people felt confined and demoralised. A Catholic mission and school presided over a society that for many years had struggled with everyday eruptions of violence, binge drinking and a deep sense of hopelessness.

In the year before filming got under way, the Innu had begun a campaign against NATO flights over their lands. Those who went out to hunt, fish and trap in the great spread of subarctic forests and lakes were dismayed by the jet fighters roaring overhead, sometimes no more than 200 feet from the ground, travelling at such high speed that sudden explosions of engine noise were the first sign they were there. For the Innu of Sheshashiu, the incursions of these planes onto their lands meant that the places they went to escape the demoralisation and violence of settlement life—places where they could live in ways that helped restore pride and health—were also being invaded.

A young Innu man, Alex Andrew, offered to work with me on the film, acting as both guide and interpreter. He suggested that we interview some men in Sheshashiu who would talk about life in the settlement; this would be the background against which the impact of the NATO planes could be understood. Alex proposed that we meet with Sebastien Pastichi, who had never before talked to outsiders about what had happened to his people. Sebastien seemed heavy with sadness, and he spoke with great care in a slow voice. This is a translation of what he said:

"They spread death everywhere. They leave an oily scum on lakes and rivers. They fly too low over our marshes and forests. They spread death on our land. They fly incredibly fast making an incredible noise. The animals can't feed properly ... Neither trees nor animals can grow as they should ... When they go over it's like an explosion. They scare the life out of me.

"They make laws against everything, especially hunting. They harass us even when we have done nothing. We have to have some pleasure, so we drink. It's the one thing we're allowed to do. Our way of life seems to be illegal. It's hard for us to be ourselves. When we hunt, we hunt for our children. We kill what we need, no one should stop us ... Even when we're hunting they interfere. So we drink. Me too, that's why I drink. The only right we have is to be drunk."

We also did several interviews with Elisabeth Penashue, one of the most articulate of the women who had led the anti-NATO campaign. She too spoke about "government," assigning to it many of the causes of Innu demoralisation. At the conclusion of these interviews, she said, fighting back tears, "I really envy my parents. They were always out on the land. It could have been the same for me, if the government had not done all these things—the church, the school, the alcohol. If only they hadn't done so much harm to the Innu."

Innu people told us, over and over, that their health and their future depended on getting out "into the country," away from the settlement, away from government, school, the Church and alcohol. Many Innu families were spending time at fishing and hunting camps during the spring and autumn. The community had set up an outpost programme, with a fund to hire float planes that could take people to the lakes where they had always most liked to spend part of the year.

The campaign against a century of missionary pressure was getting results. On our first shoot, we met two canoes being paddled by young men from spring fishing camps, far inland, back to Sheshashiu. This was the first overland crossing of ancient trails for many years, and it represented a new determination that Innu must live all through their territories. We heard many times that the old spiritual beliefs and powers of the Innu could be sustained only out on the land. The Church might dominate the settlement; but it could no longer keep people there. We realized that if our film was to capture the importance of Innu spirituality, we too would have to make a journey to "the country."

7

While working in Sheshashiu, I had met Pien Penashue. He was a tall man in his late sixties with the wiry, tough physique and weathered face of someone who had spent much of his life on the land. I had heard that he was an expert hunter and trapper, and Alex told me that Pien was also the most articulate of all the community's elders when it came to Innu spirituality. Pien suggested that we should visit him at his autumn camp; that would be the place to talk about Innu beliefs.

I set out for the camp with the other three members of the film crew. We flew in a single-engine float plane—a journey of about forty minutes or, as Alex told me, a two- to three-day walk. Every joint of the fuselage rattled as the engine snarled and backfired. The pilot had the nonchalance of those who have developed an eccentric familiarity with wild parts of the world. At one point he asked Caroline Goldie, our sound recordist, if she had ever flown a plane. No, she shouted over the incredible racket. Then have a go, shouted back the pilot, and he proceeded to leave the controls in her hands. There followed a series of lurching changes of altitude. I stared out the window and imagined us crashing. The land below was a bewildering mosaic of lakes, rivers and forested ridges. Not a house or campsite had been visible since we left Sheshashiu. I wondered how long we would be able to live down there. We were carrying sleeping bags. But did we have fishing lines or guns? I thought not. Human habitations were separated by hundreds of square miles; without Alex, we would have no real hope of finding our way. And yet I knew that every lake and hill and river had a name on an Innu map.

Caroline was not long at the controls. After a while, the plane banked and turned back on itself. There below us, almost hidden in the trees, were four white tents. The smoke of several fires and stoves rose into the air. The plane banked again, and the camp was lost to view. The pilot dropped with alarming abruptness.

We landed in the bay of a long, narrow lake. The water had not yet frozen, and it lapped against a long beach of rocks and pebbles. We clambered from the plane's floats onto shore, then unloaded our gear. The pilot, in a hurry to leave, shouted good-bye, shut the passenger door, untied the rope that moored the plane, pushed off, and climbed into his cockpit. The engine fired, the prop turned, and the plane taxied out into the lake, accelerated, and roared off over the trees.

The light had faded into dusk. There were sounds of wood being cut, the paddle of a canoe returning to shore, the shouts of children. A group of these children came over to see us and help us put up a tent. Once that was done, Alex suggested that we go and visit Pien. He would want to say hello, and we could fix a time to talk with him. We walked the short distance through the trees.

Pien sat in his tent, on a box close to the stove. He had been working on an otter skin, which was now stretched to dry on a long wooden pole. His wife was plucking a Canada goose. Another elder was visiting. Others were coming and going. The tent was warm. The people in it were busy, concentrating, quiet, peaceful. One of the children was sleeping. Pien was welcoming. He chatted with us for a while, but it was the end of the day, and he seemed tired. We suggested that we come back the next day to film an interview.

Early the following morning, in a light grey with autumn mist, we found Pien working on his otter. We filmed him as he scraped fat from the inside of the skin. He talked about the condition of animals in the area, telling us he had killed several spruce grouse that were very thin. Their thinness was caused by low-level flying, he said. He was indignant about the flights and the ways in which Innu life on the land was made ever more difficult. Pien spent as much of his life as he could out on the land, moving between many different hunting, fishing and trapping camps. He told us that he had never agreed to stay all year round in a settlement. He spoke in Innu'aimun, and Alex translated. "The government doesn't care about the animals. They want us to stop hunting. But I won't stop being a hunter. When I'm dead, that's when I will stop hunting."

A few hours later, we filmed Pien as he paddled his canoe along the lake in front of the camp, where he was catching many small trout. "The bombers cause so many problems," he said. "They overfly our camps. There were two here yesterday. There never was any government out here, on our land. We never saw them. Now the government gives people money. Welfare allowances. The government wants to own us. They behave as if we have no rights. But we have been hunters here since time immemorial. In the country everyone works together, but in the settlement people fight each other."

Through Alex, we arranged to film another interview in which we could ask Pien about Innu spirituality. Pien suggested that it be done in his tent. Later that day, as we sat in the warmth of the wood stove surrounded by members of his family, I asked Pien about Innu ways of knowing where to hunt. "We know from the drum," he said. "We can see in a drum, and with our songs, where animals are, and sometimes when to go. The drummer has visions, like dreams." He told us that it was not now the time to drum, but one evening it would be possible to watch this. He had a drum with him, in the camp.

He also spoke about the animal master, the spiritual force that expressed the concerns of the caribou and whose collaboration was necessary if hunting was to succeed. The same kinds of spirit forces were important for all hunting, for all actions on Innu land, he said. If you did not know about these, and show them respect, then you could not know about hunting or about animals.

I asked about scapulamancy: I had heard that Innu could use the shoulder blade of a caribou to learn about hunting, or even about the future.

Yes, he said, the shoulder blade of a caribou. Or a porcupine. He would show us with the shoulder blade of a porcupine. He had one here. Where was it? Yes, in a box somewhere. He searched and found it. A small, white, flat bone. He put it aside while he cut a short piece of wood, which he split at one end. He slotted the porcupine's shoulder blade into the split, so the stick became a handle.

Moving to the wood stove, Pien raked some embers together. He squatted within reach of the fire, and paused. He looked at the shoulder blade, then at us, then at the fire. With great care he held the bone in the fire, keeping hold of the wooden handle. The hot coals glowed red around the bone. After a few seconds he leaned back and drew the bone from the stove.

He examined the shoulder blade. "I see a skidoo," he said. "Someone is going to be coming on a skidoo. This means that the hunting will be good; that is why they are coming here."

He chuckled, conscious, I suspect, that this was not quite the magical kind of message we outsiders may have expected. I leaned across and looked at the shoulder blade. It did indeed have the shape of a skidoo burned into it.

Pien talked about other Innu ways of contacting spirits and getting knowledge. He spoke about the shaking tent, the most powerful of all Innu shamanic techniques. A man would go into a small, carefully made shelter covered with caribou hides. A spirit would join him there. They would struggle with each other. The tent would shake. Those watching were left in no doubt: the movements could only be those of supernatural powers. But this was almost never done now, he said. He had not seen a shaking tent for many years.

Pien went on to explain that the missionaries had done everything they could to bring an end to the use of shaking tents. They had also opposed the burning and reading of shoulder blades. The missionaries had said that these practices were devil worship, and evil, and must stop. Men who continued to use the tent were refused credit at the store, were criticised and undermined by the missionaries in every way possible. They became afraid. Only a very few Innu continued even to burn shoulder blades. Pien said that his people must overcome the missionaries' ideas. Innu strength would come back only when they rebuilt their spiritual life, their own knowledge.

After the interview with Pien, we filmed scenes around the camp, then settled back in our tent to get some rest. Late in the evening, I walked in the dark to the rocks at the edge of the lake. The air was still, and the water was calm. Ripples along the shoreline lapped with the softest of sounds. I stood for a long time, thinking about the day, wondering at the quiet beauty of hunting camps. I was awed by both the vastness of the natural world and the group of tents set among the trees, so small and seemingly vulnerable in this huge landscape. I knew that the spread of Labrador forests, lakes and ridges covered an area the size of Ireland. To the north, the land extended to the treeline and beyond. To the west, the lands of the Innu shaded into those of the northern Cree; these bordered in turn the lands of the Chippewyan, an Athabaskan people whose territories bordered those of other Athabaskans beyond Great Bear Lake—the Dogrib, the Slavey and, in the eastern foothills of the Rockies, the Dunne-za. This continuous line of hunting and gathering peoples was about 2,500 miles long; even in the 1980s, it was a line that crossed no more than two highways and cut through no towns. All of this land was named and known by the cultures that, intellectually and morally, had shaped it for thousands of years. The glow of lamps through the canvas of the Innu tents exuded a profound confidence. I was warmed by the intimate quality of those lights, but at the same time chilled by both the geography and history in which they endured.

For a time I sat on the rocks by the dark water, thinking about the mixture of chill and comfort. Then, needing to feel the reassurance of the tents, I got up and walked towards them through the trees. Suddenly I heard singing. I stood for a time and listened. First a woman's voice, then a man's joining her. It was a tune I half knew but could not identify. Then I remembered: it was a hymn. The voices stopped. Another song began. This also was familiar: it was the tune of a Quebec folk song. The isolated sound rose into the darkness and silence of the forest. After a while, the singing stopped. I made my way back to my tent.

The following morning I asked Alex about the songs. Oh yes, he said, older people often sang. Sometimes at the end of the day, after lulling children to sleep, and sometimes when they woke in the night. Pien and his wife liked to sing; Alex told me it was them I had heard. They sang like that almost every night. And the songs? Hymns, said Alex. They had books of hymns that the missionaries had taught them. Some might be tunes they had learned long, long ago. Pien's father and mother had passed on what they had been taught as children by the Catholics. The family was very religious. Religious? Yes, said Alex, all the older people were Catholics.

I was astonished. The previous day I had watched Pien's proud demonstration of scapulamancy, and had delighted in his insistence on Innu forms of knowledge and spirituality. How was it possible, then, that he would end that same day by singing Catholic hymns?

The next day we visited Pien's tent again. He was standing outside, working on his otter skin. Alex explained that I wanted to ask some more questions. Pien said he did not mind; he wondered what I had been thinking. I told him how puzzled I was: yesterday we had heard him talk about Innu spirituality, Innu beliefs, and he had read a message from a burned shoulder blade. He had spoken about how the missionaries had tried to stop these parts of Innu life. He had explained that the Innu beliefs and customs are true. Then I heard the singing of hymns. The songs that the missionaries brought. How could I make sense of these two things?

Pien listened to me, and then to Alex's translations of my

questions, with close attention. He had no difficulty giving me an answer. I wrote it down: "The Innu religion is the religion of life. Christianity is the religion of death. We have to follow Innu ways in order to get our food here on our land, to live. But we have to follow the Christians in order to get into heaven. When we die. So we need them both."

8

According to many Christian teachers, non-believers are damned—if not to hell, then at least to purgatory. This has given rise to debates about children who die in infancy. Do they go to heaven? Analogous questions have arisen about adults who have never had the chance to hear the Christian message. Are they doomed, by ignorance, to hell? That such questions can be asked, can become, indeed, matters of intense theological preoccupation, reveals the absolutist notions at the centre of the Christian message. From this absolutism missionaries take their first justification: the "heathens" must be converted, for they otherwise are doomed. If they can be led into belief in God, and be persuaded to follow the laws and customs of Christian life, then they can be "saved" from damnation.

The Christians who proselytised among hunter-gatherers spoke of spirits that appeared to have immense power. Both the missionaries and the spirits they described were illuminated by the wealth and authority of the societies from which they came. Missionaries were representatives of both a new God and the white, colonial world. Nor can the arrival of missionaries be separated from the spread of European diseases. Even if missionaries themselves did not bring the diseases, they were preceded or followed by many who did. Missionaries were the advance guard and companions of colonial processes. Often, they offered Christianity as a way to ease the pain of which they were, albeit unwittingly, an important source. They also brought medicines with which to treat the new illnesses that so often weakened, even decimated, hunter-gatherer populations. Many would be dismayed, in due course, by the destructive power of these processes, and would find in their Christian faith some powerful arguments to be made on behalf of the people who were being oppressed and dispossessed. But their religious and moral message was unequivocal: Christians had the truth, and hunter-gatherer shamans were evil liars. Missionaries worked with determination and, at times, with great personal courage to persuade hunter-gatherers to give up their shamanism and embrace Christianity. These messengers of Christianity attempted to supplant beliefs as residential schools attempted to eradicate languages. They sought to change peoples' minds. No coincidence, then, that under the guise of "education" the two projects so often went hand in hand.

In regions where shamanistic beliefs were expressed in regalia, rattles and drums, missionaries often urged people to make great bonfires on which to throw all their "heathen" possessions. Totem poles and house posts, deemed to be idols in some kind of devil worship, were similarly cut down and destroyed to demonstrate acceptance of Christianity. In some cases, missionaries pursued Christian ideals with astonishing cruelty. In Gitxsan country in northwest British Columbia, a Reverend Tomlinson insisted that men for whom he found Christian wives should, in return, work as labourers for him for a minimum of seven years. He also built a jail, in which he imprisoned those whom he charged with having broken the new, Christian laws. In neighbouring Wet'suwet'en land, Father Morice, among the first and most influential missionaries in the area, turned his first converts into informants who had the job of reporting to Morice the breaches of Christian rules among their neighbours; Morice is remembered as having instituted public floggings and imprisonment for the offenders. The record also speaks of him tying the hands of mothers who had not had Christian weddings, and keeping them tied even when a mother's child "would be crying and crying for her, and the mother's hands were tied and no one could untie them without the order of the priest."

In both the Arctic and the Subarctic, missionaries were often dependent on the goodwill of their Inuit hosts and converts. Some achieved warm relationships with the families who most readily adopted the Christian message. But even there, they came as men who knew what was true and best, beyond doubt. They were dedicated to changing the beliefs, and therefore many of the customs, of all the Athabaskan and Inuit peoples they met.

9

A feature of Inuktitut is its use of words that mean "perhaps" and "maybe," with different affixes to signal varying degrees of likelihood. Many speakers, especially elders, almost always qualify their statements with some degree of caution. "Will you be going to hunt tomorrow?" "If the weather is fine, if things are all okay, possibly." One elder I knew well would often add "if I am still alive" to anything he said about his plans.

Outsiders may experience this way of speaking as equivocal and hesitant. This is not an unreasonable assessment: hunters are aware that decisions must reflect reality, not be imposed on it. And the real, the facts, are always changing. So people must leave decisions to the last moment, eschewing anything that could be called planning. In an economic system whose success depends on accuracy, speaking anything but the truth creates real dangers. These are social or functional explanations, but I suspect there is another, deeper reason for the ambiguity in Inuit and perhaps all hunter-gatherer language.

The knowledge that marks hunter-gatherers' relationship to their territories is an intricate mixture of the real and the supernatural. There are facts about things and facts about spirits. And the wall between these two kinds of entity is not solid. People can cross from the natural into the supernatural; spirits can move into the human domain. Just as this divide between physical and metaphysical is permeable, so also is the divide between humans and animals. In this way, the boundaries around the human world are porous. This porosity is the way of seeing and understanding the world that underlies shamanism.

All hunter-gatherer societies, and many other tribal cultures, are said to be shamanistic. The term comes from "shaman," a word that the Tungus people of Siberia used to denote a person who has the power to cross from the human to the spirit world, and to make journeys in a disembodied form. Anthropology and new-age religions apply the word "shamanism" to belief systems in which spirit possession and spirit journeys are of central importance. There are many versions of this form of spiritual and religious life, each of which carries the idea of spirit possession into its own social and economic domain. The important and distinguishing feature of hunter-gatherer shamanism, however, is its mixture of flexibility, on the one hand, and firm attachment to a particular territory, on the other.

Hunter-gatherer stories reveal an array of spirits that influence what happens and are available to be influenced. These spirits are flexible and ambiguous in character. The supernatural creatures that have undertaken to make and keep the world as it is are tricksters as well as godlike. In the myths and histories of the hunter-gatherer world, there is a lack of defining line between good and bad, playful and serious. There is an instability of moral qualities, therefore, that matches the instability of identity.

A ghost becomes a boy becomes a raven becomes a feather becomes a man. A man becomes a salmon becomes a spirit becomes a woman. A girl becomes a dog becomes a seal becomes a spirit. A spirit becomes a penis that is fed to a woman who becomes a fox who becomes excrement that becomes flies that are spirits. A spirit becomes a man who has sex with a corpse who gives birth to a spirit who becomes a boy who becomes a bird.

Here are metaphors and magic, and a world where there are no material certainties. In these conditions, respect for the world is as important as knowledge itself. The spirits have to be placated, their potential for causing harm kept to a minimum. There is no limit to what the world may contain, or to how the unknown may reveal itself. Everyone must pay close attention, be careful, use every faculty to be aware of the land and all that it may hold. And this is a specific land, a territory; the power of shamanism lies in a particular area and is not a way of understanding or influencing the world as a whole.

It is easy to see how these shamanistic ideas both express and contribute to the hunter-gatherer's reliance on detailed information and great intuitiveness. Shamanism prepares the brain to work at its fullest, widest potential. There is a profound and intelligent uncertainty. No one knows what is going to happen or which decisions about any part of life will turn out to be correct. Hence the importance in Inuktitut, for example, of expressing caution and qualifications of all kinds. The analogue nature of myth mirrors a sense that the world itself defies digital ways of speaking. For those whose judgements about the real are at the centre of everyday life, subtlety of expression and habitual equivocation are ways of keeping close to the truth.

To equivocate is to refuse absolutes. People have their beliefs, but they do not claim these as the only *possible* beliefs. A people's confidence in their own territory comes from their understanding of, and reliance upon, the spirits that belong there. Their stories about this place are part of what makes it theirs, and establish how they can rely on it. Other places require different knowledge, have different stories, are influenced by spirits that they do not know. This makes a change of territory a dangerous matter; there is only one possible land. Spirits may be full of trickery, sensitive to human disrespect, but they are part of what is understood. And the knowledge of these spirits is what makes one's own territory a good place to be.

By comparison, the Genesis story of creation does not imply moral ambiguity. Instead, it can be seen as a succession of binary pairs. Nothingness : something. Water : earth. Sea : land. Day : night. Male : female. Adam : Eve. Clean : unclean. Cain : Abel. The binary nature of God's project in Genesis is symbolised in the two trees, embodiments of dichotomies that, in the divine scheme, are at the centre of human destiny. Knowledge of good : knowledge of evil. Mortality : immortality. The language of Genesis contains, as poetry and symbolism as well as in a succession of events, a set of dichotomies. It is the starting point of an intellectual and religious heritage in which conflict between two opposing forces defines what human beings should and should not do.

Dichotomised morality is so integral to the Judaeo-Christian moral idiom that those brought up in its tradition have great difficulty conceiving of any alternative. There is a profound logic to this: the language of Judaeo-Christian morality relies on judgements expressed in two kinds of verbal category, good and bad. The concept of morality is somehow inseparable from a binary ethical structure. The myth of Genesis reinforces this logical point by establishing a story in which religion itself is binary: Jews, Christians and Muslims believe in two ultimate powers, God and the Devil, with one standing for all that is good, the other all that is evil. For Judaism, Christianity and Islam, the divinity is widened by the Holy family and by prophets, and the Devil has his agents. But the existence of these additional figures does not affect the dual nature of the moral and religious message. There is no intermediate figure, no third term, sitting between God and Satan, representing some middle way between good and evil. Monotheism is the doctrine that there is, and can only be, one good God. The missionaries who affirm this monotheism therefore spread the dichotomous moral word. They start with Genesis, as it were, and continue to the events and moral lessons of the Jesus story. They wish for others, for everyone in the world, to accept these stories as ultimate truth and as guides to everyday life.

When Christian missionaries live among hunter-gatherers, they hear about a panoply of spirits. They discover that these are people who use trances, drums, dreams, dancing and even gambling games to make contact with the spirit world. Here are some men and women who can leave their bodies to make dangerous journeys through the air, into the sea and even underground, to meet with spirits. Missionaries regard these beliefs as primitive and ridiculous. They assert, instead, the "civilised" and "developed" ideas of their own creation story, their own God and Jesus, giving the account of the crucifixion and resurrection; they insist that these "real" truths exclude the possibility of shamanistic ideas. In many places, missionaries labelled hunter-gatherers' spirits as devils, miniature forces of evil, and shamans themselves as agents of Satan. Morever, they identified the whole practice of shamanism with impoverishment and disaster here on earth, linking it to poverty and to death by new diseases; and they foretold that the punishment for refusing Christianity would be to burn forever in the fires of a place called hell.

For the shamans, Christianity offered new spirits, unfamiliar but not false. They and their people heard about God and Jesus while experiencing a desperate need for remedies for new diseases that they themselves could not cure. They connected missionary authority with colonial riches and power, and feared eternal damnation—they had not heard of this new region of the supernatural, but they were easily able to add it to the supernatural possibilities they already knew about.

The central difference between Christian and shamanistic beliefs and practices was the very thing that made hunter-gatherers willing to accept, or at least find room for, Christianity. It is the *spirituality* of shamanism that has caused it to accommodate Christianity, a religion whose representatives dedicated themselves to getting rid of shamans and all that they stood for. As the Mowachaht elder Chief Jerry Jack said in 1994: "We never told the Christians that they would go to hell if they did not accept our religious beliefs. That's the difference between our spirituality and the white man's."

10

On arriving to work in Pond Inlet in the autumn of 1974, I learned that Lisa, an old friend from another Inuit settlement, was also visiting. Lisa was a distant relative of Anaviapik, and was staying in the household of one of his daughters. She had left a message with Anaviapik that I should be sure to come and visit whenever I could. That evening I went to see her. We sat at the kitchen table, enjoying the coincidence of our both being in Pond. She commented on the way Anaviapik and Ulajuk often called me *irninguak,* adoptive son; might there be some kinship term that Lisa and I could use when speaking to one another? Adoptive cousin-in-law? Others in the family drifted in and out of our conversation—much of which was in English, since my new-found relative spoke far more fluent English than I did Inuktitut.

Early on in this visit, a young man—I'll call him David—came into the house. He had hoped to persuade Lisa to come and party with him. David was one of the community's most enthusiastic users of mood-altering substances, spending much time looking for and finding local sources of alcohol and marijuana. I knew that Lisa herself had something of a reputation for partying. David was not happy to see us chatting together at the kitchen table, and with some surliness he turned down Lisa's offer of a cup of tea. Nor would he sit and join us. After a few minutes of leaning against the wall, arms folded, hostility in his every gesture, he left. Lisa said she thought he might have been drinking or smoking already.

About an hour later, David came back. Lisa and I were still at the kitchen table, chatting about mutual friends, Lisa's children, the work I was doing. The other members of the household, apart from one child who was in his bedroom at the back of the house, had gone to the store. David came up to the table. His eyes were narrow, unfocussed; he exuded tension and anger. Since the opposite qualities so prevailed in everyday Inuit life, I may have reacted with undue alarm. I was very uncomfortable, even afraid. Lisa, however, seemed to be untroubled, and she made some jocular remark to David about needing a cup of tea more now than when he had last visited the house. This was not well received. David took an aggressive step towards us. I got up from my chair and took a step away from him, at the same time suggesting that he sit down and visit with us. I must have sounded timid and ridiculous. White man, determined not to acknowledge the real, attempts to be polite. For a moment there was a tight silence in the room. David stared at us, looking from one to the other. Lisa now stood and also moved away from him. With remarkable agility and strength, David reached both hands under one side of the table and flung it upside down across the room. He made some alarming threats, then stormed out of the house.

I had never seen anyone in Pond behave like this. On the few occasions I had been with men and women who had had a few drinks, I had seen only laughter and silliness, no violence. But I had experienced bad drunks in other parts of the North, in other isolated communities. I was shocked, and I feared that David would soon be back, ready to attack us instead of the furniture. Lisa laughed at my fears. "No," she said. "He won't come back now. He'll be partying." We picked up the table and tidied up the things that David had strewn on the floor. I stayed until the others in the family returned, then went back to where I was staying. It was dark, and I walked in anticipation of David ambushing me somewhere along the way. I assumed that his fierce outrage came from Lisa seeming to have refused his attentions in favour of mine.

The next day I told Anaviapik about the episode. I think he had already heard of it from Lisa or his daughter. He listened to my description of what had taken place, but I had a sense that he was embarrassed. For Inuit, especially the elders of Anaviapik's generation, displays of anger are deemed to be signs of mental weakness in adults. Anaviapik also had a strong dislike of drunkenness. On both these counts, he may well have been ashamed on David's behalf.

But when I had finished telling him the story, he replied: "*Sakkatuinnalaurpuq.*"

The infix *tuinna* means "merely" or "only." But merely what? The "laur" indicated past tense, but the root he used was unfamiliar to me, and, as had so often happened, the events occasioned a language lesson.

Sakkajuq, Anaviapik explained to me, was what shamans used to do. Going into a trance, leaving the body, travelling in the air, invisible, out of time and out of mind, at least out of ordinary time and mind: this was to be in the shamanic trance. So David was merely trancing, in the manner of a shaman. Did Anaviapik's explanation dignify drunkenness and show disrespect for shamans? I think not. Rather, it showed a familiar skill at finding humour and making connections. It was neither a metaphor nor a joke, yet it was also, in an understated and elusive way, a bit of both. Only a bit—"*mikitukuluutuinnaumat,*" "it's only a wee little bit," as Anaviapik might have put it.

11

There are two kinds of reasons for not drinking to excess. The first concerns economics. The second has to do with pride. Someone with a family to feed and a job to keep must either stop drinking or lose both family and job. Someone who cares about how he is regarded by others may start drinking in order to impress, but he will have to stop for the same reason: chronic drunks are despised by everyone. Farmers and hunter-gatherers have very different economic lives, and they experience social status with different degrees of intensity. But both have these two reasons for not drinking—so long, that is, as they have viable economic lives and something to be proud of. It is all too easy to see how colonial processes give ever more reasons to the colonists to control their drinking, and ever fewer to those whose lands, languages and religions the colonists destroy.

Alcohol is made from cultivated plants or from the milk of domestic animals, and it does not seem to have been developed by any peoples who lived as hunter-gatherers. The agricultural frontier is also the original drunkenness frontier. Hunter-gatherers in some parts of the world used plants to transform their state of mind, and hunter-gatherers everywhere have used drumming, chanting and direct "possession" to reach states of altered consciousness and to make spirit journeys. This may help to explain the enthusiastic welcome many peoples who lived beyond the drunkenness frontier gave to alcohol.

So much is said and written (more said than written, since the anger and racism at work are often restrained or self-censored) about the drunkenness of "Indians" and "natives." Those poor people, those terrible people, those hopeless people—various adjectives consign aboriginal societies to a historical cul-de-sac, some place *without* society. To oppose these stereotypes is to uncover difficulties at the heart of the encounter between the hunter-gatherer and the farmer. To be drunk is to be without order, out of control. Or is it?

Imagine a bar in Fort St. John, the nearest such place to Dunne-za territories. A large, gloomy space, round tables crowding into the distance. Circles of drinkers sit at some of these (it is Saturday afternoon, run-up to heavy drinking time), the tabletops thick with glowing glasses, full, half full, empty. To one side is a row of pool tables. To the other the curve of a large counter, where barmen load tray after tray of half-pints. The noise in the place is harsh, raucous, sharp with both edges of dispute and shouts of laughter.

A curious feature of drinking and drunkenness in the Fort St. John bar is its defiance of the effects of alcohol. This same irony is to be seen in bars throughout the colonists' towns, be they in England, Poland or Japan. Drinkers gulp down a substance that frees them from restraints of all kinds, but they are determined not to lose control. The tough guy drinks hard and does his best to appear to stay sober. Imagine a pub in any English village. Saturday night, at the end of the evening's drinking and darts and dominoes. Men and a few women sit on benches along the walls, or at tables set close together.

The tables are littered with pint, half-pint and small glasses for spirits. At the bar itself, men sit on stools or lean side by side, holding their pint glasses or waiting for them to be refilled. Behind the bar, the publican and a barmaid pull beer and tip measures of whisky, gin, brandy into the appropriate glasses.

Now imagine the way that the drinkers at this English pub talk about their states of mind. "Give us another, Joe. Pint of your best. And one for my friend here. Maybe you can give us a couple of chasers. Time's running out. It'd take more than this to make a man drunk. Isn't that right, Joe? Yes, Joe can tell you. He's never seen me walk out of here with a wobble. Right, Joe?" Even in more elegant circles, drinkers are determined not to stagger or slur, and boast of their ability to walk a straight line or keep a steady hand despite great amounts of drink.

By comparison, the language that Inuit or Dunne-za drinkers use is far more welcoming of the effects of the alcohol. "Give us another. We're getting going here, eh? I'm getting drunk today, boy! I want to feel good. Let's feel good! And when you and I feel good, we'll just keep right on going until we feel better."

These snippets of imagined drinking talk do not do justice to the elaborate way in which everyday English speaks of drunkenness. Pissed, tight, bombed. There are many expressions that refer to and rely on the negative and repressive. "Out of his head" is perhaps one phrase that captures a hunter-gatherer view of the condition. Inuktitut also shows its flexibility when it comes to talk of drunkenness, though there is a tendency to use English—acknowledging, no doubt, the original source of the substance itself.

All drinkers seek release. For the farmer in an English rural pub this is measured, orderly, controlled. The face strains to show no sign of what is felt, no indicators of the release. A loss of balance, a slurring of words—these are weaknesses. The farmer does his best to remain himself. He merely gets drunk. For peoples who welcome transformation, the release is different in nature. It is more complete, for there is a welcome loss of self, a flight into another state of being, another kind of person. "*Sakkatuinnajuq,*" "he merely makes a spirit journey."

These are broad generalisations, of course, but they express opposite approaches to drunkenness, or to any other altered state of consciousness. This is neither to idealise nor to applaud one or another kind of drinking. Alcohol has been a destructive force for many people from both traditions. And where hunter-gatherers have endured dispossession, racism, loss of language and the corrosive effects of low self-esteem, their use of alcohol has a relentless and desperate quality. Inner breakdown compounds the culturally established readiness to lose control, until the combined forces of the old and the new create a maelstrom of personal, family and community disaster.

12

Hunter-gatherer knowledge is dependent on the most intimate possible connection with the world and with the creatures that live in it. The possibility for transformation is a metaphor for complete knowledge: the hunter and his prey move so close to one another as to cross over, the one becoming the other. Here is an intimacy that secures complete understanding. Being able to move with accuracy on the ground appears to require a parallel freedom of movement in thought—an absence of constraint, a welcoming of many states of mind, from humour to trancing to drunkenness. A fluidity of boundaries, a porousness of divisions, can be seen as useful and normal.

This ever-present possibility of transformation is both the opposite of, and an equivalent to, control. Rather than seeking to change the world, hunter-gatherers know it. They also care for it, showing respect and paying attention to its well-being. All hunter-gatherers had rules about the treatment of the animals they hunted and the plants they gathered—rules that were designed to show and perpetuate goodwill. This led to selective harvesting and forms of wildlife management. Plant gatherers left tubers in the right kind of soil to ensure that plenty of plants would be there the following season.

Some hunting peoples of the Subarctic and the Prairies set fires so that fallen trees or dense thickets of brush would not displace the berries, willow and grasses that their economies depended upon. Many fishing systems paid careful attention to the safe passage of salmon and other anadromous stocks. Western science may be able to explain the functional nature of each such practice; for hunter-gatherers themselves, a central concern has been their relationship with the creatures they harvest. In this form of management, people trust that if they do the right things, the world will stay as it should; the creatures and plants they eat will feel welcome and know they are respected, and will therefore continue to make themselves available.

The distinction between respect and control is of immense importance to an understanding of how agriculturists approach hunter-gatherers. The skills of farmers are centred not on their relationship to the world but on their ability to change it. Technical and intellectual systems are developed to achieve and maintain this as completely as possible. Farmers carry with them systems of control as well as crucial seeds and livestock. These systems constitute ways of thinking as well as bodies of information. The thinking makes use of analytical categories that are independent of any particular geography, and not expressive of any given set of facts; the achievement of abstraction and the project of control are related. In addition, the deduction of conclusions from analytical, abstract theory is dependent on a precise rational process. This process depends, of course, on reason itself; and a foundation of reason is the law of excluded middle. This is the seemingly straightforward proposition that nothing can be both true and false, or both itself and not itself, at the same time. The dichotomy of mathematical logic, therefore, combines with the dichotomies of Christianity to express the agriculturalists' fundamental intellectual and moral characteristics.

Hunter-gatherers, on the other hand, rely on a relative absence of exact or abstract categories that transcend geography and specific facts. Their knowledge is compounded of many specifics. At the same time, they believe that the boundaries around these facts, around the real, are unstable; one kind of thing can become another, for reasons that have nothing to do with everyday material events. For hunter-gatherers, the distinctions on which Western "expertise" are based are not at all clear. Hunters' success depends on the relationship with the "wild" animals in their lands, and the wildness of these animals can appear to hunters to be far less than that of cattle or horses. Hunters have a relationship with their animals; and the basis of this relationship is that each depends on the other. Similar kinds of relationship exist between hunter-gatherers and the land itself. A firm line cannot always be drawn between the animate and the inanimate. Among some Australian hunter-gatherers, for example, pregnancy is linked to the movement of the "dust" or "essence" of something on the land into a woman's womb. This something can be a tiny part of a rock or tree or the soul of an animal. Conception is not just an egg being fertilised and becoming a baby.

At the most general level, failure to take care of the land is a moral risk that brings spiritual danger. It also involves spiritual realities, and the relationship between humans and the spirits of the places where both humans and spirits live. These places cannot be controlled, and must not be reshaped. To know them means being aware of both detail and instabilities. Many Inuit would put a little fresh water in the mouth of a freshly killed seal or whale so that it, and all other seals and whales, would know that the hunter respected the willingness of the hunted to be killed and eaten. Gitxsan and Nisga'a would mark the harvest of the first salmon each season with statements of thanks to the salmon for agreeing to return to be harvested in the people's rivers. Algonquian hunters would place the bones of bears they had killed in trees, showing respect and appreciation to all bears. Bushmen of the southern Kalahari who gathered roots from deep in the desert sand would often place an offering in the hole they had dug before filling it in again, as a sign of respect and thanks to the plants. Failure to pay this kind of attention to creatures and plants risks causing them to stay away or to refuse to be found. The hunter-gatherer's underlying purpose is to ensure that the natural world remains the same.

Not all farmers are, or have been, absolutists. The Judaeo-Christian religions may well be an extreme, especially at the colonial frontiers where the agriculture of capitalism meets hunter-gatherers. The farming of the modern era has witnessed the creation of immense surpluses, and hence complex involvement in markets that go far beyond the local region, reaching often to the organisation of large nations. There are connections between this commitment to large surpluses—and thence to profits that can be used to acquire ever more land—and modern society's reliance on an enforced hierarchy to control those who are the labourers and servants to the system. In the economies of the "developed" world, farming reflects, where it does not directly influence, the social and material systems of political control and human inequality. Aggressive rationality and planning are integral to such conditions. And it is here, of course, that the farmer : hunter-gatherer forms of intellectual and spiritual life are so far apart. Where monotheistic missionaries meet the peoples of the Arctic and the Subarctic—or the Aborigines of Australia, or the San of the Kalahari—the two systems of thought are as polarised as the two societies' ways of managing human relationships and using the land. Even in imperial systems that are not monotheist—as with Hinduism in the Indian subcontinent and Confucianism (or indeed Communism) in China—the domination of indigenous peoples, especially hunter-gatherers, has had an absolutist character. The polarisation of farmers and hunters is not to be found only at the Christian frontiers.

Not all such encounters are as polarised as these, however. In some parts of the world, agricultural tribes have lived for many centuries as neighbours to hunter-gatherers, with the farmers often looking to the hunters for spiritual guidance. And there are farming societies where the creation of surplus and the related development of political institutions have continued on a small scale for long periods of time. The relentless economic expansionism, and the accompanying need to justify it, so well known on agricultural frontiers, are linked to the success of capitalism and the rival ambitions of powerful nation-states. Far from national centres and modern economic activities, there are indigenous societies that rely on agriculture in which people use spirit possession and other elements of shamanism to explain and manage misfortune, to make decisions and to foresee the future. For some of these peoples, the spirits of the land itself exist much as do the spirits of wild creatures for hunter-gatherers.

So a porous divide between the human and the non-human world is not unique to hunter-gatherers. Witchcraft, for example, is used in many indigenous agricultural societies (especially in west Africa) to intervene in people's lives. Spirit possession is common in tribal India to make contact with, and placate, gods with influence over harvests and the health of children. And in much of Central and South America, indigenous farming cultures have long relied on hallucinogens for getting into contact with spirits and receiving "spirit power." In all these cases, however, relations with spirits are tied to rituals split off from the routines of everyday economic and social life. In many cases, these relations also are integral to problems of hierarchy and social control, the concomitants of agricultural social development. It is among hunter-gatherers that shamanism, and the porosity of the divide between the natural and the supernatural, pertain to the activities of every day and to all resources. In hunter-gatherer intellectual systems, which are developed with a view to knowing the natural world rather than to shaping it or to controlling other peoples, knowledge and shamanism are inseparable.

13

Shamans, like the Dunne-za dreamers, use sleep and unconsciousness to move through the walls around reality in order to know it better.

The dreamer makes a journey into the land, although his body remains asleep in the safety of his home. The dreamer crosses the boundary between humans and animals, making contact with his prey, and may also move through the boundaries of time. All humans dream, of course, and many dreams are experienced as an escape from the boundaries of everyday life. But for hunters, the dream experience is real. The events of the dream are relied upon as a guide to trails and the location of animals. In this way, dreaming is also an actual hunt or a phase of gathering.

For many hunter-gatherers, dreams are a form of decision making. Along with other forms of insight and intuition, hunters use dreams to help them decide where to hunt, when to go there and what to hunt. These decisions can be matters of life and death; they certainly make the difference, day to day, between an adequate and an inadequate supply of essential food. Some decisions are not complicated: if a family is living in an area good for a particular kind of gathering, then the decision to do that gathering each day can be routine. And if a family has chosen to camp at a place where migratory animals move each year, then the decision to spend much of each day waiting in the best places may also be automatic. But there are more complex decisions to be made. When should the next step of the seasonal round be taken? Will the fish be running in a particular river? Will the deer be feeding in this or that part of the territory? Is it time to go inland, or to return to the coast? Will there be water at such-and-such a pan in the desert? Decisions of this kind involve large risks. Families make journeys on which they consume the few supplies they can carry, and they pass through lands where they cannot hunt or gather, trusting that something their lives depend upon—water, meat, roots—will be where they have chosen to go. The wrong decisions can result in death.

To make these decisions, hunters need knowledge. They must bear in mind all the facts that inform the choice. The patterns of other years. The difference between this year and the norm. The ways in which weather has been changing, over a period of months and in the past few days. The patterns of animal movements, the cycles of animal populations, the relationship between animals and the places they may or may not be feeding. Hunters and gatherers must draw on knowledge that they have accumulated over many years. They must also take careful note of what others say they have seen and done in the last few days. They listen to one another's accounts of hunts and journeys. They watch the sky and feel the wind.

In theory, all this knowledge could be made into a long list, in the way that anyone in a dilemma might make a list of pros and cons, reasons for doing one thing or another. But a listing of facts is not enough. Which facts are the most important? The information may not be consistent. Wind changes urge a delay; news of animal movements suggests all possible speed. The weighting of factors is another kind of knowledge; for decision making, it may be the most important. The simple truth about a difficult decision is that its difficulty comes from the irresolvable quality of knowledge. A deep dilemma arises from the way in which accumulation of facts seems not to decide the issue. So what should be done? In the end, there is a need for some other kind of knowledge, some leap of the imagination, some way of processing the facts so that they yield a conclusion. This is what dreams can do.

Dreamers are aware of the facts; their brains are full of the right kinds of knowledge. But they leave it to a final intuition to "see" the correct choice. Dreaming, like many shamanic techniques, allows a form of knowledge that in effect processes all other knowledge. Intuition is a way of paying the closest and deepest possible attention to the world.

The hunter-gatherers who rely on this form of knowledge could not achieve more reliable decisions by explicit deduction. They acknowledge, from the start, that they do not and cannot control events. Rather, their task is to understand events by moving as close to them as possible. Dreams take the dreamer not to some surreal universe in which the natural order is transcended or reversed, not to a land of fantasy, but to the place and creatures he or she knows best. Moving from the natural into the supernatural, passing through the porous walls that surround reality, the hunter arrives at essential knowledge.

With this mixture of information and shamanism, hunter-gatherers signal and accept that their world is not in their control. They prosper by knowing, not by controlling. They accumulate encyclopaedic information about their territories, yet rely also on myth and belief. They understand the world and make critical decisions about it without trusting to dichotomies of either rationality or ethics. By escaping mere facts, they discover the most important facts of all.

14

The boundaries around the hunter-gatherer world are imprecise and uncertain. This is neither a source of fear nor a cause for insecurity. On the contrary, access to the creatures and the land itself, an ability to make journeys from the human to the non-human world, are ways of being in touch with the very things people depend on. To know the world is to get beyond the confines of everyday reality. To pay real and close attention is to learn what is happening on the other side of the boundaries where everyday life begins and ends.

One boundary that is not easy to describe is that which defines the roles of men and women. Although many hunter-gatherer languages, including Inuktitut, make no distinctions according to gender and have grammars in which being male or female is irrelevant, men and women in hunter-gatherer societies do have particular social and economic roles. A common stereotype of gender roles among hunters asserts that, in the harshness of "primitive" conditions, men are men and women do what they are told. The savagery of everyday life, according to this idea, allows no room for gentle and "civilised" exchange between the sexes. Grunting and resolute, the tough male drags his would-be spouse by the hair to a cave where she becomes his obedient wife. The woman who does more than mind the home (or cave), have babies, stir the pot and stitch skins together can expect a remedial beating. In John Boorman's film *The Emerald Forest,* set in an idealised hunter-gatherer community in the depths of the Amazon rainforest, the hero's bride-to-be explains that in order to marry her, he must club her over the head and drag her off by the hair. He obliges, and a stereotype of male-female relations in the hunter-gatherer world is endorsed.

In this stereotype, dominance of men over women is a function or a stage of evolution. The caveman is at one extreme, inarticulate, club in hand, ensuring that "his woman" is compliant and servile. Then comes the early farmer, with peasant brutality but a touch of sentimental affect. And then the migrant from country to town, the labourer: though he is rough and harsh, he has a supportive wife with whom there is some potential for collaboration. Next there is the country landowner, who, if he is well educated and a "gentleman," treats his wife with affection and respect. Finally we have the pinnacle of civilisation and civility, the educated, thoughtful, literary and "modern" man, who, being more evolved than the others, understands that there must be equality between the sexes.

The theory of progress here draws on myths about each of its stages; women can be treated as servants and objects whatever the supposed level of civilisation. But in popular theories of progress, hunter-gatherers are stuck back there with the caveman. The supposed crudeness of their sexual relationships is portrayed as harsh reality, a kind of grim common sense, for a group of people who have nothing. Poverty, in this model, is something of a corollary to low human development, explaining and justifying any sort of extreme. In reality, those who have the least interest in possessions are the most open to gender equality. Hunter-gatherers do establish well-defined roles for men and women, but there is more mutual respect, a greater recognition of mutual dependency, than in any other kind of society. In a way, the stereotype reverses the truth: those who live in wild environments, or live by harvesting nature most directly, are the peoples who are least "primitive" in their attitudes to women.

The stereotype of caveman or hunter-gatherer male chauvinism is easy enough to explain. Those "primitive" people are supposed to be our earliest ancestors—the original versions of ourselves. What they are tells us who we are, deep down, at our core. So the stereotype of the brutish, hairy man who drags his woman to a cave for sex and servitude is our version of the origin or essence of ourselves. It has nothing to do with any real peoples. In fact, early anthropology set out a theory of development from matriarchal origins to patriarchal revolution. But as a mythic notion, the caveman is a projection of the modern onto the ancient. The grunting misogynist is a crude and idealised version of modern man. If we were but our natural and real selves, thus it would be.

Among hunter-gatherers, there is a widespread division of labour between men and women, with the men often devoting themselves to the pursuit of large animals across great distances. The men, reading weather and tracks, depending on physical stamina, set out to kill creatures sometimes larger and stronger than themselves. The value of this kind of hunting is obvious: one ungulate can yield between three hundred and one thousand pounds of meat, and one sea mammal can provide even more. Yet many studies of hunter-gatherer nutrition show that the hunting of large animals by men produces less food than the hunting and gathering done by women. In fact, both levels of harvest are important: the men have a chance of making a kill that delivers a bonanza of protein—but their rate of success is not high enough to be relied on. The women harvest smaller sources of foodstuff, but do so with great regularity and reliability. In the case of salmon-fishing people, where men pull fish from the river, it is the work of women—cleaning, filleting and preparing fish for smoking and drying—that sets the level of the harvest. In the Plateau societies of the northwestern United States and the interior mountains of British Columbia, at least 30 per cent of nutrients comes from plants, most of which are harvested and processed by women. The well-being of these societies relies on the economic activities of both sexes. Each is the complement—and is *acknowledged* to be the complement—of the other. So division of labour does not result in either sex achieving supremacy over, or independence of, the other. This is a clue to the porosity of boundaries between the genders.

Among the Innu, as in most hunter-gatherer societies, shamans could be men or women. At a more mundane level, men and women are often willing to do one another's tasks. Role reversal is not greeted as either comic or perverse. Inuit men are quite ready to put on a woman's *amautik,* the jacket shaped for carrying an infant on the mother's back. And many Inuit women are happy to join in all parts of the hunt. At least one woman I knew well had lived as a hunter rather than as a mother—though she had also raised several children, relying on the domestic skills of three successive husbands. In Northwest Coast cultures, hunting for large animals and most salmon fishing have been seen as men's work, but the territories and fishing places where the men go are inherited through the female line and have often been held by women chiefs.

The egalitarian individualism of hunter-gatherers has implications for the status of women: these are communities in which women tend to have been treated with respect. A relative lack of formal institutions and ceremonies for marriage may also have connections with both the relative equality between the sexes and mutual respect within the family. It may be that the absence of ritual or rite of passage for couples who decide to live and have children together is a signpost of equality. The more elaborate the rites of passage, the more concerned a society is to tie gender to role and to establish rules governing the assets and behaviour of the married. The absence of such rituals does not mean that hunter-gatherer systems are paradigms of gender equality, or that hunter-gatherer women never have to fear the domination or aggression of violent husbands. It is important to notice, however, that ritualised marriage ceremonies correlate with hierarchy and inequality between the sexes. The *institution* of marriage is closely linked to concerns about property and inheritance, and these are sources of discrimination against women in most agricultural societies.

In many places, women have been powerful defenders of their people's rights. Mary Johnson was not the only female chief who played an important role in the case of *Delgamuukw v. the Queen.* Anaviapik's friend Utuva was more articulate than any of the men we interviewed when it came to describing the nature and form of police and government discrimination against the Inuit. Another Pond Inlet woman, Inuja Kumangapik, spoke with great clarity and forthrightness about the impact of biologists on the snow goose populations of Bylot Island: "There were many, many geese over there, nesting there. Then the goose experts came. They caught them in nets and put rings on their legs. They frightened them all away. They used to be like snow on the ground, the geese over there. But now there are very few. They say that we Inuit did that; they say that we killed them. But it's not true. The goose experts have chased them away."

Among the Innu, women have played an assertive and central role in modern politics. The NATO air force base at Goose Bay was built on Innu land. Many Innu have expressed anger and dismay about the British, German, Dutch and American planes that take off from this base to practise "super low-level" flying over places where Innu families live when they are away from Sheshashiu. Others have complained about canoeing accidents and trauma in their children that they say result from the sudden appearance of planes travelling at 600 m.p.h. right over their heads. More generally, they have argued that this invasion of Innu land is a violation of their rights to live, hunt and travel where and when they like.

In the 1980s, when Innu complaints appeared to make no difference to the NATO flights, the women of Sheshashiu launched a campaign of direct action. They began with a declaration that the men, so many of whom had slid into despondency and drunkenness, had failed to achieve anything, and so the women would now take the political lead. Women had continued to fight for the culture, they said. They were the ones who most often insisted on taking families out onto the land, away from the forces that undermined Innu knowledge and self-respect. In recent years, women had done more harvesting of Innu resources than had men, and they had kept the knowledge for doing this alive. This was not an argument about a modern as opposed to a "traditional" role—Innu women have always hunted and gathered. Their point in the present, in the aftermath of so much loss and despair, was that they were the ones who made sure their sons and daughters learned the skills of the land.

Innu women organised a series of invasions of the Goose Bay airstrip. They walked onto the base, young and old, and sat down in front of the planes. They were arrested, charged, fined. Base commanders arranged for a high wire fence to be built around the runways and set up an armed guard post at the only entry point to the base. The Innu, again led by women, broke through the fence and occupied the runways. More of them were arrested. Some went to jail.

Elisabeth Penashue played an important part in these protests against the NATO presence on Innu lands. When we were filming *Hunters and Bombers,* she was arrested. On her way from police custody to a hearing, she stopped outside the court, turned to the policeman escorting her, and insisted on speaking. At first Elisabeth asked for a translator from among the Innu who were demonstrating against the arrests, but she eventually spoke in Innu'aimun. She said: "The government treats us as if we don't exist. As if we were invisible. It's only when we protest and go on the runways that they take any notice. We've been arrested for defending our land. We won't give up now. No matter how hard it is for us, we won't give up."

The effectiveness of women in so many campaigns against colonial incursions into their lands and lives is inseparable from characteristics deep within hunter-gatherer societies. Although there are clear role divisions, these are systems of family and economic life in which men and women are aware of the daily interdependence of the sexes. Also, these are societies where refusal of rigid boundaries between any categories—whether between humans and animals, men and women, or the natural and the supernatural—has been integral to a way of being in the world. For hunter-gatherers, people's knowledge, health and even survival have depended on being able to move from one kind of reality to another.

15

Innu actions against NATO flying exercises led to a series of court cases. As the result of an appeal process, the Innu had a chance to bring forward evidence that the flights were indeed doing damage to their lands. Lawyers representing the Innu wanted to lead with evidence about the impact of low-level flying on wildlife. They put forward Pien Penashue as an expert witness.

The lawyers for the Crown protested. Pien Penashue was a hunter, they said, not an expert. He did not have any of the qualifications by virtue of which expertise can be recognised by the courts. The Innu's lawyers argued that Pien was the most relevant kind of expert. The Innu defendants and those who had come to watch the trial were outraged by the suggestion that their own expertise was somehow less than that of white academics. How could intellectual work by outsiders achieve the detailed appreciation of the animals, birds and fish in Innu territory that would be accumulated by a man of Pien's life experience? How could anyone other than Innu know best what was taking place on their lands? How could a court reach a just conclusion if it denied Pien the right to be heard on equal terms with other experts?

The court agreed with the Crown lawyers. The judge ruled that Pien lacked the training and degrees that establish expertise. He could give evidence, but not as an expert. The Innu responded by refusing to speak English in the court and by withdrawing interpreter services. The court then had the task of finding alternative interpreters. No white academics were willing, and no court workers were expert enough in the Innu'aimun language to take over the interpreting work. There are Innu communities in Quebec, along the north shore of the St. Lawrence River, as well as in Labrador. These groups speak mutually intelligible dialects. But the court failed to find even one Innu who would help: the people closed ranks. If Pien Penashue was not to be given expert status, the Innu would not offer their linguistic expertise to let the case continue. No more evidence could be heard.

16

The expertise of Pien Penashue is very unlike that of those schooled in the academic natural sciences. The hunter-gatherer is eager to hear what others have learned; he no more insists that he knows everything than he asserts a doctrinaire rejection of Christianity. Knowledge of the animals and of his land is personal and intimate; and the most important judgements rely on that elusive balance of knowledge best called intuition. The hunter-gatherer is a true expert when it comes to naming and knowing the things he or she depends upon. Hunter-gatherers believe in the spiritual link between the creatures they hunt, the places they travel and themselves. European scientists would decry this as a failure to distinguish between material and religious discourse.

In either case, the language of science is opposed to the language of shamanism. Biology relies on division of the natural world into genera and families of creatures. It seeks always for more complete—and potentially ultimate—explanations of how things have come to be what they are; the possibility of a discourse between species is far outside the discipline's parameters. Shamanic forms of expertise are based on a multitude of specific facts; they do not arrange the natural world into hierarchies of families and genera. And shamanic knowledge to some extent depends on the possibility of communication between humans and animals. It accepts without difficulty analogous communication between animal species. Shamans embrace mystery not as a temporary failure of explanation, but as an integral way of apprehending the world around them. Pien Penashue's expertise is indeed very different from that of the men and women whom courts are prepared to qualify as experts.

Hunter-gatherer knowledge is not dependent on absolutism or dichotomies. It is inductive and intuitive; its conclusions emerge by allowing all that has been learned to process itself. Reasoning is subliminal, and therefore has the potential to be more sophisticated, more a matter of assigning weight to factors, than can be the case with linear logic. It is a way of gaining and using knowledge that also seeks for continuity and renewal. It is not tied to attempts to control or change the world. Defining and apportioning land, measuring yields, trading surplus, planning further changes, establishing military forces, organising conquests, dividing spoils—these are activities integral to agriculture that respond well to applications of deductive reasoning and mathematical expression. The difference between farmers and hunters is a matter of thought itself.

The aggression of colonists and settlers has often been fuelled by their absolutism. A dichotomised account of the world and its processes is of immense help to those who advance by conquest and expropriation. It dispenses with negotiation, compromise and reflection on relative truth. Those peoples who do not accept the newcomers' views are evil; those who lead different kinds of lives must be transformed. This set of beliefs and ideas serves the underlying purpose of agricultural settlement: those who occupy lands that the newcomers need can be removed.

It is all the more remarkable, given the wealth and power of many agricultural societies, that the spirituality, stories and beliefs of some hunter-gatherers have endured. Small groups have kept important parts of their heritage alive, both defying and accommodating to new religions. Their example, and their beliefs, have an importance that is out of all proportion to their numbers. They have inspired and influenced people throughout the world who are not hunter-gatherers, but who nonetheless know the need in themselves for many different kinds of god, and know also the importance to humanity of resistance to absolutism of all kinds.

SIX: **MIND**

1

In 1973, after a year in the Arctic, early one morning I visited Anaviapik's house in Pond Inlet. Anaviapik and Ulajuk were at their small kitchen table, drinking tea. Everyone else was asleep. I went in, helped myself to a mug and sat down with them. As always, they said how pleased they were to be visited. "*Pulartiarit*," said Anaviapik, meaning literally "Visit well," "Be welcome." Then he asked: "*Isumassaqarpit?*" "Do you have the material for thought?", meaning "Is there something on your mind?" "No," I said. "I'm just visiting."

"That's good," said Anaviapik, "because I have a thought, and I have a question for you."

His question, which seemed to arise from a conversation that he and Ulajuk had been having just before I came into their house, was this: "*Qanuingmat tassumanik Qallunaat isumaqattalaursimajuit Inunnit isumarqanigitunit iila isumarqijugut Inuulluta? Taima isumangmata tusaumavit?*" "How is it that in the old days the Qallunaat always thought that the Inuit had no thoughts and that we Inuit were mindless? Is that what you have heard?"

As I thought about what to say, Ulajuk and Anaviapik smiled at me. Then both of them burst out laughing. I recognised their laughter as a way of removing any possible risk that they may might sound aggressive. This was just a question, a puzzle, something they wanted to talk to me about.

2

In 1979, late in the year but before the snows had come, I arrived in Fort St. John to discover that Thomas Hunter, the Dunne-za elder, was in hospital. Thomas was now in his eighties, a small, tough man whose face had become a maze of wrinkles but whose hands were still strong and whose eyes were still bright with curiosity and ideas. He had been ill for a while, coughing a great deal and suffering periods of sudden weakness.

A few weeks earlier, Thomas's family, along with several others, had moved from the Halfway River Reserve to a moose-hunting area a few miles to the north. There they had been preparing dry meat. Moose were plentiful, and the hunters had killed several. The meat had been brought back to the family tents for the women to slice into long, paper-thin sheets. They hung these on wooden frames beside fires, where the meat part dried and part smoked. After a few days, the slices were almost weightless and perfectly preserved. They were a source of concentrated protein that could be stored for many months and packed into bags that were easy to carry. Thomas had not gone to the dry-meat camp that year. He had become very weak; before the others left, he had been taken to the Fort St. John hospital.

I found him lying on his bed in a small public ward, dressed in jeans and a vest. As soon as he saw me, he said, as he had so often before when I arrived back in Dunne-za country: "I know you here. Time to hunt. We better go someplace, look for moose." I thought he was making a joke, then realised he was serious. "But can you leave the hospital?" I asked. "You see doctor," said Thomas. "Then we go to dry-meat camp."

Thomas's English was fluent, but it came from his fur-trading days. He used it to make direct statements or to ask straightforward questions. His real language was Dunne-za Athabaskan. So we did not have a discussion about just how he was going to manage in the camp. He had made himself clear, and I went to find his doctor.

The doctor was a tall young man in a hurry. He said that Thomas was very sick. His lungs were "gone." I asked what the actual diagnosis was. "TB," he said. "Maybe cancer as well." He paused, then added: "Nothing we can do." So was it all right to take him out of the hospital? "Fine," said the doctor. "I just hope that you take him somewhere he can get some care." Were there medicines I should take with us? Was there any treatment he should come back to the hospital for? The doctor shrugged. "There's nothing much anyone can do," he said. "He's old and his lungs are gone." I could take Thomas whenever he wanted to go.

I was upset. There was a certain callousness about the way the diagnosis had been given, and something eerie about the lack of medical prescription. Was Thomas about to die? I went back to Thomas's bed and sat down beside him. He looked at me. He must have known that I had seen the doctor. He didn't say anything. I looked at his eyes, which seemed to have sunk back into his skull. He had lost weight. "The doctor says you can leave here when you want. But I guess you could stay, too, if that seemed better." I paused. "You would be cared for here."

Thomas did not hesitate. "Okay we leave now?"

"Yes," I said. "When you want."

"Now," he said. He sat up on his bed, swung his legs over the side and stood up. We found his shoes and jacket and walked out.

Thomas wanted to go straight to the dry-meat camp. He did not say much, and I noticed that he was breathing with some difficulty. I was worried. Maybe we should go to the reserve, where he would have a bed and be within a few hundred yards of the clinic; at least we would have access to a phone. He could call for emergency help if he needed it, or see the nurse who came each week. No, said Thomas, he wanted to be in the camp.

When we got to the camp, everyone else was away. I asked Thomas again if he would like to go to the reserve, at least to find other members of his family. "No," he said. "Better finish here." He went to his family's tent and sat down at its entrance. He was tired and weak. I made a fire for him and put a kettle on to boil. I fetched a few pieces of dry meat so that he could eat whenever he was hungry. He sat very still, looking around, not saying much.

As the fire blazed up and the kettle began to steam, Thomas said: "See the horses." I looked up and saw that the family's horses were all standing close to the little group of tents. They had their heads low to the ground and seemed to be staring at us. I was astonished: the horses always kept their distance from the camp, moving far into the forest, doing their best not to be caught and ridden or loaded with packs. Every morning they had to be trailed into the woods and, often with some difficulty, herded back to the tents, where they could be tethered for use. Now they stood there, close, unmoving.

"Horses they know," he said. "In my mind."

I waited for a while, making sure Thomas got his tea and was comfortable. The horses stayed there, watching. Thomas died a few weeks later.

3

In 1988, during the filming of *Hunters and Bombers,* we interviewed Mary Adele Andrew, the mother of Alex Andrew, my Innu guide and interpreter. She was a large, energetic woman who had brought up a big family as a single parent—her husband had died when the children were still young. She had a strong inner warmth, a generosity of spirit that insisted anyone who came to her house must sit and eat whatever she happened to be cooking. I had been with her at a summer camp, and I knew how much she loved to be far away from the settlement, out on the land. I knew she would have a great deal to say about what had happened to her family, and to all the Innu she knew, as a result of having spent so much time stuck in Sheshashiu.

Mary Adele sat at her kitchen table and talked to Alex. She spoke slowly, carefully, with great force. "These houses were built to trap us," she said. "They told us, 'Stay here, you'll get a house.' But it was a trick to get our children to go to school and to make sure we stayed in one place. It was a lie, so we wouldn't see our land being destroyed. They hoped we wouldn't say anything. They said when our children leave school, they'll get good jobs. But nothing happened."

Instead, the people had lost their real wealth, their real homes. Their land had been taken. The Church and the school, the priests and the government, had joined forces to do this. In one of our interviews, Elisabeth Penashue told us: "In the old days we used to revere the priests. They were powerful and said, 'Don't go out on the land. Send the kids to school or you will lose your family allowance.' Parents were afraid. They made the children go to school. When a priest told us to do something, we did it. We listened to him as if he was Jesus."

In defiance of these intense pressures, Mary Adele Andrew and Elisabeth Penashue, as well as some others, continued to spend as much time as they possibly could on their lands with their families. They were teaching the young how to live there, as Innu. It had been hard, and some had lost a great deal, but there was hope—so long as they could keep going onto the land.

One day at a summer camp, where the women were baking bread in ovens they had scooped out of sand heated with large fires of driftwood, Mary Adele said: "On the land we are ourselves. In the settlement we are lost. That was the way they made our minds weak."

4

European "discovery" of the New World, those great adventures to the Americas as well as to southern Africa and Australasia, led to a set of theories about the peoples who lived in these lands. The theories, which disregarded hunter-gatherer economic systems, languages and belief, were underpinned by the idea that hunter-gatherers were not quite human beings at all.

Articulate colonists of southern Africa in the sixteenth century declared that the Khoisan peoples they encountered at the Cape of Good Hope were "the very reverse of humankind ... so that if there's any medium between a rational animal and a beast, the Hotantot [*sic*] lays the fairest claim to that species." In Australia, Aborigines were classified as being at a midpoint on the evolutionary ladder, more a species of animal than human. When William Lanney, the last Aborigine of Tasmania, died in 1869, a struggle to get possession of his bones was fuelled by the belief that "he represented a last living link between man and ape."

In sixteenth-century Spain, the question arose as to whether or not the original inhabitants of the colonies in the Americas were "natural slaves." The notion came from Aristotle, whom the Spanish monk Juan Gines de Sepulveda relied on for judging the rights of "the Indians." A "natural slave," according to Sepulveda's interpretation of Aristotle, was a person whose inferiority and ignorance were such that only through servitude could the necessary human development be achieved. In Aristotle, this idea arose as part of a justification of slavery in Greek society; he sought to show that the slave's opportunity to work in the master's household was a chance to experience and learn the arts of civilisation. Applying this to Spanish colonial rule, Sepulveda saw slavery as a necessary opportunity. The enslavement of the Indians of the newly conquered territories in South America, who lived far beyond Christian influence and teaching, was "natural"—and it would bring them into the "natural" joys and benedictions of the Christian Church. But slaves did not have the right to own property; those who did the enslaving had the right to take all the land and compel its former inhabitants to work for them.

Had it not been for King Philip of Spain, this quasi-Aristotelian justification for the dispossession of South American Indian peoples would have been long forgotten, a small footnote in colonial rationalisations of conquest and theft. In 1550, alerted by his legal experts to disagreements about the question of Indian rights to their lands, the Spanish king authorised a formal public inquiry—the first Royal Commission to deal with indigenous peoples. The inquiry took the form of a debate. On the one side was Sepulveda, very much the theorist: he had never set foot in the new colonies. He advanced the idea that these new peoples were "natural slaves" and therefore had no rights that could restrict the claims of Spain's conquerors and settlers. The other side of the argument was entrusted to Bartolomé de Las Casas, a monk who had spent thirty years in South America and lived close to indigenous peoples there. He had been among agricultural, imperial societies—not the hunter-gatherers of Amazonia—and he had been surprised by complex and, in many ways, familiar kinds of social institutions. Las Casas had written extensively about his experiences, arguing that the Indians of the colonies had systems of law and administration, as well as ideas of property and morality, that should be respected. The debate lasted almost two years; the king of Spain then took eighteen years to decide which argument he would accept. In 1568 he gave his support to Las Casas's point of view. Meanwhile, of course, Spanish policy and practice had been ruthless: Indians were deemed to have neither land rights nor souls. Some were enslaved, many were killed, and their lands were appropriated. Neither the sophistication of the Indians Las Casas described nor the vigour of the public debate in Seville moderated the conviction, shared by colonists and Christian missionaries alike, that these people had not yet reached the evolutionary level of real human beings.

On the settlement frontiers of the North American colonies, the question of the Indians' humanity was also raised. If the indigenous occupants of the lands to which settlers were moving were not humans, but roamed, rather, "as beasts of the field," then they had no right to resist the new Americans' "manifest destiny" to take and use all new-found lands.

It is easy to see that an insistence on people's being something other than, or less than, human has been inseparable from the wish to occupy their lands. Doubts about an unquestionable right to appropriate land did effect some challenge to the theory of colonial expansion: legal justifications were sought to defend claims to new territories. At the frontiers themselves, however, colonists at times killed hunter-gatherers as if they were animals. Colonial governments licensed these murders by turning a blind eye, sometimes even condoning them as a suitable response to the "primitives" and "savages," whose want of humanity made them a threat to settlement. History records Aborigine hunts in Australia, the killings of Bushman families throughout southern Africa, attacks on the tribes of Amazonia, and relentless campaigns against the Indians of the American West. Early in the nineteenth century, the philosopher Hegel observed that the development of the American West had cost some two million Indian lives—an outcome he deemed to be a necessary part of progress.

5

The assertion that some humans are not human—or are not human enough to have rights—is absurd as well as brutal.

The Spanish *conquistadores* destroyed communities with elaborate agricultural and urban systems. They insisted that the Aztec were barbarous and undeserving of compassion, for they practised human sacrifice and cannibalism. In reality, the Spanish had arrived among people whose material wealth they wanted to plunder. The idea that the victims of this plunder were "savages" was belied by the very thing the Spanish most wanted from them—the elaborate and magnificent creations of Meso-American artists, including worked silver and fine jewellery.

In reality, explorers and adventurers arrived in the lands of societies that were at least equal to their own. Hunter-gatherers did not display abundant material goods or the technology of warfare, and they did not have the knowledge of mathematics, astronomy, engineering and textiles that was to be found in some indigenous societies of South America. But the Europeans who came ashore from those ships of exploration—dirty, malnourished and ill clad—encountered hunters and gatherers who showed all the signs of being well fed and healthy. The societies of these people were more stable and more secure than those of the explorers; they were also societies in which private and public well-being intertwined to ensure much fairer distribution of resources and greater social justice than the newcomers had ever experienced. But the representatives of "civilisation" did not hesitate to condemn as "savages" the people who provided the food that kept them alive.

These encounters between unhealthy newcomers and vigorous tribal communities began in the 1400s and continued until the beginning of the twentieth century. Another such paradox is to be seen in the history of European ideas. While describing the peoples they were discovering as inferior beings and claiming their lands for themselves, European travellers and intellectuals also began to extol the moral superiority of the peoples who were being conquered, enslaved and dispossessed.

Columbus's description of the very first Carib hunter-gatherers he met includes a recognition that they are "so generous with all that they possess, that no one would believe it who has not seen it ... and [they] display as much love as if they would give their hearts." Social philosophers made intellectual use of what they saw as life that existed in pure nature. Hobbes's famous reference to life as "nasty, brutish, and short" came from his imagining a society "with no place for Industry ... no Culture of the Earth ... no Arts; no Letters; no Society." He wrote this in 1650, a critique not so much of the hunter-gatherer world as of the early capitalism of the England in which he lived. In subsequent centuries, philosophers as diverse as Vico, Montesquieu, Rousseau, Hegel, Engels and Marx made use of the idea of a "natural" human condition both to criticise the societies of their day and to celebrate "innocence" and "simplicity"—a core of deep human goodness—in peoples whom we would now call hunter-gatherers. In one context, these newly discovered tribes were savages whose conditions gave them no rights to life, liberty or property. In another, they became symbols for the human potential for goodness, equality and freedom.

These paradoxes point to the cynicism of the colonial project. The explorer Ralph Standish wrote in 1612 about the "savages" he had met at the Cape of Good Hope. Given their want of human achievements, it was, he said, "a great pittie that such creatures as they bee should injoy so sweett a country." The countries of the new worlds were indeed "sweett" to the farmers who wanted to make them their own. Inca and Aztec silver provided wealth for the impecunious monarchy of seventeenth-century Spain; extensive grasslands and forest offered at least a hope of wealth to the landless settlers who had reached the expanding edges of European empires. In each place colonists claimed as their own, they concocted justifications for dispossessing people and taking their territories. To say that those they encountered were not human was the most general—and, in its way, the simplest—device for depriving hundreds of cultures and millions of human beings of their rights.

6

When Anaviapik and Ulajuk raised the question of how southerners have seen the Inuit, they expressed a concern that must have puzzled, alarmed and at times oppressed many, if not all, hunter-gatherers. They have experienced the attitudes of settlers and settler governments towards them. They have felt the consequences of the judgement that they, the original inhabitants, are not entitled to their own lands, languages and ways of life. They know that this judgement is somehow connected to a refusal by the colonists to see that hunter-gatherers have minds—a refusal that sits at the centre of the history of colonial misrepresentation.

The indigenous peoples of Canada have been forced to respond to a strong implication in modern legal theory that they do not qualify as fully human. This version of frontier racism has arisen in legal arguments about "aboriginal title," the rights indigenous peoples may or may not have to their own systems, territories and resources. The legal and political actions that lie behind these contests over title are referred to in Canada as "land claims." For a century, various aboriginal peoples have had to make these claims. After his election in 1968, Prime Minister Pierre Elliott Trudeau maintained a long colonial tradition by insisting that the nation's Indians should be assimilated into the mainstream of national life. In 1971, Trudeau announced a change of policy from adamant rejection to circumspect acceptance of a land-claims process. This shift resulted from a judgement in the Supreme Court of Canada in what is known as the Calder case—an action brought by the Nisga'a people, who argued that they had aboriginal title to their territories throughout the Nass Valley. Although the judges were split on the decision 3-3, the message was clear: aboriginal title did, after all, have some legal reality. Since then, arguments over who has which rights have been full of legal complexity. Land claims have at times appeared to define the nation.

The term "land claim" is itself an anomaly, implying that the onus should be on the original occupants to claim their homes, resources and territories from the colonists. This is a reversal of common sense; the burden of proof should lie with the newcomers. This reverse sense is a first indication of the onerous task indigenous groups have had to undertake. Their elders, historians and lawyers must find ways of satisfying criteria set by the Canadian courts for testing whether or not a claim can indeed be said to amount to a claim to aboriginal title. Litigants from hunter-gatherer and fishing societies have had to prove that:

- they use and occupy a definite territory to the exclusion of all other peoples;
- they have used and occupied the territory "since time immemorial";
- they are "an organised society."

These tests have arisen from accumulated precedents in cases that reach back to the nineteenth century. But each of the particular requirements for evidence has been confirmed by Canadian courts in the 1970–90 era. The problems inherent in proving the first two requirements—exclusive use and occupancy—are severe. In many hunter-gatherer systems, there are imprecisions and overlaps of territory that unsettle the demand for boundaries and boundary maintenance that the colonial model requires. For oral cultures to prove continuity of land use beyond the present generation, to the satisfaction of courts that rely above all on first-hand experience and written documents, is also a daunting undertaking. The imposition of legal process and rules on the peoples the colonists have sought to dispossess makes it hard for hunter-gatherers to give evidence of their own kind in their own way. But the difficulties that arise with questions about use and occupation of land do not challenge the hunter-gatherers' humanity. It is in the "organised society" test that the deepest prejudices reveal themselves.

What are the qualities by which society is judged to be organised? Rules and conventions of behaviour, shared economic practices, common religious beliefs and customs—these things *are* society. To speak of society is to imply organisation. People who live without a shared set of values and rules cannot live as a people. The human condition is composed of social realities. And the central indication of this is language.

Language and society are inseparable. Each is a necessary condition for the other. Children who grow up without any form of society do not learn to speak; human societies do not exist without language. The organisation of the human mind requires a community of fellows who speak to one another, sharing and thereby teaching the words and rules that constitute the language. We know what words mean because they are used by a group of people to mean things. Society has an existence because it is composed of people who share these meanings and who use them to share everything else that makes human life possible.

To suggest that there are human beings who might live without "organised society," therefore, is to say that there are human beings who live without society at all. To suggest that a people might live without society is to imply that they are living without language. And to imagine a people without language is to suggest humans who are not quite human.

Think of an elder on the witness stand being quizzed about whether her society is "organised." Are there laws? the lawyers ask. Do you have rules about your ways of using land? Do you collaborate? Do you live in anything that we can call a society? To experience this kind of interrogation is to endure skepticism about the obvious. The suggestion is that your people have not lived long on their lands, that they have not lived there in the belief these lands are indeed theirs and no one else's, that they are not attached to these lands in any profound way, and that they do not have customs or beliefs uniquely their own. To answer these kinds of questions is thus to respond to insults. Aboriginal people who take the witness stand in land-claims cases often have an intense feeling of not existing; their history, their homes, the integrity of their grandparents are all contested. To be obliged to prove that which defines you is to have a sense that your very humanity is in question.

7

Imagine the crowded, roaring bar of the George on 96th Street, Edmonton, a Prairie city in the Canadian midwest. The George is a rundown beer parlour in a part of town where no one goes. No one, that is, except drifters, down-and-outs and hard drinkers. There are Indians, some hookers, winos from many backgrounds. It is a place to have friends who don't ask questions, providing warmth that has nothing to do with family or home and a chance to lose any sense of weakness. The tables are crowded; the noise is a shrill mixture of shrieking laughter and shouted conversation; fights break out. A tough place. But for all its toughness, the men and women who come here are far more often generous to one another than they are belligerent. Everyone has almost nothing; people give out drinks, cigarettes, small change and advice about where to bum a hot meal. Everyone belongs because they have all chosen, for a while at least—maybe a week or maybe a year—to belong nowhere.

I went to the George every day for a few months in the spring and summer of 1969. It was my first field work in Canada and my first encounter, therefore, with "Indians." In this and other bars, on street corners, in abandoned shacks and at the edges of a park, I talked and drank and laughed with men and women whose lives reached into the Plains of the Midwest, to the Pacific Coast and to the forests of the Subarctic. They spoke to me of spending time "in the bush," away from "the white man." Sometimes, at night, when a slight drunkenness had not yet given way to incoherence, when sadness rather than defiance or humour was the mood of the moment, a few of those I knew best would tell me they would soon be heading home. But I was never sure what home meant or just how many of them would be able to get there.

One of the first people I met in Edmonton was Harry. A tall, heavily built man of about fifty, with the strong features of a Plains Indian, he was leaning on the wall outside the George, playing a harmonica. The music was beautiful—a bubbling of sounds, fast and rich, with a blend of tunes that I learned later was part Scottish and part Cree. It was the music of the fur trade, of the encounter between those who manned the trading posts and bought the furs and those who trapped and sold them. I stopped to listen. Harry watched me as he played, then paused and asked if I had a quarter to spare. "Sure," I said. I handed him fifty cents. He looked at the two coins, as if measuring their possibilities. "Good," he said. "Now you and me can go buy a few beers."

Harry became my best friend on skid row. He was a good friend to have. Everyone knew him, and many took pride in his musical skill. He played the fiddle as well as the harmonica, and knew where he could go to find both cups of tea and instruments that he could borrow. He made money by busking on street corners, but only enough to fund more rounds of beer and the occasional bottle of cheap sherry. I suppose he was an alcoholic, but he never got rotten drunk, and he somehow placed himself at a distance from those who did. He was courteous and helpful, and a bit of a street-person social worker: he would take care of those who were most destitute, and he did what he could to prevent fights from drawing real blood.

One Sunday, when the bars were closed and the skid row community broke into little clusters of people on street corners and vacant lots, Harry told me to come and hear some real music. He took me to a bootlegger who sold us a bottle of the cheapest sherry, then led me to an abandoned house a few blocks from the George. It appeared to be boarded up. But Harry knew a way in through a broken door to its basement. Down there, in the gloom, a group of men and women sat in a circle. A few bottles stood around, and one man was unconscious in a corner.

We joined the group. No one said much, and then two or three people began to sing. Then another few people; a different song. It was Algonquian music, with drumbeat rhythms tapped out on a broken chair and the floor, the voices high-pitched, the words a strange and haunting wail. The singers sat with their bodies hunched a little forward and their eyes shut tight. They strained to get the sounds right, to keep the rhythm, to take and keep themselves elsewhere. This was no longer skid row—or was it actually the very heart of skid row, where those who lived as marginals could be themselves at the centre of the white man's city?

The songs were separated by quiet pauses, a hand passing the bottle, a shifting of bodies, but no one spoke.

After one of the songs, a young woman broke the silence and said to me: "Now you hear our minds, in our song. How come the white man says we have no mind? When they hear a Cree song, I guess they think it's a coyote howling."

Everyone laughed.

8

Twentieth-century cameras have allowed us to look into the eyes of the wildest of animals. There is no corner of nature, however remote or small or dangerous, that does not appear in intimate proximity on television screens. At the same time, researchers into animal behaviour and its genetic sources are developing ever more sophisticated techniques for seeing the mechanisms and achievements of the natural world. There is a new kind of relationship, based on technical sophistication and the knowledge of experts, between humans and the rest of nature. Photographers and scientists are the wizards, if not the shamans, of our age, making revelatory journeys into places where, in the course of ordinary life, the rest of us cannot go. We rely on them to show us the world that is not human.

This ever-increasing closeness to the natural world influences our sense of the dividing line between human and animal. We discover an unexpected complexity of animal systems and learn about the intricate links between one animal and another. We find that there are divisions of labour, with one part of an animal community raising newborns, others getting food, and others defending the group. We are shown the sophisticated behaviour required to capture or evade capture, elaborate forms of courtship, and myriad forms of communication. We discover that leaf-cutter ants make gardens and harvest crops; that gannets can recognise their own nests among a hundred thousand others; that humpback whales coordinate their fishing; that in some species, monkeys can warn one another about several different kinds of impending danger; that male fruit flies sing elaborate love songs to court females. These are occupations and purposes that we understand. They depend upon qualities, characteristics and motivations that humans also possess. Detailed portraits of the natural world again and again reveal similarities between humans and other beings. The evidence of DNA, with its apparent overlap of gene sequences between mushrooms and people, is the newest way in which the lines between us and the rest of nature can seem to be uncertain.

This apparent blurring of the divide could also be seen as an echo of shamanism. Perhaps the boundaries around the human are less definite, more porous, than most scientists and many farmers have tended to suppose. Animal rights advocates often make this kind of argument; they point out that the use of animals for experiments, or even as food, depends on human beings keeping animals in a separate and inferior moral category. The promise of Genesis is integral to the domestication of animals and to their use as a resource. The exiles from Eden go forth and have dominion.

In fact, real shamans—as opposed to those who provide wildlife programmes for television or make new-age forms of argument—say that humans and animals exist in separate domains. In the shamanic myths of very ancient times, humans and animals lived in the same circumstances, able to speak and procreate with one another. But these myths also tell of how ancient times yielded to an era in which the divide between humans and other creatures became much clearer. Indeed, the need for shamans, or for spirit possession, derives from the periodic need to cross this divide. Hunter-gatherers and their shamans insist that life depends on maintaining the right kind of relationship with the natural world, and the negotiations necessary to sustain this relationship are difficult. The power of the shaman centres on this difficulty. By overcoming it, the journey can be made from human to animal and back again. This is the power that comes from transformation.

When we look into the eyes of animals, be they pets or creatures that we hunt, cows and horses on the farm, dogs in our homes, or the subjects of wildlife documentaries, we seem to see thought. Animals watch, wait, appear to ponder. They look as if they are assessing one another's movements in order to make sure that their own are safe or effective. They give many signs that they are thinking.

In 1999, British television broadcast a wildlife programme made with a new level of film technology: it had become possible to film in almost complete darkness. The footage included sequences of lions stalking and killing their prey. They moved with great stealth, peering, watching, calculating, manoeuvring. Several individuals collaborated to approach and surround their prey; two remained still, one continued to move closer. Then, with great skill and precision, they made their attack. Were they thinking? Surely they must have been. To calculate in this way, to make decisions about how best to carry out the kill, must require some form of thought. Or is this so? It is very hard to imagine thought without language. And these lions do not speak; they have no more than the most minimal form of vocal communication. Their brains work without words.

This wordlessness is integral to how we see animals. It arouses in us a form of gentle sympathy, an anthropomorphic kind of pity. It also means that, for all their ferocious killing of weaker species, we see animals as innocent. Those lions do not lie, because to tell a lie requires speech; they cannot be condemned for the cruelty of their ways, because morality arises only with articulate thought, in words. The lions show a kind of purity of judgement, rather like pure emotion. There can be no process of the kind that depends on thinking as a silent form of speech. Lions do not talk to themselves. When we look at them we see, rather, the strange dumbness of the animal as it thinks without thoughts. We see feelings that we recognise, of course: fear, excitement, even pride. These feelings also arouse our sympathies. But animals are untainted by the ambiguities and distractions and complexities of what humans know to be the heartland or even defining features of thought.

9

"What, then, is the difference between brute and man? What is it that man can do, and of which we find no signs, no rudiments, in the whole brute world? I answer without hesitation: the one great barrier between the brute and man is language." These are the words of

Max Müller, among the first theorists of language, writing in 1875. Müller's view has been shared by many others, including late-twentieth-century scientists who have sought to identify the part of the brain where the potential for language resides. Language theorist Derek Bickerton observes that language is probably "the antecedent of most or even all of the other characteristics that differentiate us even from our closest relatives among the apes." The miracle of this ability to understand, use and make language is the miracle of being human. "Simply by making noises with our mouths, we can reliably cause precise new combinations of ideas to arise in each other's minds."

There has always been a popular view, endorsed to some extent by Darwin's ideas about language, that places the calls of birds and the cries of animals on a single spectrum of communication. This is to say that humans do the same as animals, except that they do it more and better. Yet the attempts of primatologists who spend years teaching a chimpanzee to recognise and in some way make use of four or five words, or the intensive efforts of marine biologists to decipher the communication systems of dolphins, are projects that seem to confound any claim that even the most intelligent of other mammals have anything like language. They communicate, but they do not have language. There is no equivalent of grammar or syntax; no parallel, therefore, to the way human children learn to speak. Animals do not do what humans call thinking. They may exist in a Zen-like state, where the brain works without self-consciousness (consciousness is also inseparable from words), but they are not thinking. This is why animals are outside moral judgements and why, also, they inspire such a sense of puzzlement. To look into their eyes is to see a creature with a brain. We see facial expressions and even gestures that are very like our own. Yet something is missing.

We humans may be able to get a sense of animal "thinking" from our remarkable capacity to make quite complex decisions without thought. The driver of a car who suddenly has to deal with an emergency is capable of making a quick set of decisions—changing gears and speed and direction—without any apparent thinking. Similarly, drivers of cars often have the experience of picking the route home without being aware of doing so. Actions of this and many other kinds are said to be unconscious, in that the thinking takes place somewhere other than where we are using—or are aware of using—language. The terms "mindless" and "thoughtless" indicate the significance of action that fails to proceed from the necessary mental processes: such behaviour is wrong, in either a moral or a practical sense, precisely because we did not think—that is, we acted without the benefit of words.

In mythology and literature, creatures that are almost human are often monsters. Their animal characteristics are exaggerated in a symbolic manner: they represent the frightening power of the animal in ourselves. These creatures are somehow primitive or savage; they take human shape, yet are outside human culture. And their condition draws its most poignant qualities from their lack of articulate language. Prospero's Caliban, Beauty's Beast, King Kong, Marian Engel's Bear. The stories in which such characters have greatest effect are those in which they are discovered to be innocent despite their apparent beastliness. They are without the guile of language.

When human beings began to use language, their brain structure made an evolutionary leap of huge importance. The physiological difference between those who spoke and those who did not may have been very small, a tiny fraction of the total brain. But once it was there, a divide opened up between human beings and all other animals, a divide that had immense and ever-expanding consequences. Language allowed human evolution to take a very different and much more elaborate path.

In the absence of language, inheritance is limited to the gene pool. Parents pass on to their offspring a bundle of genes and very little else. But with language, they can pass on vast bodies of knowledge, moral codes, forms of social arrangement. And with language, it is possible to think. With thought, it is possible for each generation to transform knowledge and ideas, which are then passed on to the generation that follows. Once it had language, the human species spread throughout the .world, from environment to environment, each group with its own ways of occupying territory, knowing about their land and ordering their lives in a particular region. In this way humans came to live in cultures—that is, in many kinds of articulate and organised societies.

The human mind is this combination of language, thought and culture. The capacity to be human is inseparable from the capacity to think, be articulate and change life through words. The best-trained chimpanzee and the least-educated human being are far, far apart in linguistic skills. The one has nothing more than a tiny number of words it can use in restricted conditions. The other has grammar, syntax and hundreds of words that he or she can use in any circumstances. The human mind, at its least, is rich with potential that makes human evolution unlike anything else in history. All the mind's capacities are shared by all humans, irrespective of any other consideration. Each culture may give rise to its own kind of mind. But mind is what gives rise to culture itself.

10

No one knows when human beings first used language. *Homo erectus,* the human ancestor who lived about two million years ago, used tools and has been declared by some archaeologists to have had many qualities that are more human than animal. But the kind of tools *erectus* used seem to have remained unchanged for about a million years, and the structure of its upper body suggests that it did not have the breathing system necessary for elaborate speech. As one scholar has put it: "If these ancient people were talking to each other, they were saying the same thing over and over again." The evidence that does suggest language, where the tool kit is complex and changing and where the physiology of the upper body is consistent with the use of speech, comes from about 800,000 years ago.

The dating of the acquisition of language, however, is less important in this context than the nature of the process. Did language appear through a long and gradual evolution, with lower levels of linguistic achievement giving way to higher levels? Or was the ability to use language a more sudden, cataclysmic event, or set of events, which meant that humanoids became humans in a revolutionary change to the mind? The importance of this issue is very great. A slow process means that different peoples could have been at different levels of linguistic ability, or that languages with different forms of linguistic sophistication could have emerged and disappeared. A revolution means that a single evolutionary development created the mind of the original *Homo sapiens,* the ancestor of all modern human beings.

Various kinds of evidence suggest an answer to this question. There is, first of all, what seems to have been a sudden explosion of human culture—the great spread of hunter-gatherer systems around the world, each with its own sophisticated and specialised technology. Then there is the evidence of language itself. What has been called "the language instinct" turns out to be fundamental to the activity of all human minds, even those denied the normal means to develop language. Every child acquires or uses grammar, irrespective of the circumstances. And given even minimal language-learning opportunities, children begin to employ grammatical techniques with astonishing speed, soon applying linguistic rules and making their own sentences—ones that they could never have heard before. These findings suggest that the capacity for language developed in a short period of time and by means of a specific evolutionary leap—the ability to acquire and employ grammar.

Some experts have argued that the ability to learn a language is "hard-wired" into the human brain. By this they posit the existence of a faculty that serves as the language-learning element in the human mind. In other words, this is the capacity to be a human that all humans inherit; and it is this capacity each society then relies upon to build its particular array of knowledge, skills and norms. The immense evolutionary advantage that came with this hard-wired faculty lay in the way humans could think, know and distribute resources in collaboration with one another, in ever-changing ways. The social dimension of language is thus intrinsic to both its form and its opportunities.

These pieces in the jigsaw of human prehistory can be assembled to show a picture of human beings living in groups that use language, and therefore thought, to deal with all their concerns. The universal qualities of mind mean that humans are able to learn one another's languages. One person knows more or less than another. One person is more eloquent than the next. But all of us share the faculty that makes eloquence possible.

Languages rely, of course, on the sounds that people make. Linguists have identified some 140 separate pieces of sound—the total for all the world's ways of speaking. English uses about 40 of them; Norwegian, the most elaborate vocal system of all Indo-European languages, uses about 60; Inuktitut uses about 50; there are Bushman languages in southern Africa that use about 120. As one of the world's most eminent linguists has said, the Bushman is "the acrobat of the mouth." This wide range of sounds does not suggest that the sophistication of a particular culture has any links with the outward complexity of its language or languages. Nor are there differences in grammar that indicate any one language is more or less evolved than another. The popular idea that some languages are "primitive" is false. Each language has its own sophistication, but all share a basic level of intellectual achievement.

The Khoisan languages, which may have the most direct links to the birthplace of language itself, have been despised by Europeans as "the chattering of monkeys." In reality, the Khoisan use about 85 per cent of all language sounds. This is not to argue that their languages are *more* complex or can therefore achieve greater intellectual heights than those of other peoples. The point is that they are not *less* complex. There is no relative simplicity of language. The underlying faculty for language, the hard-wired component of the human brain, is universal: an ultimate equality of opportunity.

11

The brain struggles when it comes to thinking about thinking. Being able to speak, however, may have much in common with other kinds of human natural potential—to have arms, for example, or a particular arrangement of nerves. Noam Chomsky, the most influential modern theorist of language, has described an important implication of the similarity between the capacity for speech and other capacities:

> No one would take seriously the proposal that the human organism learns through experience to have arms rather than wings, or that the basic structure of particular organs results from accidental experience. Rather it is taken for granted that the physical structure of the organism is genetically determined.

Chomsky makes this self-evident observation to support his proposal that the underlying feature of language—the thing in every human brain that makes language possible—is also a physical structure that is genetically determined. Chomsky situates the source of language, the piece or pieces of the brain that make language possible, alongside other faculties that are inherited rather than learned. In the same context, he goes on to ask: "Why, then, should we not study the acquisition of a cognitive structure such as language more or less as we study some complex bodily organ?"

According to Chomsky and other researchers, the speed at which a child's vocabulary grows, and a child's ability to use grammatical rules, are not influenced by teaching. Studies have found that attempts to correct children's grammatical errors by repeating the correct form back to them are unsuccessful. Researchers looking at cases where parents corrected their children's English found that this correcting "had no effect—if anything, it had an adverse effect—on the child's subsequent development." This research, and the discovery that children do not respond to being taught correct grammar, is discussed in Steven Pinker's remarkable book *Words and Rules,* which builds on many of Chomsky's original insights.

Pinker focuses attention on irregular verbs as indicators of how the brain acquires, builds and uses grammar. In this context, he looks at studies of identical twins, noting that "vocabulary growth, the first word combinations, and the rate of making past-tense errors are all in tighter lockstep in identical twins than in fraternal twins." Pinker's striking conclusion to a chapter centred on the relation between nature and nurture in language learning is clear enough: "Every bit of content is learned, but the system doing the learning works by a logic innately specified." This account of language learning establishes that every child inherits a fundamental set of characteristics prior to and independent of culture. These create the possibility of language and also set the limits to what can be taught.

Findings from the modern heartland of linguistic theory are consistent with hunter-gatherer ideas of child raising and education, in which children are expected to develop in their own ways, at their own pace. Nature is relied upon to do its part in the business; the mind is expected to grow very much on its own. Pedagogy is viewed as of limited benefit at best, and as counterproductive at worst. Children learn when and what they are ready to learn.

In Inuktitut, there is a linguistic indication of this faith in the extent to which human potential is hard-wired. When a person experiences extreme grief, he might say "*Isumaga asiujuq.*" "My *isuma* is lost; I am out of my mind." When Anaviapik's son Inukuluk was talking about his experience of adult education and his struggle to "be a white man," he began what he said with "*Isumanguar,*" "appearing or pretending to think," which I translated as "I sort of thought." And in many conversations I heard "*Isumatuinnarpunga,*" "I just thought," a caveat that conveys the sense of the English words "It's only my opinion." The root *isuma* has many uses; they show it to be something that also has an independent existence. In Inuktitut, thought is tightly linked to the capacity for thought.

When a child misbehaves or misunderstands, adults are likely to say "*Isumaqijuq*," meaning that she lacks *isuma*, is without the necessary thought. It is striking that in this use of *isuma*, the child is no more blamed for this lack of thought, for having an undeveloped *isuma*, than she could be criticised for having short legs or a large nose. Everyone is aware of a child's development, in body as well as mind. Comments are made noting progress. These comments are not judgemental, though they will, of course, have their effect. Social pressure comes from what a child hears said about her. Adults do influence the way a child learns, shaping aspects of character and affecting the rate at which many kinds of learning take place. But the Inuit trust that individual development comes from what goes on in the child, not from any systematic pedagogy. *Isuma* grows at its own pace. A child has the potential, and she will grow in her own good time. Once the capacity has developed, and there is a mind and language that expresses mind, then some children will learn better than others. External factors come into play, just as they do with the strength of arms and legs. But for hunter-gatherers, the individual mind is the thing that must choose to learn, develop, make decisions. Pressures from others on that mind, according to the deep beliefs and social conventions of hunting peoples, are more destructive than instructive. The mind has the capacity to learn, and, left to develop on its own, it will do so.

The place of thinking in Inuit ideas of psychological development, and the related ways in which the word *isuma* is used, offer many clues about Inuit society. Parents identify children with respected elders, trust children to know what they need, do not seek to manipulate who children are or what children say they want. This way of treating children tends to secure confidence and mental health. And Inuit child raising is inseparable from many aspects of interpersonal behaviour. Adults respect one another as separate but equal. This is the basis for cooperation—by respecting individual skills, judgements and knowledge, the strengths of the economy and the social order are shared. *Isuma* is the notion that underlies and unites many of these features of Inuit life, for it affirms that in crucial ways the development of *isuma* is independent of social manipulation and control. Embedded in this use of the word for mind is a view of mind itself.

The work of Chomsky and Pinker and the Inuit use of *isuma* reveal the same truth: there is some logical and physiological antecedent to the cultural specifics of learning. In the hunter-gatherer reliance on individual egalitarianism lies the freedom for everyone to be themselves, and a confidence that the integrity of society—the respect that hunter-gatherers show to one another as well as to the natural world around them—will achieve the best results for both individuals and the group. The egalitarian individualism of hunter-gatherers is of a piece with a compelling theory of mind.

12

The linguistic theories of Noam Chomsky have been closely associated with the school of thought, or the theory of thought itself, known as structuralism. Structuralist theory originates with ideas about grammar and its relation to the structure of the mind, and it relies on the view that grammar has at its heart a logical principle akin to the law of excluded middle, the principle that nothing can be both X and not X at the same time. The merits of this approach to grammar may well be inseparable from the insights of Chomsky and others into universal grammar. But anthropology made structuralism its own with a series of assertions about a seemingly fundamental dichotomy between culture and nature. Beginning with these ideas as a way of looking at ritual and myth—seeing rituals and myths as expressing or mediating the need for humans to establish their cultures in defiance of nature—structuralist anthropology then proceeded to explore many other kinds of dichotomies, some of which resonated with the culture : nature paradigm. Man : woman. Dark : light. Raw : cooked. Upstream : downstream. Sky : earth. Sun : moon. Human : animal.

The anthropologist who made the most elaborate play with pairs of opposites, and who originated the claim to see in them a clue to the nature of the human mind itself, was Claude Lévi-Strauss. Lévi-Strauss's followers in the English-speaking world were led by Edmund Leach, who for many years was professor of anthropology at the University of Cambridge. Their views became an orthodoxy. No ritual or myth, no piece of social life or sacred text was safe from a grid of interpretive dichotomies. This "structure" was set out as if it were an explanatory reduction of social and intellectual life, a scheme that laid bare some primary meaning.

When I first read Lévi-Strauss in 1967, I was impressed by the originality and apparent insight of his way of writing about both tribal and European cultures. In *Tristes Tropiques* and *The Savage Mind,* and then in his ever-burgeoning *Structural Anthropology,* Lévi-Strauss carried a generation of intellectuals on a wave of ideas that began with Cartesian philosophy and proceeded to overtake all branches of sociology and philosophy. This was work that addressed central questions about reason and mind, while taking the reader to tribe after tribe. Lévi-Strauss's description of the Nambikwara, hunter-gatherers of the Amazon, was both compelling and poignant. Here were people whose material simplicity was matched by extraordinary cultural beauty, and yet they were disappearing from the world. His account of the myth of Asdiwal, collated from Nisga'a stories set in the Nass Valley, with its hero figure moving up and down the river and between the earth and the sky, was analysed into binary pairs as if it were underlain by a kind of mathematics of human consciousness. To read this work was to experience a sense of intellectual wonder, to feel that one was being led on a journey of remarkable discovery. Yet the journey was oddly frustrating, as if it passed through fabulous landscapes but never reached a destination. The ideas floated high above reality, circling; and many of us circled up there with them, not sure where we were or where we were going, in awe of the height, feeling uneasy, but not quite daring to land. Over the years, the magic of the journey faded; a sense of dissatisfaction remained.

In retrospect, the trouble with the intellectual claims of structuralism is not hard to discern. To say that a tribal myth contains opposing elements, and that its structure is demonstrated by laying these out in a formal manner, revealing at the same time dichotomies that are otherwise obscure, is to explain nothing. The analysis may well claim that a reduction of these pairs shows them all to be expressions of the fundamental opposition of culture to nature. But what is the explanation being made of the myth? The myth expresses the core issue of all societies: to establish how human life seeks to separate itself from natural life. Or the myth expresses the deep nature of human mind, the structure of mind itself, where dichotomies lie at the heart of us all. But these are not explanations of anything. To say that a myth expresses the core of society and the nature of mind is to say very little. Myths and ritual are the products of language and society; it would be strange indeed if they did not give expression to them.

Structuralism in anthropology is an elaborate and at times fascinating game. It looks scientific, for it has a direct link with scientific linguistic theory and follows scientific methods of exposition, deducing an underlying reality from social, verbal and textual surfaces. But there is no explanatory achievement that goes beyond an a priori assertion about the structure of mind and a somewhat tautological process of deduction: the theory knows what it is going to find, then relies on a thorough but predictable exposition to find it. It is very striking that when Edmund Leach wrote about the Bible, and gave himself the task of explaining the nature of much of the text, he showed how a particular complexity of the stories is in the interest of a priestly class. Here *is* an explanation, but it is to do with function, not structure.

A further objection to structuralism in anthropology is more down-to-earth. The use of binary pairs to create an analytical grid is at odds with the way in which indigenous cultures, starting with hunter-gatherers, achieve so much by avoiding dichotomies. Hunter-gatherers also reject any complete reliance on deductive reasoning. So the structuralist analysis that commits from its outset to display the inner workings of dichotomies is in a perplexing, and somewhat imperial, relationship to its subject matter. The anthropologist's ways of thinking here occupy the intellectual territory, obscuring and ousting the people's own modes of thought and discourse. An irony of Lévi-Strauss's achievement is that it yields much more insight into his own culture, so centred on binary logic and attempts to create rational order, than into those of the tribes he examined.

Postmodern work in literature and social science has emerged in part from frustration with this structuralist reliance on dichotomies and its attendant pseudo-scientific qualities. These new approaches to history, culture and knowledge centre on meaning rather than on mind. And they pay close attention to the ways in which meanings themselves are constructed by the would-be analysers. Thus post-modernists deconstruct the accounts, be they myths or theories of myths, allowing presuppositions, intentions and colonial purposes to disclose themselves in whatever array of complexity or contradiction may emerge. It is easy to imagine the intertwining puzzle that this exercise in scrutiny can yield.

The postmodern task is the analysis of analyses—an approach that can indeed yield insightful theories about theory. Its problems, like those of structuralism, stem from a failure to describe a world whose reality would be recognised by those who spend their days living in it. The deconstruction of hunter-gatherers has contributed to the view that they do not exist at all; they become, instead, a myth of colonial theory, a part of someone else's ideology, or, at best, an edge of some other way of life. To those who live, or whose ancestors have lived, by hunting and gathering, this deconstruction must come as a surprise.

13

Let me return for a moment to dichotomies, this time to the pair of terms that I have relied upon not only for the writing of this book but in much of my thinking about the world. In my notes and letters, places where writing is not laundered for fear of critics, I have long used the pair of abbreviations h-g : p-g. These stand for hunter-gatherers and potato growers. This pairing began as a small and rather obscure joke about the romantic appeal of hunter-gatherers in their vast territories opposed to the confined, harsh lives of peasants in their fields of potatoes. The joke—if it can claim to be more than a piece of grim and private humour—needs some explanation.

I began anthropological work in the west of Ireland. I spent long periods of time living and working with men and women whose lives centred on small gardens and a few fields. This was an experience of peasant life. And all the notions and tensions of what it means to be a farmer in Ireland turned my attention to potatoes. The history of the families I knew, and the stories of their farms, had been shaped by the potato. Peoples' attachment to their fields, their occupation of the hillsides and bogs and islets of the west coast, were made possible by potatoes. They could live on ever smaller areas of ground and still feed large numbers of children thanks to the potato's remarkable productivity and nutritional value. The shattering of this system, the breaking of the cycle whereby each family would divide its holdings to allow the sons to inherit land, and in which sons and daughters would marry young and begin their own households, was the result of the potato famine of 1846–51.

The Ireland I knew had its roots in the changes after the famine. Not that the potato was gone; rather, it could no longer be the basis for an ever-increasing farming population. Modern Ireland was born of potato blight and a population that was declining fast. The language of modern Irish writing (some of James Joyce, Samuel Beckett, Sean O'Casey, Patrick Kavanagh, even Seamus Heaney) reflects all this, though I know there is a harking back to the "real" Irishness of pre-famine days in Yeats and Synge. These Irish "traditions" were of short duration, but the potato was always there. It shaped the harshness of the work and became the condition of both attachment to the farm and the inevitable emigration of so many to other places. The history of the potato in Ireland provided a chance to see, in stark and clear form, the reasons for and consequences of the exile from Eden. Here were peoples who were indeed enduring the curses of the God of biblical creation. They lived by the sweat of their brows, gave birth to many children, then had to go forth and multiply elsewhere.

After living in and writing about Ireland, I travelled to the lands, lives and stories of hunter-gatherers. I have already described the intense feelings this gave me of making a journey to a very different kind of human condition. The immense landscapes were matched by human beings who seemed to be free and at home. I had made a move from the realm of troubled exiles to the other side of Eden. Through this move I was to discover a contrast that was both personal and anthropological: I felt liberated from the anxious inner landscapes of middle-class Europe, and I was able to experience the warmth and ease of people who have the deepest possible sense of being in the best of all possible places. So the h-g's came to oppose the p-g's, the hunter-gatherers the potato-growers.

One absurdity of this construct is obvious enough. Potatoes came from the Americas, and were domesticated there, I would imagine, by peoples more in the hunter-gatherer mould than in any other. There must be a better agricultural abbreviation, one that refers to some resource other than potatoes. The obvious candidate is g-g, grain-growers. At least this makes a childlike pun on the existence of the horse (gee gee), though this is a creature that can leap across the divide—as, of course, it did when Indians in the Plains and the Northwest got hold of horses in the 1600s. In fact, the h-g : p-g abbreviation, along with its alternatives, should be put aside so that other weaknesses of this dichotomy may be confessed.

Hunter-gatherers rely on wild animals and plants. But there are many indigenous societies that depend on domesticated animals and crops—tribal peoples who are not hunter-gatherers. Relations between these two forms of indigenous economy are often marked by mutual suspicion and animosity, with tribal agriculturalists tending to despise their hunter-gatherer neighbours. James Woodburn, the British academic who has made some of the most important contributions to hunter-gather anthropology, summarised the kind of discrimination that hunter-gatherers have suffered from their pastoralist neighbours in southern Africa, noting that they tend to be described as "dirty, disgusting, gluttonous, ignorant, stupid, primitive, backward, incestuous, lacking a proper culture and language and even as animal-like, not fully human." He goes on to explain how these attitudes are often accompanied by deep fears about the apparent exotic powers of hunter-gatherers. The important point here, however, is that the divide between agriculturalists/pastoralists on the one hand and hunter-gatherers on the other appears between indigenous societies as well as between hunter-gatherers and European settlers. Both the sociology and the colonial histories of Australia, North America and southern Africa justify, and can be illuminated by, some version of the h-g : p-g distinction.

In the Americas, some indigenous agriculturalists developed the dense populations, severe inequalities and military aggression that I have linked to settler farming. Similarly, to reflect on the ancient history of Europe, where farmers occupied hunters' lands millennia ago, is to contemplate the possibility of many mixed economic systems. Farming peoples moving into new lands would have relied on herds and hunting while they established fields and waited for crops to grow. Some hunter-gatherer societies, having exhausted their supplies of wild game, either adopted a form of agriculture or pastoralism or entered into complex dependence on neighbours who were farmers or herders. And hunters in many regions may well have tried to farm or make use of domestic animals as adjuncts to their hunting systems. The evidence of language, as I have said, argues that the farmers overwhelmed the hunters. But this does not mean that farmers were not also hunters or that the hunters, before being overwhelmed, did not attempt some farming.

Anthropological accuracy requires, therefore, a great deal of caution about the hunter : farmer dichotomy. In reality, there is a possible spectrum of economic systems—with hunters at one end, farmers at the other, and many kinds of mixture in between—rather than two exclusive categories, some pair of opposites that between them include all possible human societies. In this respect, the hunter-gatherer : farmer divide is itself a form of myth.

Nonetheless, I believe that within this distinction lies a set of intellectual and imaginative opportunities. Thinking about the place of hunting peoples in the human story offers an insight into the history of the world. It provides a parallel insight into the nature of the human mind. The destiny of the hunter-gatherer is both an external and an internal process, an issue for societies and for individuals.

14

The hunter-gatherer mind is humanity's most sophisticated combination of detailed knowledge and intuition. It is where direct experience and metaphor unite in a joint concern to know and use the truth. The agricultural mind is a result of specialised, intense development of specific systems of intellectual order, with many kinds of analytical category and exacting uses of deductive reasoning. The hunter-gatherer seeks a relationship with all parts of the world that will be in both personal and material balance. The spirits are the evidence and the metaphors for this relationship. If they are treated well, and are known in the right way, and are therefore at peace with human beings, then people will find the things they need. The farmer has the task of controlling and shaping the world, making it yield the produce upon which agricultural life depends. If this is done well, then crops will grow. Discovery by discovery, change by change, field by field, control is increased and produce is more secure. The dichotomies of good and evil, right and wrong express this farming project: control comes with separating manipulable resources from the rest of the environment and working with determination and consistency against all that might undermine this endeavour.

The differences between hunter-gatherers living before agriculture developed or beyond the later farming frontier, and small indigenous societies based on a mix of farming, herding, hunting and gathering, may not best be understood as issues of mind. As noted, ideas about spirituality and the boundaries between the physical and the metaphysical are shared by many indigenous societies, both hunter-gatherers and small-scale farmers. However, all agriculture depends on controlling and remaking the natural world, and farmers have the task of both defending their fields and finding new ones. These are social and economic reasons for relatively high levels of organisation and aggression. It is no coincidence that in so many parts of the world, including regions where different indigenous systems live alongside one another, agriculturalists despise hunter-gatherers for being "primitive" and hunter-gatherers complain that farmers are belligerent. In the colonial era of the past five hundred years, "developed" agricultural societies have launched themselves with particular ferocity against all other peoples, and have, in particular, sought new land in vast territories occupied by hunter-gatherers. However complex the overlap between different kinds of indigenous societies, the dichotomy of hunter-gatherer : farmer says a great deal about how the world and the mind have been shaped.

Two ways of being in the world yield two kinds of human condition, each with its own set of circumstances. History reveals that exponents of the one have made war on the other, and the world has changed accordingly. Yet these different ways of thought are, as potential, within everyone. Human beings can reach into themselves and find two versions of life, two ways of speaking and knowing. Internally, many people are torn between these two ways. Individuals are born into one or the other society, and therefore learn its particular skills and disposition; but nobody is born to *be* either. The potential for language, and therefore for thought itself, is a shared human characteristic. The specifics of the language that are learned—not language itself—are what embody the intellectual and personal characteristics of one or another kind of mind and society.

What makes us who we are? Things we inherit, be they aspects of body or the hard-wiring of the mind. But language means that much of who we are does not lie *within* us as individuals so much as *between* us. The child is shaped by the society she lives in as a result of how others speak and behave towards her. All of us learn and live in relationships with one another: much of our reality lies in how these relationships take shape, function and maybe fail. Much of who people are comes from events and processes that are more than just internal and personal. And this is where we can see a particular importance of hunter-gatherer societies: they have established and relied upon respect for children, other adults and the resources on which people depend. If these relationships are not respectful, then everything will go wrong. The sickness of particular individuals, the failure of the hunt, the weather itself—these are all expressed in terms of relationship. The egalitarian individualism of hunter-gatherer societies, arguably their greatest achievement and their most compelling lesson for other peoples, relies on many kinds of respect.

The hunter-gatherer achievement, however, is not a matter of mutually exclusive qualities. Every healthy human being has the potential for all human qualities; nobody develops one kind of strength to the complete exclusion of its opposite. To this extent we are all hunters *and* farmers. The differences between one kind of society and the other are therefore to do with balance. And the imbalance has arisen because farmers have achieved such complete domination over hunter-gatherers.

Many hunter-gatherer societies have made accommodations to

farmers and herders. In many parts of southern Africa and South America, hunter-gatherers have created gardens and become shepherds or farmworkers or suppliers of pots and spears. In North America, many became cowboys, and some worked as domestic servants. Modern hunter-gatherers have taken advantage of farmers to supplement their own resources, or they have looked to the new wealth of farmers to help them deal with the loss of land and the destruction of wild animals that the farmers' arrival has caused.

But in many places, in many ways, hunter-gatherers are not at ease with farming and herding ways of life. Again and again, the farmers, while using their hunter-gatherer neighbours as casual and cheap labour, complain about their "unreliability." The hunters want to go hunting; gatherers like to gather. Hunter-gatherers tend not to plan and manage surplus. They need food or money now, not in several weeks' time. In the modern world, the hunter-gatherer often appears to be restless as well as poor.

The genius of hunter-gatherers is not rooted in their readiness to learn from or to work for others, however widespread these attributes may be. The compelling expression of hunter-gatherer culture lies in the balance of need with resources; the reliance on a blend of the dreamer's intuition with the naturalist's love of detailed knowledge; and the commitment to respectful relationships between people.

The hunter-gatherer within the modern, urban world is not extinct. Remaining hunter-gatherer societies continue to exhibit the qualities they have always possessed. And there are also hunter-gatherer points of view, beliefs and habits of mind within the farmers' world, inside the nations and towns that the exiles from Eden have established and from which they continue to press outwards with a nomadic imperative. Some anthropologists have pointed to the presence of people who forage at the urban community's social and economic margins—men and women, even some families, who make do on a day-to-day basis, relying on resources that are found here and there rather than earned from the routines of daily labour. This portrayal of the hunter-gatherer in the modern city accords with widespread images of hunter-gatherer destitution and landlessness. It speaks to loss rather than to achievement.

But there are more optimistic views of the hunter-gatherer in the urban setting. Indeed, there are eruptions of the hunter-gatherer mind in the farmers' world, as evidenced in the many voices raised in opposition to the unquestioned dominance of the agricultural way: protests against repressive order, bureaucratic planning, chronic inequality, patriarchal conceit, poisonous pedagogy and the denial of all that is essential to art. These are arenas in which a rival mind seeks expression and longs for its particular forms of freedom. Throughout the western world, there are men and women who consciously choose low levels of material comfort and small numbers of children to avoid the need for large incomes, thereby pursuing lives in which they may survive without regular jobs and devote themselves instead to creative work and family life. This way of being encompasses a concern about the destruction of the natural world by the ever-growing pressures to reshape it in the interests of surplus and profit. And dissident voices within mainstream culture have long criticised the use of repression and violence both in maintaining social order and in raising children. In all these we can hear echoes of hunter-gatherer ideas and practices.

Men and women galloping on horses across the countryside in pursuit of foxes; men with shotguns who fire at pheasants, grouse or partridges driven towards them by a line of beaters; those whose wealth allows them to trawl the seas for game fish in their powerful boats—these people may claim to represent the hunter-gatherer within us all. Yet their habits and minds are fixed firm to the farming condition. Their hunting, shooting and fishing is evidence of the very characteristics that agricultural development has exaggerated, with the help of capitalist and industrial developments, to an extreme. They are suppressors rather than exemplifiers of the hunter-gatherer. They live by the systems of privilege and organisation that are hallmarks of the agricultural mind.

No: the hunter-gatherers in the heartland of the exiles, living in the nation-states of farmers and in the cities farmers have built, are opponents of the dominant order. They oppose hierarchy and challenge the need to control both other people and the land itself. Consciously or not, they are radicals in their own lives. At the least, they experience the tension in themselves that comes from a longing not to plan and not to acquiesce in plans; at most, they use a mixture of knowledge and dreams to express their vision. It is artists, speculative scientists and those whose journeys in life depend on not quite knowing the destination who are close to hunter-gatherers, who rely upon a hunter-gatherer mind.

The visionaries in society are always there, and they are perhaps a part of us all. The agriculturalist mind and its economic order never quite obscure evidence of the hunters. Many people feel the strain of a way of life and a mind-set that disallow all forms of improvisation and intuition. The controlling features of a life that has no place for the hunter-gatherer mind create a longing for spirituality and underpin many forms of protest, from Quaker ideals of equality to the call for deschooling society, from new-age mysticism to concern about rainforests.

There is a common experience of something being wrong that may receive real illumination from a much more direct acknowledgement of rival forms of mind. Rival forms of mind are, of course, reducible to rival forms of society—and, in the end, to the displacement of one kind of economy by another.

Hunter-gatherers may well represent a need in all peoples to experience a profound form of freedom. But they are also within the social universe in a literal way: there are hunter-gatherer societies, along with other indigenous peoples, whose demands for cultural survival and actual territories are no less vociferous than they ever were. In every part of the colonial world, the issue of aboriginal rights is alive as an issue whose urgency and poignancy are augmented by the prospect of a final destruction.

In Australia, Aborigines contest government and industrial invasions of their lands. There have been struggles in the courts, in political campaigns and on the land itself. Settlers still want to eliminate hunter-gatherer claims to the Australian outback. Aborigine organisations protest, and in some crucial places win, their right to their own way of life, their own heritage, their own place on the earth. In 1992, the Australian High Court established a common law principle of native title to native lands.

In South Africa, the ‡Khomani San, the remaining Bushmen of the southern Kalahari, have managed to survive their complete displacement and dispossession, after fifty years of subsistence at the most destitute margins of South African farms and townships. With the fall of the apartheid system, these few survivors began a campaign for the return of their lands and for the right to live in whatever combination of social and economic systems they choose for themselves. In March 1999, the South African government accepted their claim and agreed to return some 50,000 hectares of original ‡Khomani San territory. In Amazonia, Colombian hunter-gatherers have been mapping their territory as part of new negotiations with the government, which has at last moved towards recognition of tribal rights to tribal lands.

In Canada, the land-claims era that began in 1971 has led to a number of settlements that recognise hunter-gatherer heritage and lands. The Cree of the James Bay area in northern Quebec have managed to secure the basis for a hunter-gatherer economy in a large and complex settlement of their dispute over the flooding of much of their lands to create hydroelectric dams. The Inuvialuit of the western Arctic have negotiated a settlement of their claim to aboriginal title in a large area in the Mackenzie Delta and adjacent islands. The Gitxsan and Witsuwit'en have won much greater basis for their rights in the Supreme Court of Canada. The Nisga'a have reached an agreement with Canada that gives them a core of territory and extensive rights of self-government. And the Inuit of the Northwest Territories have secured from the Canadian government a territory and jurisdiction that is based on the spread of their hunting territories in the eastern Arctic. This is the new territory of Nunavut—the most ambitious attempt in the history of the encounter between colonists and hunter-gatherers to secure coexistence at the frontier.

These victories are all the result of modern political processes. They represent a small proportion of the struggles and claims that continue, in South America and Australasia, in Siberia and the United States, in Australia, Africa and Canada. In all these places, there are indigenous groups who fight for their survival. The exiles from Eden, the nomads who roam the world looking for new places to settle, transform and control, must make common cause with these struggles.

15

Human activities have shaped the surface of much of the world. Farmers have effected the greatest changes, turning forests, deserts and swamps into fields and pastures. Hunter-gatherers have also shaped the world, but more subtly, intent on keeping it a place where wild plants, animals and fish can thrive. The displacement of hunters by farmers has meant less "land" and more "countryside." These different ways of moving on the earth reflect the economies and societies of different kinds of human being.

The world is also shaped by stories. What people feel, know and need to pass on from generation to generation has existed in words: words that speak of how the world began, of how humans emerged in it and found both places to live and ways to deal with one another. Words are entitlement to these places. The stories of farmers, including the Creation as described in Genesis, give meaning to their ways of life. The stories of hunter-gatherers give meaning to theirs. These are different meanings, different kinds of stories. The history of the one has dominated and, to a large extent, silenced the other.

Many hunter-gatherer ways of knowing the world have

disappeared, along with hunter-gatherer languages. These are rich and unique parts of human history that cannot be recovered. If the words are gone, so are the stories. A particular shape is lost forever. There are fears that hundreds more languages—many of them those of hunter-gatherers—will have disappeared within another generation. Each such case represents a harm that is inestimable: the cumulative loss of language constitutes a diminution in the range of what it means to be human.

There is, however, a story about these stories that has its own importance. The encounter between hunter-gatherers and farmers is a history of loss to one kind of society, gain to another. For hunter-gatherers, some losses cannot be compensated. But the story itself is one form of compensation. If the world can acknowledge who hunter-gatherers are, how they know and own their lands, what the encounter with farmers and colonists has meant, then some restitution can be made. An inquiry into the fate as well as the achievements of hunter-gatherers is, in this regard, part of a story that hunter-gatherers need to tell and have told. Without the hunter-gatherers, humanity is diminished and cursed; with them, we can achieve a more complete version of ourselves.

NOTES

Full publication information for the sources cited in these notes is given in the bibliography. Numbers in the left-hand column refer to page numbers in the text.

In addition to the specific sources I refer to in the notes that follow, I have received help and inspiration from a wide range of Arctic literature. Some of this has been popular and anecdotal, as in the case of the work of Farley Mowat (which some would say is more literary than anecdotal). I have drawn on detailed ethnography, especially in relation to the central and eastern areas of the Arctic. No field work proceeds without a background of writing by others that sometimes shapes and sometimes contradicts one's own work.

There is, of course, an extensive literature dealing with many aspects of Inuit archaeology, ethnography and social organisation. The work that gives a particularly wide-ranging but detailed account of Inuit life across the Arctic is that of Knud Rasmussen, who, as the child of missionaries in Greenland, grew up speaking Inuktitut, and who therefore, when he became an explorer and ethnographer, had the rare ability to speak to the peoples he met in their own language. The results of his epic expedition across the Arctic were published in the *Report on the Fifth Thule Expedition*. The work runs to several volumes; Kaj Birket-Smith wrote some, Rasmussen others. I have always found volumes 4, 6, 7, 8 and 9 most compelling; they include shamanic stories, poems and vivid accounts of Inuit elders as well as a wealth of ethnographic detail. (It should be said, however, that Rasmussen has been criticised for taking liberties with some of his sources and oddly failing to give Inuktitut versions of all the stories he includes in the report.)

Prior to Rasmussen's Thule expedition, two other Arctic ethnographers of great importance were Vilhjalmur Stefansson and Diamond Jenness. Stefansson wrote a great deal, but his book *My Life with the Eskimo* draws together much of his work in a very readable form. Also, his *Not by Bread Alone* is a remarkable exploration of Inuit (and other people's) diets. Jenness's most important ethnography is *The People of the Twilight*, in which he describes his stay with the Copper Eskimo of the central Arctic. Jenness subsequently became an influential member of the Canadian administration of northern affairs, and was associated with plans for relocating Inuit farther south, where, he proposed, they would be able to take advantage of employment opportunities arising with new activity on the subarctic frontier. (Plans of this kind for the most part failed very quickly.)

In the 1950s and 1960s, the Canadian anthropologist Asen Balikci did extensive field work with the Netsilingmiut of the central Arctic. His book *The Netsilik Eskimo* sets out his findings and reflections. It also led to a set of twelve films, ultimately edited and distributed by the National Film Board of Canada, under the title *Netsilik Eskimos: The People of the Seal*. These films, intimate and painstaking, follow the year of a Netsilingmiut family who agreed to re-create their "traditional" life. The Netsilik films are arguably the most complete documentary of hunter-gatherer life ever made, and they provide a unique window onto the material life, economic activities and human qualities of the Inuit. Their success was the result of a collaboration between Balikci and the Oblate missionary Guy Mary-Rousselière. (Mary-Rousselière himself is the author of *Qitdlarssuaq*, a book about an Inuit migration between northeast Canada and Greenland in the late 1800s, in which it is possible to glimpse aspects of shamanism and to experience some of the language used by an Oblate intellectual, even when writing as a well-informed ethnographer, about Inuit customs.)

During the early 1970s, the Northern Science Research Group, the academic branch of the Canadian Department of Indian and Northern Affairs of which I was fortunate to be a member from 1971 to 1974, published a series of monographs, making available the outstanding work, in particular, of Nelson Graburn (in Arctic Quebec), Peter Usher (on Banks Island) and Derek Smith (in the Mackenzie Delta). At the same time, a number of Area Economic Surveys were undertaken and then published by the Department of Indian and Northern Affairs. It must be said that these surveys were rather crude, and somewhat informed by an implicit, if not explicit, anticipation of a final transformation of hunter-gatherer economics and society into some form of frontier capitalism, with the Inuit as minimally qualified labourers.

In the 1970s, a comprehensive study was carried out, in collaboration with the newly created Inuit organisation the Inuit Tapirisat of Canada, to detail all the places throughout the Canadian Northwest Territories where Inuit had hunted, gathered, fished, camped or had sites of cultural and historical importance. This large and, on the whole, very successful undertaking was overseen by Milton Freeman, though regional studies were carried out under the aegis of (among others) Peter Usher, Bill Kemp, Don Farquharson, Carol Brice-Bennett, Tony Welland, Rick Riewe and myself. The results were published in three volumes as *The Inuit Land Use and Occupancy Project*. An important finding of this work lay in the extent to which it revealed that Inuit hunting and gathering had increased, rather than decreased, as a result of the modernisation of techniques. And the Inuit interviewed in depth about their system emphasised over and over again that, in their view, their health—at both the individual and community levels—depended on use and occupation of their lands. A similar study carried out in Labrador was directed by Carol Brice-Bennett, and its results were published in *Our Footsteps Are Everywhere*. Labrador is remarkable for the way settler families have lived alongside Inuit for some 150 years. They constitute the only group of Europeans to make lives for themselves in North America more on the terms of an indigenous population than as representatives of a colonial enterprise. Brice-Bennett's study includes the settler story.

In Arctic Quebec, fascinating work has been done among Inuit since the early 1970s by Bernard Saladin D'Anglures and associates at the Université de Laval. Volume 5 of the *Handbook of North American Indians*, edited by David Damas, is a compendium of much Arctic scholarship.

In the 1980s and '90s, Inuit studies have been dominated to a considerable and necessary extent by questions of sovereignty and decolonisation. In this regard, Inuit families' relocation from Arctic Quebec to north Baffin Island has generated both historical and theoretical debate. Relating to this, the work of Peter Kulchyski and Frank Tester, especially their book *Tammarnit*, is of importance, as is the work of Alan Marcus. One person who has written at the end of a long career as both explorer and administrator is Graham Rowley. His book *Cold Comfort* offers many remarkable insights into the key changes in Inuit circumstances during the colonial period.

ONE: **INUKTITUT**

11 **Inuit** Inuit is the ethnonym of many of the Eskimoan people of the Arctic. It replaced "Eskimo" in the Canadian central and eastern Arctic in the early 1970s, and since then it has been adopted by all Canadian Inuit, as well as by some Greenlanders. In Alaska, the parallel term Inupiaq is used, although there has been a much stronger tendency to hold on to the word Eskimo. In Siberia, the Eskimoan people of the Chukotsk peninsula refer to themselves as Yuit, though they now tend to be grouped with the Eskimoan peoples of southwest Alaska as Yupik. The peoples of the Aleutian Islands of the Bering Strait call themselves, and have long been referred to as, Aleut.

Inuit is a plural form that translates best as "persons" rather than "people." Its singular form, *inuk,* means "person," though it can be used to refer to a human being. A dual form, *inuuk,* "two persons," creates some difficulties. For the sake of simplicity I have avoided it here.

The word *Inuktitut* is based on *inuk,* to which is added the infix *-titut,* meaning "in the manner of." A rule of Inuktitut morphology governing the adding of infixes to stems requires than when a final *k* meets a *t,* the *k* disappears and the *t* is doubled. Hence the correct spelling of the word is Inu-tt-itut. In Canadian popular writing, however, this rule is ignored, and Inuktitut has come to be the standard spelling.

In almost all the renderings here of Inuktitut words, I do my best to use the Roman orthography devised by Raymond Gagné (published as *Tentative Standard Orthography for Canadian Eskimos*) and taught to me by Mick Mallon. The exceptions (along with "Inuktitut") are the names of places and people, where I use the form that has been adopted by cartographers or by the people themselves. All the names for modern Inuit settlements have both an English and an Inuktitut form. Hence Pond Inlet is known to the Inuit as Mitimatalik, and Arctic Bay as either Ikpiarjuk or Tununiarusirq. In most places, I have used the English names.

14 ***Miut* and *miutaq*** Inuit identify family or hunting groups with particular places, using the affix *miutaq* (singular) or *miut* (plural). Hence the Tununirmiut are the people of Tununiq, and a person from there is a Tununirmiutaq. In fact, people can be *miut* of different places, just as other cultures have different kinds of community membership—a person may be an American, and also a New Yorker, and also a resident of the Bronx. Similarly, an Inuk can be a *miutaq* of Baffin Island, Frobisher Bay and a core hunting area that his or her family looks to as an original home.

The anthropology of *miut* groups, however, is of great importance, for it provides a clue to systems of land use and harvesting for distinct hunting and gathering communities, often in the form of extended families with ties to a set of camps that constitutes their seasonal round in a large territory. Therefore, the most important Inuit membership—the key *miut* group—is the one that designates this territory as a whole. A good account of this is given by Thomas C. Correll in his essay "Language and Location in Traditional Inuit Societies."

The etymology of *miut* and *miutaq* is revealing. Acting as an infix or affix, it is composed of separable phonemes: the *mi* means "in"; the *u* indicates existence, as in sentences such as "*Suna-u-na?*" ("What is that?") or "*Inu-u-vunga*" ("I am an Inuk"); and the *ta* suggests ownership, as in "*Suna-ta-qarpit?*" ("What have you have got?"). This last element is missing in the usual plural, though a full version of the plural, *miutait,* is also used as an affix.

15 **"This girl growing …"** The first years of Inuit childhood are spent in close contact with the mother and are marked by a child's being given all that is asked for until weaning happens. As Jean Briggs has described so well in *Never in Anger,* there is then a relatively harsh change in how the child is treated: demands for the breast and continuous attention are firmly rebuffed, with the consequent crying often being ignored. Briggs also observes the ways in which teasing and similar forms of humour are used to convey to children where the limits of behaviour lie.

In fact, the permissive and gentle quality of Inuit child raising, especially the lack of corporal punishment, was remarked on by many early explorers and missionaries in the Arctic. For examples, see G. F. Lyon's *Private Journal,* written when his ship was stuck in north Hudson Bay in the early 1880s; Father Roger Buliard's *Inuk;* and Arminius Young's *One Hundred Years of Mission Work in the Wilds of Labrador.*

18 **The precision of Inuktitut** Inuktitut achieves its precision by making use of many infixes—small building blocks of meaning set into verbs and nouns—which allow careful specificity and also create possibilities for all manner of playfulness. The elements in the words of the drill with coloured bricks, for example, are: *aupaluk,* meaning "red" (as of blood) + *tumik,* signifying the accusative case, followed by (though it could equally be preceded by) *tigusi,* indicating "take hold of," + *gavit,* indicating conditional second-person singular, "if or when you."

A more complex example is *nguaq,* the root of the word for imitating or pretending, which can also be used as an infix. So: *nguartut* means "they are pretending," *pisunguartut* means "they are pretending to walk," *isumanguartuq* means "he/she is seeming to think," while *nguarnguartuq,* "pretending to pretend," is often used to mean "he/she is acting." Interestingly, the words for adoptive son and adoptive daughter are *irninguaq* and *paninguaq.*

Infixes can be built into Inuktitut words to achieve irony and self-deprecating overtones. The person who says "*Isumanguarpunga,*" literally "I am seeming to think," is likely to be affecting a modest tone, translatable as "I just think." By combining infixes that undercut and play with one another's meanings, a speaker of Inuktitut can infuse what is said with subtleties and jokes that are not easy to translate. Thus Anaviapik's son Inukuluk, speaking about learning English at the adult education centre, said, as a sort of final judgement on a long series of educational experiences, with a world-weary sigh and a glimpse of a smile, "*Isumanguarllarilitainaktukuluuvungali,*" which breaks down into: thought / seemed to / to my surprise / once again / at length / sweet little, wee / am / I / and a final *li* that denotes lightheartedness. How would we put this in English? Something like "Ah well, poor little me finally surprised myself by realising …" And he went on, "*Qadlunangumarijjaijunirpunga,*" white man / be / real / never will / I. Or, to set this out in English word order: "I'll never be a real white man."

The logic of Inuktitut works in reverse phoneme order, with each thing moderating everything that has preceded it, as in mathematical expressions that rely on a series of brackets, of the form $x(y[z\{w\}\ ])$, meaning that y is modified by x and z by xy, and so on. I had been learning Inuktitut for almost two years before I recognised this feature of its logic and the play of ideas that it makes possible.

The logical quality of Inuktitut word formation also means that there are virtually no irregular verbs or exceptions to the ways in which the basic building blocks of the language are assembled. It may be that the absence of irregulars is a result of the isolation of the language; it appears that no surviving languages have had any significant impact on Inuktitut. The linguistic archaeology of impact, to be seen in variants and irregularities, may tell us, therefore, that Inuktitut's isolation has been more or less complete for a long period of time. However, there are signs of pre-Inuktitut words in some place names; and it is undeniable that Inuit in many regions have had long, albeit uneasy, relations with their subarctic Athabaskan and Algonquian neighbours. It could be, however, that Inuktitut's agglutinative quality explains more about the language's regularity than about its isolation. As Steven Pinker shows in *Words and Rules,* the coining of words causes their grammar to be regular. Pinker gives examples of how English nouns and verbs that are contrived tend to be given regular endings, even where their roots evoke common irregular forms (p. 162ff.). This may be what takes place in Inuktitut more or less all the time.

The play of infixes allows Inuit men and women to ensure that what they say will be truthful, on the one hand, and not likely to cause offence, on the other; or that the facts are set out but self-glorifying claims are avoided. Inuit often choose to put themselves into what is said, showing that something is speculative, or definite, or dependent on fortune, making sure that they are heard to be both careful with truth and modest. This is a profoundly important feature of Inuktitut culture and psychology and is also, I suspect, a feature of all hunter-gatherers. To express these social and individual qualities, the speaker must be able to use all these subtleties of a language.

27 **Skidoos and sledges** In the 1970s the snowmobile, which came to be known in much of North America as the skidoo, arrived in the Arctic. The skidoo has a motorcycle type of seat, long enough for two, mounted on a frame, and an engine that drives a single, wide track. Skidoos very quickly displaced dogteams, for they could pull a loaded sledge at considerable speed—anything up to fifty miles an hour. At first these machines were somewhat unreliable, having being designed for wintertime leisure activity much farther south, but the Inuit are excellent mechanics and managed to keep them going. Over the years, models have become tougher and faster.

Although skidoos have largely replaced dogs as the method of pulling sledges, sledges themselves have continued with relatively minor adjustments. The sledge's two runners are separated by a row of slats that are tied on rather than nailed on, to allow flexibility as the sledge meets varied ice and snow surfaces. Although modern runners are "shoed" with iron strips rather than with whalebone or mud, they are still coated with a thin layer of water that freezes to create the smoothest possible contact between sledge and ice.

31 **Qallunaat** Qallunaat (singular Qallunaaq) is the Inuktitut term for people who are neither Inuit nor the Inuit's Athabaskan and Algonquian neighbours (for whom the term *Allait* is often used). It means anyone who comes to the North from the south. But the

derivation of Qallunaat is obscure. Since *qalluk* means "eyebrow," some commentators have speculated that Inuit were surprised by the prominence or hairiness of European eyebrows. However, the first use of Qallunaat appears to come from Greenland, and in Greenlandic Eskimo there is a root *qalluq*, pertaining to the south. Since some of the earliest encounters between Inuit and Europeans took place on the Greenland coasts, this may explain the word's origin.

Readers familiar with Arctic literature will notice that the word has often been spelled "Kabloona" (see Frank Vallee's *Kabloona and Eskimo in the Central Keewatin* as an example). This spelling results from a difficulty with the guttural *q*, which travellers writing about the North often write as *k;* and from a regional dialect variation in which the sound "dl" in the eastern Arctic becomes a "bl" in the central Arctic. In some dialects, the sound becomes a fricative, with the sound of the "Ll" at the beginning of many Welsh place names (Llangeitho, Llangyrig, Llandudno). Many scholars (following the Gagné orthography) avoid this problem by using "ll" to cover these and other related dialect variations. Thus the "ll" in *qalluk* is the same consonant as the "dl" in the widely used spelling of Qadlunaat (or as the "bl" in Kabloona).

35 **Knowledge of the land** The affirmation of an existential bond between self and place is very striking among hunter-gatherers. The ties to a place and its resources may be more clearly affirmed than ties to history, community or even family. Some anthropologists have noted a freedom from social and kin ties that secures the necessary relationship to land and resources, thanks to which hunting and gathering are flexible and productive. The work of the British anthropologist James Woodburn is of great importance in relation to these issues, especially his essays "Minimal Politics" and "African Hunter-Gatherer Social Organisation."

The French anthropologist Claude Meillasoux has also addressed the extent to which hunter-gatherers are disengaged from social constraints with which other peoples are all too familiar. See, for example, Meillasoux's essay "On the Mode of Production of the Hunting Band." In his book about the Ik of northern Uganda, *The Mountain People,* Colin Turnbull devotes much of his introduction to summarising the more remarkable features of hunter-gatherer systems. His sense of these comes from his work among Pygmies of the Congo (described in his book *The Forest People*), and he notes that "Just as the Mbuti Pygmies in their lush tropical rain forest regard the forest as a benevolent deity, so do the Ik, in their rocky mountain stronghold, think of the mountains as being peculiarly and specially theirs" (pp. 24–25).

Nonetheless, I am aware of a tendency on my part to exoticise the Inuit. I was learning the Inuit language and many things about how to travel by dogteam on spring ice. But it was *possible* to learn these things. Dan Sperber, in his essay "Apparently Irrational Beliefs," has made the telling point, when discussing cultural relativism, that the experience of anthropologists is "the best evidence against relativism," whereas their writings are the evidence for it. In a striking summary of this paradox, he writes: "It seems that, in retracing their steps, anthropologists transform into unfathomable gaps the shallow and irregular boundaries that they had found not so difficult to cross, thereby protecting their own sense of identity, and providing their philosophical and lay audiences with just what they want to hear." Sperber is quoted in Glenn Bowman's fascinating essay "Radical Empiricism"; see especially pp. 1–10.

40 **Syllabics** One reason syllabics are so useful for writing Inuktitut comes from the way the language relies on consonants and vowel sounds, not tones. Although finals and doubled consonants modify vowel sounds, this modification can be shown by using the syllabic symbols as diacriticals—putting a miniature < (pa) above the full-size symbol, for example, to indicate that the consonant alone is sounded, not the whole syllable. This makes it possible to show all final and doubled consonants. It must be said, however, that many Inuit do not bother to use diacriticals when writing, relying on context to resolve possible homoscripts.

42 **Words for fear** In Inuktitut, roots for verbs cannot stand alone. The root *kappia*, for example, if said by itself, is ungrammatical. To become a word, it must be modified by an infix that gives it sense, then an ending that gives it person. Thus a very simple word based on the root *kappia* is *kappiasukpunga*, "I feel *kappia*," i.e., "I am afraid." Where a word is not a word, so to speak, because it lacks the necessary additional suffixes and modifiers, I indicate this with a hyphen (-).

45 **Tiredness** The Inuktitut words for tiredness have *takka* as their root. Anaviapik would often say to me: "*Takkavit?*" "Are you tired?" But when I asked him if *he* were tired, he would almost always reply by saying he was not *takka* but weary or sleepy, using words based on the root *uingu*. I came to think that adults do not feel, or do not admit to feeling, simple tiredness. This seemed to be a word reserved for children—along with anthropologists.

46 **Words for hello and good-bye** In some regions the word "hello" is used, though it can be extended to mean a person from the south (as in *helloraluk*, literally "the great or large hello," to mean a policeman). *Saimo* is used in much of Arctic Quebec for "hello"; its origin is obscure to me. In the eastern High Arctic the term *qujannamik* is used to mean "thank you." Its etymology is puzzling; the root, *qujannu*, is a widely used idiomatic term meaning "it doesn't matter" or "not to worry." In the same dialect area, the term *tavvautik* is used to mean "good-bye." The root for this is *tavva*, which means "there" or "thus." Words widely used to say sorry are based on *mammiuna*, a term that refers to intentionality. The expression *ilaaniungituq*, used in the Ungava dialect area to mean "sorry," translates literally as "it is not intended."

When Anaviapik and other elders greeted someone they were especially delighted to see, they would exclaim, often very loudly and repeatedly, the words *puinna* and *aliunnai*, which are expressions of joy. (Oddly, the word *aliunnai* in the north Baffin dialect is *aliunna* in the Ungava and Arctic Quebec region: the difference in the final "a" and "ai" signify opposites, so that in one place the word means "happiness" and in the other it means "unhappiness.")

When thinking about English ways of saying "please" and "thank you," it is important to note that words to express social politenesses do not necessarily indicate social inequality. Some English politenesses have their origins in prayer ("Bless you," for example). And historically, everyday politenesses have sometimes signalled equality and respect rather than hierarchy. But the underlying, cumulative quality of such terms, and the way they function in everyday life, is closely tied to inequality. Hence some teenagers' reluctance to use such terms, opposed as they often are to hierarchical habits and norms of all kinds. Hence, also, the importance Quakers attached to the ways in which everyday language can express and enforce social inequities. A wonderful account of Quaker concerns with what they termed Adamic language—the vocabulary that would have been used by Adam and Eve in the

Garden of Eden, before corrupt, unequal society had worked its changes on human spirit and vocabulary—is to be found in Richard Bauman's *Let Your Words Be Few.*

46 **Feelings and *isuma*** A full discussion of Inuit attitudes to expressions of feeling can be found in Jean Briggs's book *Never in Anger.* Briggs also wrote a report in the late 1960s for the Canadian Department of Indian and Northern Affairs that was published in the Northern Science Research Group series under the title *Utkuhikhalingmiut Eskimo Emotional Expression.* The insights set out in these two publications are based on participant-observation in an Inuit family in a remote area of the central Arctic. Briggs's work is remarkable for the intimacy of its portrait of Inuit life. But it also gives the reader a rare opportunity to follow the way in which an anthropologist works and to consider how the insights achieved may be closely linked to problems and conflict as well as to seemingly neutral observation. These are two of the few detailed studies of Inuit society carried out by a woman—and the results show insight that men typically would not achieve.

47 **Measuring time of travel** When writing about journeys in the spring, it is awkward to refer to the times we slept as "days." In fact, Inuit count the length of a journey by speaking of the "number of sleeps," asking the question "*Qassinik sinilaurpit?*" (literally "How many did you sleep?"). This conveniently avoids the issue of whether sleeping takes place during the day or at night.

49 **Fulmar** The fulmar or northern fulmar *(Fulmarus glacialis)* is a circumpolar species with populations on all Arctic coasts, including Alaska, Siberia and the Barents Sea. Both adults and young spit foul-smelling oil to defend themselves at nesting sites. The fulmar is a species whose numbers have risen dramatically. In Iceland, for example, there was one breeding colony in 1640; by 1950, there were 155 colonies. The increase is thought to be the result of commercial fishing—fulmars are excellent scavengers.

49 **Words for snow** The history of the "vocabulary hoax" is explored by Geoffrey K. Pullum, linguistics professor and essayist, in his book *The Great Eskimo Vocabulary Hoax* (pp. 159–74). Pullum's central point is that there is a remarkable resistance, even within the heartland of academia, to the real facts about Inuktitut words for snow. He begins with the observation: "Once the public has decided to accept something as an interesting fact, it becomes almost impossible to get the acceptance rescinded. The persistent interestingness and symbolic usefulness overrides any lack of factuality."

Pullum describes the difficulty experienced by anthropologist Laura Martin when "attempting to slay the constantly changing, self-regenerating myth of Eskimo snow terminology, like a Sigourney Weaver fighting alone against the hideous space creature in the movie *Alien* ... You may recall that the creature seemed to spring up everywhere once it got loose on the spaceship, and was very difficult to kill" (pp. 161–62). Pullum's book does not seek to resolve the question of what is and is not the truth about Eskimo words for snow, but rather encourages students to challenge their teachers when teachers reproduce a potential myth without the benefit of authentic data. He offers the imprecation: "Don't be a coward like me. Stand up and tell the speaker this: C. W. Schultz-Lorentzen's *Dictionary of the West Greenland Eskimo Language* (1927) gives just two possible relevant roots: *qanik,* meaning 'snow in the air' or 'snowflake,' and *aput,* meaning 'snow on the ground.' Then add that you would be interested to know if the speaker can cite any more" (p. 167).

In fact, other dictionaries do cite more—see, for example Lucien Schneider's two-volume *Dictionnaire Esquimau-Français du parler de l'Ungava.* But the issue is not how many words or roots Inuktitut has for speaking of snow, but rather what the relationship might be between distinctiveness of language and distinctiveness of ways of seeing the world, if not of mind itself. Although I shall suggest in part 6 of this book that differences of mind are connected to differences of society, I have found evidence in the snow example of a considerable number of Inuktitut words for snow but no overwhelming problem of inter-translatability.

Laura Martin describes her experience of the snow issue in her essay "Eskimo Words for Snow."

49 **Language, meaning and Wittgenstein** The debate about how language may shape reality, and thus have both specific and profound consequences for human behaviour, was relaunched by Benjamin Lee Whorff in the late 1930s. Whorff took as a starting point the views of his former teacher Edward Sapir, who had noted that "the 'real' world is to a large extent unconsciously built up on the language habits of the group." Whorff then elaborated the ways in which both the meanings of words and the forms of grammar can shape how people think and behave. He sets out his views very clearly in his essay "The Relation of Habitual Thought and Behaviour to Language."

The question at the centre of Whorff's reflections is: Does language shape mind or does mind shape language? It is a question that has exercised many, from Wordsworth and Coleridge to postmodern linguists. But it seems to me that the choice implied by this question is inappropriate—the more we learn about cultures and brains, the more we see that there are both fundamental reasons for saying mind shapes language (as in the case of the so-called hard-wired capacity for language and grammar) and fundamental reasons for saying language shapes mind (as in the evidence that Whorff sets out). There are important differences between the languages of hunter-gatherers and those of, for example, modern agriculturalists. But I do not think that looking at language is the best way to see and understand the differences between peoples.

The argument about language and meaning is found, in its twentieth-century form, in the work of Ludwig Wittgenstein. He explores the problem in the first pages of his *Philosophical Investigations* (paragraphs 6–15, p. 6ff.). It arises also in his *The Blue and the Brown Books* (on pp. 27–28 he sets out the problem of communicating definition). In a striking passage in *Lectures and Conversations on Aesthetics, Psychology and Religious Belief,* Wittgenstein makes the observation, when speaking of "appreciation"—a term like the example I give of "worship," where definition is especially problematic: "It is not only difficult to describe what appreciation consists in, but impossible. To describe what it consists in we would have to describe the whole environment" (p. 31).

In my view, when it comes to the words for snow, along with all the other words an anthropologist must learn, the task does depend on understanding a great deal about "the environment" and may at times be daunting or indeed impossible. But it *is* possible to gain enough knowledge to be able to do a great deal of good translating. Hence my view that Anaviapik and Wittgenstein are fundamentally in agreement: both are aware of the problem and the limiting cases (of which the words for snow are not an example).

54 **Igloolik** The spelling of Inuit place names and personal names can be confusing. The village of Igloolik, an island at the northwest corner of Hudson Bay, is a good example. The

word means "where there are dwellings," with the root being *illu* ("dwelling" or "interior space") plus the affix *-lik* ("there is" or "there are"). So the correct spelling—following Gagné—should be Illulik. But because of southerners' spelling of "igloo," to mean a snow-house, they wrote the name as Igloolik, which has become the name on the map.

At least Igloolik is a name that originates with the people themselves. In most cases, Inuit communities and geographical features have been given names by southerners that disregard, and can come to displace, the people's own names. Thus Pond Inlet, which should be called Mitimatalik ("the place where birds land"), or Tununiq ("an area that is in the shade"), and Rankin Inlet, whose Inuktitut name is Qanalliniq, "the inlet." The evidence of British imperial exploration of the Arctic is to be seen in many of the names on the maps for features in the Pond Inlet area, including Admiralty Sound, Prince Regent Inlet, Prince of Wales Island, King William Island and Victoria Island. These are all places with Inuktitut names that are in danger of being lost.

The kinds of names Inuit use for geographical features are often surprisingly simple: Lake, Big Lake, Looks Like a Lake, Big River. But there are also places that got their names from either appearance (Red Cliff, Profile Headland, Fast Narrows) or stories (Place Where People Starved, Pisspot, Where Qallunaat Are Buried, Shipwreck and so on.)

59 **Dogs dying** A stereotype of starvation in the Arctic includes the eating of dogs as a last resort. In reality, dogs provide very little nutrition, and dogs that are themselves starving even less. Inuit have told me that if you are starving and then eat meat that has no fat (arctic hare, for example, or starving dogs), your starvation is accelerated rather than checked. This is because the body metabolises its own fats and, when there is no fat left, its own muscle. Eating pure protein causes the rate of self-consumption to speed up. Those who are starving must be able to eat fats as well as, or rather than, pure proteins.

60 **Anger** The way in which Inuit and other hunter-gatherers manage feelings of anger is indeed impressive, and can be said to constitute a social or even a moral achievement. However, it should not be understood as a virtue in and of itself. The qualities in hunter-gatherers that inspire and impress those from other kinds of societies exist for good, material reasons. The Inuit are not pacifists. There are many accounts, including those Inuit themselves narrate, of both internecine and intertribal violence. If the apparent virtues of restraint, avoidance and flexibility were not sufficient to deal with a threat from an Inuk or a Qallunaak, the person causing the threat was often killed. In the early days of colonial presence, this resulted in some dramatic impositions of southern law, as in the case of the Janes murder and the subsequent trial at Pond Inlet in the 1920s (see Qanguk's story in part 5 of this book) and the killing of two missionaries in the Coppermine area in 1914. A good account of the problematic relationship between Inuit and southern law is given by the Dutch anthropologist G. Van Den Steenhoven in his doctoral dissertation *Leadership and Law among the Eskimos of the Keewatin District, Northwest Territories.* He was fortunate to spend time among the Inuit of the Barren Lands west of Hudson Bay when shamanism was still much in evidence.

Accounts of hunter-gatherer life among the !Kung of the Kalahari, the peoples of the North Pacific Coast, the Andaman Islands, the Inuvialuit of the central Arctic, and the Nez Perce and Plains societies of the western United States all reveal that hunter-gatherers are capable of high levels of individual and, at times, collective aggression, and will use murder to achieve their ends. It is the coexistence of so many non-violent and socially harmonious characteristics *along with* a capacity for violence that is remarkable.

61 ***Muttuk*** Following the Gagné orthography, *muttuk* should probably be spelled *mattak*. But this word appears so much in Arctic literature as *muktuk* or *muktuck* as to make the technically correct *mattak* esoteric and confusing. In fact, the double "t" of *mattak* has a similarly ambiguous sound to the double "t" in Inuttitut. It is also striking that the initial "a" (pronounced and stressed like the *a* in "matter," followed by a hold on the double "t") of *mattak* becomes the "u" sound of the English "cluck." Europeans have difficulty with the opening syllables of Inuktitut words. Thus the word for sledge, *qamutik,* is widely used by English speakers but is rendered as "komatik."

Mattak is remarkable for its nutritional qualities. Whale skin is extremely rich in both protein and vitamin C—an example of how the mixed-diet theories of Euro-American dieticians can find a challenge in the extent to which the diets of almost pure meat enjoyed by some northern hunter-gatherers can provide all the essential vitamins.

62 ***Igunaaq*** On returning to Pond Inlet, I took suddenly and rather violently sick. This lasted no more than a few hours, but rumour immediately got going to say that it was a result of my eating cached meat. In fact, I probably had a reaction to gorging steak and apple pie given to me by Collie Scullion, the ever-hospitable wife of the friendly settlement manager, some two hours after getting back, on a very empty stomach. I was told later that there is a condition called acidosis, whose symptoms are what I suffered, that afflicts those who break long fasts with sudden overeating. I do not think that *igunaaq* has any equivalent consequences even for those who are not used to it, any more than would cheese in analogous circumstances.

TWO: **CREATION**

69 **Ben Shahn** Some of the images of Ben Shahn that delight by their juxtaposition of Hebrew and English letters and words are reproduced in his *Love and Joy about Letters.*

70 ***Aleph-beis*** There are many ideas in Jewish theology that converge on the meaning and status of the *aleph-beis.* As a whole, this convergence creates a many-layered mystery around Hebrew language. For an exploration of aspects of the mystery, and for the quotes in the text here, see Rabbi Michael L. Munk's *Wisdom in the Hebrew Alphabet,* p. 21. Munk also discusses the way in which letters can be understood as eternal (p. 18ff.).

Rabbi Munk quotes the words of Rabbi Dov Ber: "It is known in Cabbalistic literature that the letters of the Aleph-Beis [the Hebrew alphabet] were created first of all. Thereafter, by use of the letters, the Holy One, Blessed is He, created all the worlds. This is the hidden meaning of the first phrase in the Torah, '*In the beginning God created* **אֶת**—that is, God's first act was to create the letters from **א** to **ת**" (p. 19). Munk's discussion of language, letters and Genesis is on p. 16ff.

71 **"The world is only that which we say and think and know"** In the 1960s, when I reached Oxford, the hegemony of linguistic philosophy was at its most complete: the theory of knowledge had been reduced to, or become concentrated in, questions about logic and language. Furious arguments broke out among us bewildered students: how could anyone propose, with a straight face, that nothing exists other than words about sense impressions? What were we to make of the suggestion that things existed for sure only

because God was always looking at them? How could we acquiesce in a course of study that centred on eighteenth-century theories of reality put forward by Hume, Locke or Bishop Berkeley? On top of these came the thoughts of Bertrand Russell and A. J. Ayer. It seemed that nothing that could not be tested in some kind of laboratory had real meaning. Questions about ethics and the supernatural were dubbed "meaningless." Wittgenstein came to the rescue: here at last was a thinker who seemed to reach for language itself. His work was strange and often obscure, but a central insight prevailed: meaning comes from how words are used, and the way words are used comes from both a set of rules (the things that anyone can learn by observation) and an environment (the thing that a group of people who share language live in, experience and know). This insight made sense of both the obvious and the mysterious nature of language; and it pushed well to one side those arcane arguments about sense impressions: if you look at the meanings of the terms of those debates, the debate disappears.

But Wittgenstein and his ideas about language still did not address most of the mysteries of the larger world. It was still possible to study philosophy and never look at the nature of the human condition or at the meanings of myths. Creation was left to the speculative edge of the natural sciences or to art. So this last and great *cheder* did not interfere with the status of Genesis. I participated in the arguments of philosophy, but with some unease, a hint of bad faith. I wanted to be modern, radical, a materialist. I had difficulties of many kinds with the writings of those English individualists. Yet I found the underlying ideas somehow familiar and even reassuring: after all, I had felt, at my own intellectual and imaginative origins, that there is no reality but the mind, and that words lie at the heart of things and of ourselves.

71 **Myth** I use that strange word "myth" knowing that in European culture it denotes a story of special importance that is not true, yet contains great truths about the human condition. The word presents special difficulties when it is used to refer to the "myths" of indigenous peoples. Ethnography labels indigenous peoples' accounts of creation and the past as "myths" and "stories," whereas these exist in indigenous cultures as accounts of real events and processes. A book that deals extensively with the question of what myth is for different peoples is Robert Bringhurst's *A Story As Sharp As a Knife*. In it he speaks of both myth and stories and reveals, I believe, many of the reasons for the power of myth. He writes: "A myth is a story, and it is a story that insistently recurs: a piece of timelessness caught like an eddy in narrative time. Once the story is known, a single image or even a single word can evoke it. But only a linked sequence of images, words or gestures can *tell* it" (p. 47). In another part of the book, Bringhurst speaks of the way in which mythtelling "lifts a statement out of the realm of history or experience and drops it into one of two realms (the timeless and true, the persistent and false) we now call myth" (p. 113). And in yet another passage he notes that "The mythteller's calling differs little from the scientist's. It is to elucidate the structure and the working of the world. Myths are stories that investigate the nature of the world (whereas novels, for example, more often look at questions of proprietary interest to human beings alone). A genuine mythology is ... a kind of science in narrative form" (p. 288). Bringhurst's insights are drawn from reflections on the telling of myths as well as their actual qualities (their "meanings").

71 **Translation of Genesis** I have relied a great deal on Robert Alter's *Genesis, Translation*

and Commentary, a work of comprehensive scholarship with many indications of the problems and choices that face translation of this kind.

71 **Opening lines of Genesis** In fact, the Hebrew of the first sentence of Genesis is particularly difficult to translate. The key words, usually rendered as "without form and void," are *tohu vavohu.* Robert Alter notes that this Hebrew expression "occurs only here and in two later Biblical texts that are clearly alluding to this one. The second word of the pair looks like a nonce term coined to rhyme with the first and to reinforce it." He adds: "*Tohu* by itself means emptiness or futility, and in some contexts is associated with the trackless vacancy of the desert" (p. 3). This explains the meaning I give here as "nothing and futility."

75 ***Ta'awah*** In his comments on the translation of the Hebrew *ta'awah,* Robert Alter points to semantic connections between desire as keen longing for something and desire as lust. He also notes: "Eyes have just been mentioned in the serpent's promise that they will be wondrously opened; now they are linked to intense desire" (p. 12).

76 **The avenging of Abel's murder** The story of Cain and Abel ends, in verses 23 and 24, with a poem that announces murder and revenge: "For I have slain a man to my wounding, and a young man to my hurt. / If Cain shall be avenged sevenfold, truly Lamech seventy and sevenfold." Genesis then tells us that Adam and Eve have another son, in place of the murdered Abel, who is Seth, who has a son called Enosh, and that only now human beings call God by his sacred name of Yahweh, translated in the King James Bible as "the Lord." With the death of Abel, the triumph of Cain, and the means for invoking Yahweh the Creator, we can see the two sides of Creation: a human lineage, cursed but successful, and the ultimate mysteries that can be invoked through the power of the word.

Robert Alter points out that the "warrior's triumphal song" in the Hebrew follows the very precise forms and stresses of biblical poetry. He also observes that it was "cast as a boast to his wives." On the basis of this triumphal form, Alter speculates that "what Lamech [Noah's father] is saying (quite barbarically) is that not only has he killed a man for wounding him, he has not hesitated to kill a mere boy for hurting him" (p. 20). This speculation adds to the evidence of Genesis's acceptance, if not endorsement, of murder. My point can be stated at its most bald: whatever agriculture needs is implicitly, if not explicitly, condoned by the curses or events of Genesis.

78 **Canaan** Robert Alter makes the point that in chapter 9 of Genesis "the horizon of the story is the national history of Israel"; the curse of Canaan "is made to justify the—merely hoped for—subject status of the Canaanites in relation to the descendants of Shem, the Israelites." Alter then goes on to comment on the obscure nature of just what it is that Ham does to Noah. He connects the justification of Canaan's subjugation with the curse earned by some kind of sexual act, observing that "the Hebrews associated the Canaanites with lasciviousness (see, for example, the rape of Dinah, Genesis 34). Lot's daughters, of course, take advantage of his drunkenness to have sex with him." Alter also notes, however, that there could well have been some taboo against seeing a father naked that earned the curse (p. 40).

81 **The poetry of Genesis** I have summarized the first eleven chapters of Genesis with a view to answering questions about whom Genesis creates, and what human condition it establishes or prophesies. It is not to gainsay the beauty of the story. I have looked for the curses and for the qualities of the agricultural life they affirm. Nonetheless, the events and landscapes of the first part of Genesis are often dreamlike, and they are among the most

powerful and enduring images of the Judaeo-Christian intellectual and artistic heritage. Genesis 1 to 11 sets out the human condition in which Jews and Christians—and perhaps all other similar kinds of people—live, work and die; the world that the God of the Bible made; the human destiny that this God appears to insist upon. It not *just* a poem.

In fact, many creation stories are an intriguing mixture of poetry and specific information. A very striking example is a Maori creation poem that was published in an 1870 book about New Zealand:

1
From the conception the increase.
From the increase the swelling.
From the swelling the thought.
From the thought the remembrance.
From the remembrance the desire.

2
The word became fruitful:
It dwelt with the feeble glimmering:
It brought forth the night:
The great night, the long night,
The lowest night, the highest night,
The thick night to be felt,
The night to be touched, the night unseen.
The night following on,
The night ending in death.

3
From the nothing the begetting:
From the nothing the increase:
From the nothing the abundance:
The power of the increasing, the living breath
It dwelt with the empty space,
It produced the firmament which is above us.

4
The atmosphere which floats above the earth.
The great firmament above us, the spread-out space dwelt with the early dawn.
Then the moon sprang forth.
The atmosphere above dwelt with the glowing sky.
Then the sun sprang forth.
They were thrown up above as the chief eyes of heaven.
Then the sky became light.
The early dawn, the early day.
The midday. The blaze of day from the sky.

I am grateful to Stephen Grosz for showing me this poem. Maori society is an agricultural system that predates by some centuries the arrival of European settlers in what was to become New Zealand; but the tone of this creation myth almost certainly owes much to images and ideas that missionaries would have brought to Maori society by the time the poem was recorded by William Macintosh in *Te Ika Maui: New Zealand and Its Inhabitants* (pp. 109–10).

82 **The family farm** The discussion here of the family farm first appeared, in a slightly different version, as "Nomads and Settlers," an essay in *Town and Country,* edited by Anthony Barnett and Roger Scruton. I thank the editors, and the members of the Town and Country Forum, for the help they gave me in formulating these ideas.

82 **Literature that idealises farming** The examples from English are innumerable. It is striking, however, that many collections of such stories include examples from other cultures. *A Child's Book of Stories: Best-Known and Best-Loved Tales from around the World* gives examples from Arabic, French and German. *The Orchard Book of Magical Tales,* retold by Margaret May, has examples from Persia, Japan, the Caribbean, India and China. In some collections from countries where the encounter between farmers and hunter-gatherers is relatively recent, there are stories that speak to qualities of life and society not centred on farming (though these tend to idealise a particular kind of family life), and children are invited to imagine other kinds of childhood. In Canada, there are many children's books that make small expeditions into the Arctic and the adventures of Inuit and Indians. And in South Africa, collections of stories often include glimpses of hunting, albeit often by pastoralists. One example is *Stories South of the Sun,* compiled by Christel and Hans Bodenstein and Linda Rode, in which a story entitled "The Great Hunter" is, in fact, about the need to deal with a leopard that has been preying on sheep (p. 63).

86 **Ireland and "the flood of human movement"** K. H. Connell's *Population of Ireland, 1750–1845* gives the history of the growth of population up to the eve of the famine, and it is full of insight relevant to the argument here. In my book *Inishkillane,* I describe the changes in population in the aftermath of the famine and on into the twentieth century, looking at some of the ways in which attitudes to family farms can slow down or speed up the process of migration.

In much of Europe, selected sons and daughters were offered the chance of going into a religious community or taking holy orders. This was another device for dealing with agricultural oversupply of children; it is also an example of farm families seeking to increase their financial and social strength by exporting children to positions or places from which they might be able to provide ongoing help to the home farm. It is notable that in campaigns in India and China to effect reduction in family size, resistance—even by the poorest of farm families—has been intense and sustained. The need for labour is one explanation for this; another is the importance to many households of creating at least the possibility of earnings by children who will later migrate to places from which they will send money back home, where it can be used to keep the farm family going. Surplus population thus can be seen as a strategy for maintaining agricultural life despite "overpopulation" and poverty.

87 **"This success is built on opposites"** The point can be set out more fully by noting that attachment to home is no more and no less than the consequence of several generations of labour and a fear of the insecurity that comes with any loss of family lands. But if

the change is from a small farm to a large one, or from marginal land to prime land, then the farm family moves and feels that all is well, just as a city dweller is ready to change jobs, city or even country if a new opportunity is good enough.

Behind all versions of the farm family system, a tough-minded economic sense dictates who does and does not stay, and whether or not the farm itself should be sold. By similar measures of economic rationality, farmers are reluctant to trust to change—for they have something to lose. But if the deal is good enough, then change they will. The threat that the kulaks appeared to pose to Stalinist ideals of Soviet authority, for example, came from their particular independence and security; the absurdity of collectivisation in the interests of the peasants comes from its being a plan that can only reduce the power and productivity, and therefore the security, of the farm family. But even successful peasants will exchange poor land for good, a marginal farm for a productive one. Clinging to a family farm in defiance of the need or chance to move can be poignant. Reluctance to move can speak to every kind of attachment to home and heritage. But overall patterns of movement, like most individual decisions to move or not to move, have a rationale that is economic.

87 **Commitment to profit** There is an important issue of degree here. Agriculture and pastoralism both allow people to transport their economic system, either with the help of techniques for reshaping the environment, so that it will grow crops and sustain domestic animals, or by moving herds of animals. But as I have pointed out in my discussion of the farm family, the need and wish to be settled is powerful. The rise of capitalism has meant a parallel, and inseparable, rise in the force and norm of a profit motive. The match between capitalist formation and agropastoralism is a matter of degree, not entirely of kind. Capitalism depends on great flexibility and many kinds of nomadism; agropastoralism has ensured that an underlying readiness and ability to move is widespread. Agriculture also, of course, generates the population surplus in the countryside that becomes the labour resource of manufacturing industries.

88 **Loyalty to home** Any generalisation about the agriculturalist attachment to place rather than to opportunity or potential profit must take account of the differences between the regions of the world where patriarchy, dense populations and large political organisations prevail and those regions where patriarchy is less dominant, populations are more scattered and communities are politically distinct. In geopolitical terms, this speaks to a set of differences between what is sometimes referred to as "the patriarchal belt," comprising Europe, the Middle East, the Indian subcontinent, China and Japan, and the farmers or pastoralists in sub-Saharan Africa and native America.

88 **Celebrating and mourning with song and poetry** An example of musical contrasts is to be seen by comparing the famous words of two traditional songs: "Oh give me a home where the buffalo roam …" and "For all these great powers he's wishful, like me, / To be back where dark Mourne sweeps down to the sea." A more literary example of work that expresses both sides of the contradiction lies in some of Yeats's poetry, along with much Irish writing and folk music.

91 **The *Disappearing World* series** The film in Pond Inlet was made possible by Brian Moser's support and determination, along with Mike Grigsby's remarkable aesthetic and his skills as a documentary director. The cameraman, Ivan Strasberg, and the sound technician, Mike McDuffy, also played important roles in ensuring that the film was made in close col-

laboration with the Inuit of Pond Inlet. The crew was called on to work under difficult conditions and to contribute in all ways, from cooking to hauling sledges across the ice. Patty Winter, the researcher on the film, continued, thereafter, to be supportive. Her tragic death in 1983 was a terrible shock to all who had known her, including Anaviapik, who relied on Patty, especially when he was in England, for unfailing generosity and help.

THREE: **TIME**

105 **Dunne-za** The *u* in Dunne-za is pronounced midway between the "u" in fun and the "e" in den. The *e* is pronounced almost like the French "*é*" or word "*et*" ("and"). So Dunne-za almost rhymes with the beginning of a statement about the French painter: "Manet's a ..."

Many northern Athabaskan groups have ethnonyms with some form of the word for "a person" as their root. Hence Dene, pronounced to rhyme with "men eh," the "people" or "humans," of the communities of the Mackenzie Valley in the western Subarctic; the Wet'suwet'en, "the people of the interior river basin," referring to the Bulkley River system of northwest British Columbia; the Na-cho Nyak Dun, "the people of the big river" of the Yukon; and so on.

111 **The Thomas Hunter family** I spent a period of about eighteen months working in northeast British Columbia. For much of that time I was based at the Halfway River Reserve, where Thomas Hunter and his family lived. The work also took me on short visits to the other Dunne-za and Cree reserves of the region. Martin Weinstein did parallel work in the Fort Nelson area.

111 **Dunne-za territories** The mapping work was done as part of an extensive study set up in 1977 by the Union of British Columbia Indian Chiefs. Martin Weinstein, Dinah Schooner, Richard Overstall and Jim Harper all made immensely important contributions to the study, which looked at colonial and indigenous occupation of the area. Some results of the work are included in my book *Maps and Dreams.* For the hunter-gatherer land-use maps, see pp. 160–172.

The northeast corner of British Columbia and the area covered by the 1901 Treaty Number 8 appear to have been occupied by Athabaskan hunter-gatherers for several thousand years. However, in the course of the fur trade, beginning in the late eighteenth and continuing through the nineteenth centuries, Algonquian-speaking Cree families moved into the region. Treaty 8 set up three distinct communities, at East Moberley Lake, Blueberry and Doig Reserves, where Dunne-za and Cree families and territories have strong links.

It is difficult to estimate total areas of the Dunne-za and Cree territories of northeast British Columbia. Territories overlap, and some of the outer boundaries have been shown quite imprecisely. I give very rough estimates in the text. The map of the Halfway River Reserve hunting and gathering areas on p. 169 of *Maps and Dreams* indicates the total Dunne-za area of land use centred on the Halfway River Reserve and the Alaska Highway; it shows a distance north to south of about a hundred miles, with an east-west spread of about sixty miles. The territories of families living at the Doig, Blueberry, West Moberley and Fort Nelson Reserves reach beyond the Halfway hunters' areas (*Maps and Dreams,* pp. 154–72).

115 **Peoples classed as hunter-gatherers** The term "hunter-gatherer" applies, by definition, to those who do not cultivate the earth but rely instead on economies that have varying methods of using "wild" resources. As I discuss in this book, the definition of

hunter-gatherers by anthropology has been much influenced by peoples who live in relatively unproductive and harsh environments. It may well also have been influenced by the extent to which many of these peoples have lived, often for long periods, within, or with a degree of dependency on, other kinds of society. The combination of relative poverty and subordination to others may have shaped both how others see hunter-gatherers and how they see themselves. It is in relation to the complex links between hunter-gatherers and their neighbours that there is debate about whether or not the San (or Bushmen) of the Kalahari have ever lived in ways that have come to represent "typical" and pristine San hunter-gatherers. (This question is raised in the so-called Kalahari debate and by the issue of "revisionism"—a term used to refer to those who reject the objective, or even the possibility, of giving an account of "pristine" hunter-gatherers.)

Before the development of farming and pastoralism, however, ex hypothesi, all peoples must have relied entirely on hunting (including fishing) and gathering. Many, if not most, of these peoples, however, would have lived in the best possible environments. This fact means that anthropology must be very conscious, when speaking of hunter-gatherer characteristics, to bear in mind the kinds of differences that may well exist between peoples with abundant resources and those whose resources are widely distributed or relatively unpredictable. A model of differences among hunter-gatherer societies could well use a spectrum, with people of the North Pacific Coast, with their abundant runs of salmon and rich marine environment, at one end, and people like the San of the central Kalahari at the other. Between these theoretical extremes, one can imagine an immense range of economic and social systems—all of which may be called "hunter-gatherer."

118 **Sharing and collective territory** Hunter-gatherers are not the only peoples to institutionalise and rely on sharing. Gifts, and many levels and forms of distribution, are to be found in many if not all societies, hence the continuing relevance and importance of Marcel Mauss's essay *The Gift*. What is striking about hunter-gatherer systems, as opposed to agricultural systems, is the automatic or routine quality to the sharing.

An important variation in the basis for material equality, however, is whether or not hunting, fishing and gathering areas constitute formal "territories." Many hunter-gatherer groups have varying degrees of precision when it comes to identifying the boundaries of hunting and gathering areas, and parallel ways of identifying who is in each group. Among northern hunters, such as Athabaskan and Inuit groups, territories are defined by repeated use, not by a formal process. This was modified by the introduction of trapline registration in the 1920s and thereafter, when many extended-family territories were set down on official maps and often assigned to a single person. Among the peoples of the Northwest Coast and their neighbours in the Interior, territories are known and passed on with quite high degrees of precision. Knowledge of boundaries is part of the responsibility of chieftainship. These are the salmon-fishing and marine-hunting societies who developed the potlatch, at which territory is agreed upon, inherited and repeatedly defined. It has been said that these are not so much hunter-gatherer systems as a mesolithic form of society, transitional, as it were, between hunting and farming. But there are also San (or Bushmen) groups in the Kalahari who precisely define the extent of family territories, known to them as *n!ore*.

It is significant that in systems where territories are clearly defined, and where concern with "trespass" is highly developed even within the community, access to productive and important areas is much more open than the detailed rules governing it would appear to suggest. There are many exceptions and more or less flexible forms of permission. This means that even in hunter-gatherer systems demonstrating apparently strong preoccupation with territory, a considerable degree of sharing takes place. And this sharing means that equality of material well-being is secured. The example of the Wet'suwet'en, the Athabaskan group with a territory system somewhat on the model of the neighbouring Gitxsan of the North Pacific Coast, is described in Antonia Mills's book *Eagle Down Is Our Law.* Richard Daly's evidence in the case of *Delgamuukw v. the Queen* gives many insights into this issue.

119 **Small families and infanticide** The ways in which hunter-gatherers limited family size are not obvious. Contraception and abstention played a part, as did the inhibition of conception by breast-feeding. There is some evidence that the latter method does work where the mother has a high-protein, low-carbohydrate diet (typical of many hunting peoples). Also, children who were born "too soon" were often adopted by another family.

Infanticide among hunter-gatherer peoples is difficult to assess. There is evidence that Inuit of the central Canadian and Greenlandic Arctic practised infanticide on a relatively frequent basis. (See Rolf Gilberg's essay "Polar Eskimo" for some details of this in northwest Greenland.) On the other hand, the Inuvialuit of the Mackenzie Delta appear to have practised infanticide very rarely. (See Derek Smith's essay "Mackenzie Delta Eskimo.") I have heard that infanticide was practised, where conditions were especially difficult, in the early part of the twentieth century and as recently as the 1940s. (Asen Balikci has said that the Netsilik Inuit practised female infanticide "to a very high degree." See p. 424 of his essay "Netsilik"; see also David Riches's article "The Netsilik Eskimo.") And there are reports of this being done in other parts of the North. However, many of the first detailed accounts of Inuit life came from missionaries, who were concerned to show the need for Christianity. Infanticide shocked these men, and they were quick to express their conviction that this was one of several reasons for the Inuit to give up their old ways and become Christians.

Infanticide, like cannibalism and human sacrifice, is an act that Europeans are inclined to see as evidence of primitivism and savagery, and to use as justification for European colonialism. An excellent discussion of how human sacrifice can be condemned by those who practise equivalent, and far more extensive, versions of the same thing is to be found in Felix Padel's *Sacrifice of Human Being.*

Killing of children as part of a complex culture of sacrifice in Aztec society is described by Inga Clendinnen in her book *Aztecs.* Clendinnen considers the evidence and concludes that the degree of attachment that the parents had to these victims, who were kept by priests for several weeks before being killed, was enough to give rise to intense grieving (pp. 98–99). It is important to note, however, that Aztec society was very distant from hunter-gatherer modes of life, and that the ritual sacrifice of both children and adults was—as Clendinnen argues—closely linked to issues of hierarchy and to complex political institutions, both internal and imperial.

120 **Perceived differences between hunter-gatherers and others** In the legal evidence pertaining to the case of Dunne-za and Cree families who lost an important part of their territory in the early years of the twentieth century to veterans of the First World War, there are references to the pitiable destitution of the Indians—a destitution that was linked, by contemporary observers, to a lack of good clothing, horses and tents. A more

recent, and therefore more surprising, example of this same form of misconception is to be found in the judgement of British Columbia's Chief Justice Allan McEachern in the case of *Delgamuukw v. the Queen,* which I discuss in part 4 of this book. In fact, the litigants in the Delgamuukw case were hunter-gatherers of the North Pacific Coast—people who accumulated surplus produce that they distributed in complex ritual and trading arrangements, built large houses of cedar planks and carved elaborate house posts and totem poles.

Examples of prejudice and preconceptions of this kind are to be found in all places where colonists have settled in hunter-gatherer territories, be it in the Amazon Basin, Australia or Japan, where the Ainu, indigenous hunter-gatherers now confined to Hokkaido, were expunged from the official Japanese record. By comparison, the Jomon, apparently another hunter-gatherer culture, began to receive a degree of recognition as part of Japanese heritage and history, but only when the Jomon were discovered to have had high degrees of "civilisation," in the form of elaborate housing, storage pits and massive stone monuments, some of which appear to go back at least 15,000 years. In reality, the Jomon and the Ainu may well be the same peoples—the former constituting a hunter-gatherer system prior to the arrival of agricultural settlers and colonists, the latter being the group that managed to continue, in difficult conditions and circumstances, at the geographical and political margins of Japan. I am grateful to Simon Kaner, of the Archaeology Office of Cambridgeshire County Council, for his account of hunter-gatherer data in Japan. An account of the history, art and artifacts of Japanese hunter-gatherers can be found in *Ainu,* edited by William Fitzhugh and Chisato O. Dubreuil.

121 **Discriminatory attitudes of farming societies to hunter-gatherers** The case of the Vaniallaatto people of Sri Lanka has been brought to the world's attention by Jessica Wright and Andrew Pagon. The attitudes of South American indigenous agriculturalists to hunter-gatherers may be glimpsed in accounts of Maya, Inca and Aztec relationships with their subject neighbours. (One wonders, for example, how many hunter-gatherers became sacrificial victims over the course of Aztec hegemony.) Throughout Amazonia, hunter-gatherers have experienced many forms of discrimination and prejudice from neighbouring agriculturalists. In "Non-Pristine Hunter-Gatherers or Theoretical Strait-Jackets," Manuel Arroyo-Kalin shows how relations between a hunter-gatherer society and neighbouring farming peoples, even while dependent on mutually beneficial reciprocity, are remarkable for the self-proclaimed superiority of the farming peoples. The Forest Peoples Project has documented the scale of discrimination and death among the Twa or Pygmies of Rwanda: of all the peoples who suffered in the Rwanda genocide of the 1990s, the Twa are now known to have endured the highest rate of murder (some 25 per cent of total population) of any ethnic group. An excellent overview of negative attitudes towards hunter-gatherers in southern Africa is to be found in James Woodburn's article "Indigenous Discrimination."

122 **The anthropology of hunter-gatherers** It was not until the 1960s that anthropology departments around the world identified hunter-gatherers as a topic of particular or separate importance. The terms "primitive societies" or "native peoples"—to give examples of two of the categories once much in use—were used to designate a wide range of societies, including small-scale agriculturalists, herders and hunter-gatherers. The impact of the 1968 Man the Hunter conference must be seen against this background. Anthropologists Richard Lee and Irven De Vore played central roles in the conference and in the

resulting set of papers, published under the same name. In 1972 Marshall Sahlins published *Stone Age Economics*, in which appeared his seminal essay "The Original Affluent Society." This work, and the ground-breaking work of James Woodburn, offered a fundamental challenge to the depiction of hunter-gatherers as people without achievements. The most telling example centres on attitudes towards material goods and monuments: hunter-gatherers eschew these in order to maintain their highly effective systems of harvesting, yet those who consider material goods and monuments to be the indicators of progress and civilisation have taken their absence among hunter-gatherers as evidence of the primitive.

The results of James Woodburn's work among the Hadza of Tanzania were first published (and seen on film) in the 1960s; since then, he has offered insight into hunter-gatherer forms of economics, politics and equality and, in the 1990s, shed light on questions about both the nature of hunter-gatherer "encapsulation" and the low status hunter-gatherers are accorded by their agropastoralist neighbours. Woodburn is one of the few anthropologists who have sustained a long career encompassing both scholarly work and direct engagement with the people he has studied and lived among.

From the pioneering work of Lee, Woodburn and Sahlins has grown the extensive hunter-gatherer literature, including the two-volume collection of essays *Hunters and Gatherers*, edited by Tim Ingold, David Riches and James Woodburn, and most recently *The Cambridge Encyclopedia of Hunters and Gatherers*, edited by Richard Lee and Richard Daly. These are compendia of both ethnographic specifics and comparative overviews.

In the area of Arctic anthropology, Stefansson, Jenness and Rasmussen stand out as early examples of ethnographers who saw the coherence of a hunter-gatherer system (though they were writing before this had become an accepted category for anthropological study). Their insights and enthusiasm were marred by their somewhat colonial view of the future of the far north and the destiny of its indigenous populations. In recent times, the work of Jean Briggs, especially her book *Never in Anger*, shows a keen, unsentimental appreciation for the realities of a hunter-gatherer system.

Significantly, the thinking that went into and emerged from the Man the Hunter conference failed to pay adequate attention to gathering, and therefore to the importance of women's work in hunter-gatherer economies. Subsequent analyses have shown that women's gathering in many systems contributes a large proportion of essential nutrients. Their work also makes the harvest secure through the processing of meat, fish and nuts into forms that can be stored. In the 1980s and '90s, anthropologists have focussed a great deal of attention on a "discourse" about hunter-gatherers that is widely seen to have begun with the Man the Hunter conference, and continues to seek "traditional" or pristine examples of hunter-gatherers living in relative isolation. The distortion deemed to result from this search may add to other potential distortions, including a tendency to see hunting and gathering as timeless, unchanging and socially homogenous. Nonetheless, the central insights of the anthropologists who contributed to Man the Hunter continue to be of importance, especially in relation to attitudes of other societies towards hunter-gatherers.

25 **Anaviapik's sense of his moment in history** Some hunter-gatherers who have lived for long periods of time in close connection with neighbouring farmers or pastoralists are not as clear as are the Inuit about their own right to a place in the present. The British anthropologist Olivia Harris reports that the surviving Urn, fishing peoples of highland

Bolivia, have internalised the pastoralist myth of the conquering Aymana that the Urn represent a previous and doomed era (personal communication, April 2000).

128 **Inuit seasons** The Inuit seasons are *aujak*, summer; *ukiassak*, literally "material for small winter"; *ukiak*, small winter; *ukiuk*, winter; *upingaussak*, literally "material for spring"; and *upingaak*, spring. I have written about the meteorological and biological events that the Inuit associate with each of these seasons in an introduction to *Seasons of the Arctic*, a book of photographs by Paul Nicklen.

135 **Dreams and trails** The most powerful dreamers have also dreamed the trails to heaven. This process has been wonderfully described and analysed in the work of Robin Ridington, formerly of the University of British Columbia. See especially *Swan People*. I describe hearing about these powers when working in northeast British Columbia in *Maps and Dreams*, p. 266ff.

144 **Despair and anger** See Octave Mannoni's *Prospero and Caliban* and Frantz Fanon's *Wretched of the Earth* and *Black Skin, White Masks*. In the first chapter of *Black Skin, White Masks*, Fanon describes the importance that black families can attach to speaking French rather than Creole, and he gives "Hoquet" as an example of the kind of ridicule that mothers will direct at their children to make the point. It goes:

> My mother wanting a son to keep in mind
> if you do not know your history lesson
> you will not go to mass on Sunday in
> your Sunday clothes
> that child will be a disgrace to the family
> that child will be our curse
> shut up I told you you must speak French
> the French of France
> the Frenchman's French
> French French

"Hoquet" is from Leopold S.-Senghor's *Anthologie de la nouvelle poésie negre at malgache*, pp. 15–17.

145 **Faith in development as a religion** The intensity of *belief*—along with the trust in a particular kind of future, in defiance of facts about both the past and the present—is what gives faith in development its religious character. However, there are aspects of this faith that are striking for the way they often portray other societies (those *not* going to benefit from "development") as ideal but doomed. This is to deploy a cynical form of romanticism. There is also a narcissistic aspect to such faith. The position upheld or implied is that others can benefit only by becoming like "us," the "developed." (One wonders which part of "us" is thus idealised: the reality is that those who are most insistently urged to emulate and join the "developed" world are peoples whose position and resources in the dominant society will be the lowest and the poorest.)

147 **Violence** There is evidence in the Arctic of murders of Inuit by Athabaskans and vice versa. Similarly, there is no doubt that many hunter-gatherers have been as ready to kill enemies as is any other human group. The nature of egalitarian individualism may mean that people within a community are for the most part open and generous towards one another.

150 **Archaeology and language** The account of Sir William Jones's discoveries, and of the nature of Indo-European languages, is to be found in Colin Renfrew's *Archaeology and Language*, pp. 9–10. I draw extensively on this wonderful book here and in the material dealing with the spread of Indo-European languages and the links between this spread and the history of agriculture in Europe. For the direct quotes I cite, see pp. 149–50.

In fact, scholars have been pondering the antiquity, original site and subsequent movements of Indo-European languages for some time. Mario Pei's *Story of Language* tentatively concludes on the basis of shared vocabulary that Indo-European was a single, original language in about 2500 B.C., on the shores of the Baltic (pp. 22–23). Renfrew refers to much earlier consideration of the subject, however, including the work of Adolphe Pictet, published in 1877, and that of Gustav Kossina, published in 1902 (*Archaeology and Language*, pp. 13–15). For Renfrew's discussion of Tocharian, see pp. 65–67. For a summary family tree of Indo-European groups of languages, see p. 74.

155 **Farmers accepting other ways of life** Agricultural frontiers are relentless and aggressive in relation to hunter-gatherers insofar as the farmers wish to establish their crops or their herds on hunter-gatherer lands. Despite the strong tendency among agriculturalists and pastoralists to see and treat hunter-gatherers as inferiors, and to dispossess them as the need for land dictates, there are many examples of farming and herding peoples entering into relationships with hunter-gatherers that are of great material and spiritual importance to the farmers and herders. This applies to Twa in the Congo and Rwanda, some San or Basarwa in southern Africa, and Maku and others in South America. It is striking that the Xhosa, a herding people of South Africa, have incorporated the spiritual beliefs and language of the San, whom they presumably made use of before they displaced them. It is the abundance of these examples that has led anthropologists to question the separable identity of hunter-gatherers altogether. A good summary of this issue is to be found in Thomas Widlock's *Living on Mangetti*.

In the broad historical frame, the challenge is incoherent. It also fails to accommodate the extent to which hunter-gatherers *do* have a strong sense of their own distinct characteristics, which *are* linked to distinctive socioeconomic systems. Nor does it account for the extent to which the dominating farmers and herders agree that the hunter-gatherers they have links with are indeed different (and inferior) to themselves for reasons to do with a particular hunter-gatherer relationship to resources, time, knowledge and beliefs. The suggestion that *all* peoples seeming or claiming to be hunter-gatherers are in reality "incorporated" into farmers' and herders' systems—albeit as marginal, exploited and despised peoples—has force in relation to complex histories of interaction. In 1963, in his *Structural Anthropology*, Claude Lévi-Strauss asked: "Can we ever speak, then, of true hunters and gatherers in South America?" He even observed that it was not possible to know, in the case of the hunter-gatherers who appeared to be distinctive, whether or not they might be "horticulturalists who regressed" (pp. 109–10). Of course, Lévi-Strauss could have asked the people themselves, or paid due attention to their oral traditions.

The argument continues. Meanwhile, it is indeed important to recognise how much farmers and pastoralists have made use of hunter-gatherer resources and knowledge, from furs and herbs and medicines to spirit power and expertise in locating water.

57 **The spread of infectious diseases** In *Guns, Germs and Steel*, Jared Diamond documents

the ways in which agriculture and its reliance on domestic animals gave rise to many infectious diseases. See especially the chapter entitled "Lethal Gift of Livestock." The links between specific diseases and specific animals are set out on p. 207.

159 **Manifest destiny** The idea of manifest destiny is often said to be as old as America itself, having sailed into the New World with Columbus. It has been called "the philosophy that created the nation." In fact, the term was coined in the 1840s, although the argument it affirms was used to justify each stage of American conquest of Indian territories as well as claims against rival European colonists. It entered legal theory through the judgements of Chief Justice Marshall. Information about the origin of the concept, and a set of thoughts about the place of manifest destiny in American history, are given in a paper by Michael T. Lubragge that can be found at http://www.let.rug.nl/alfal/.

161 **"The best hunter"** The difficulty in imagining any equivalent concern with a person who might be "the best gatherer" is revealing. This points to the predominant focus on male hunters, despite the critical importance of women in gathering food. Unfortunately, the understatement of women's importance in hunter-gatherer life may be echoed, if not reinforced, by this book. Experience of a society is determined to some considerable extent by gender: I was a young man living in communities where it was much easier and more acceptable for me to spend time with other men. Older women often were wonderful teachers of language and tellers of stories, but when it came to being out on the land, I went with hunters—mostly men. Of course, in either a settlement or a camp, I lived with families, and I was given many opportunities to see how both women and men related to children, and indeed to one another.

163 **The sentencing of the accused in Jimmy Field's murder** The Fort St. John Crown counsel (formerly known as the Crown prosecutor) found this to be a difficult case. The witnesses were friends of the accused and not very cooperative. It was the confessions—though these may well have been incomplete—that enabled plea bargaining, which in turn helps to explain the sentences. When asked what he thought the motive was, the Crown counsel is reported to have said: "Robbery or, more likely, just viciousness" (Leslie Hall Pinder, personal communication, June 2000).

FOUR: **WORDS**

169 **Grease and potlatches** There is a difficulty here with my generalisations about hunter-gatherers as non-hierarchical. I have characterised the Nisga'a, Gitxsan and other Northwest Coast societies as hunter-gatherers. This is true in the sense that they do not harvest the resources of their lands by cultivating gardens or herding domestic animals. However, they are remarkable for the complexity of their social institutions, notably the potlatch. They are also famous as warriors who attacked neighbouring peoples, taking prisoners who would then become slaves in the captors' families. The resulting inequality between slaves and others was made more complicated by a further distinction between chiefly families and commoners. Here is a hunter-gatherer society that *is* hierarchical, and in which there is considerable competition between rival families for status—hence the use of the potlatch to demonstrate wealth and achieve symbolic victories over rivals. For these and related reasons, some anthropologists have characterised Northwest Coast cultures as intermediate between palaeolithic hunters and neolithic farmers.

Drawing attention to the hierarchical features of the societies of the Northwest Coast is useful. However, it should also be noted that the potlatch, being centred on a person's or clan's maintaining status by giving away as much as possible, contributes to general material equality. Also, the receipt of gifts at a potlatch places on the guests the burden of having to return at least as much in the near future. Failure to do so means a serious loss of status. Thus the societies using the potlatch secure, through a degree of competitiveness and related hierarchy, results that are typical of hunter-gatherers: distribution and redistribution of produce means that everyone lives at more or less the same material level.

Northwest Coast cultures are far more concerned with measuring and recording exchange than is the case with other hunter-gatherer peoples. Thus they have relatively sophisticated counting systems, with numbers going up to 1,000 rather than the more typical hunter-gatherer ceiling of 10, 5 or even less. The giving at potlatches is remembered in some detail. Also, it is probably no coincidence that Northwest Coast cultures are far more committed to oratory, with emphasis on the effect of talking on others, than are hunter-gatherers for whom oral culture is expressed in a far more domestic, intimate mode of storytelling.

173 **Film crew** The crew on a documentary film are of immense importance, not only for their technical skills in recording high-quality sounds and images, but also for their readiness to collaborate in every stage of a very demanding process. For *Time Immemorial,* Kirk Tougas was the cameraman, Caroline Goldie and Michael McGee were the sound recordists, Rob Simpson was the camera assistant, Betsy Carson was the production manager and Haida Paul was the editor.

182 **John Milloy** The Canadian historian John Milloy began his work on the history of Canada's residential schools as a project within the Royal Commission on Aboriginal Peoples of the 1990s. He undertook the task of trawling the federal-government archives for all documentation of the schools. The results of this work are as thorough as they are chilling. His findings were attached to the commission's final report and subsequently published as *A National Crime.*

In addition to giving many details about how residential-school policy was administered and documenting the dismay that was expressed (and for the most part ignored) about the policy's consequences at the time, Milloy also explores the underlying rationale for such a policy. I draw here on both his findings and his reflections on what the findings imply.

184 **Alice Miller** The book by Alice Miller that sets out her theory of "poisonous pedagogy" is *For Your Own Good.* Although this is a work of psychology centred on the ways individual parents do harm to individual children, Miller is remarkable for her ability to explore the links between psychology and social systems. She argues that the most damaging kind of child raising is to be found in a particular tradition of family life and attitudes to children. She identifies Germany as a society where this social and psychological pattern reached extremes, and in this regard includes a chapter in *For Your Own Good* that speculates about the character of Hitler and the characters of those on whom he relied to achieve his purposes, linking the Nazi project and process with the "poisonous pedagogy" that had damaged the personalities of both its leaders and its followers (pp. 142–95). Miller summarises her views on what has happened "not only in fascism but in other ideologies as well" with the observation: "The scorn and abuse directed at the helpless child as well as the suppression of

vitality, creativity, and feeling in the child and in oneself permeate so many areas of our life that we hardly notice it any more" (p. 58). It is easy to see the relevance of her views to any thinking about other societies where ways of raising children, and conducting oneself as an adult, do not include more or less routine scorn and abuse.

186 **Colonialism and language change** The British administration in India did not, of course, establish itself over a web of societies that was in any sense "primitive." Regional languages were already part of both complex administration and strong traditions of literacy. The British (like the Romans in much of their empire) had the task of dominating principalities and fiefdoms. Acknowledgement and adoption of *their* languages therefore had a strategic character—*realpolitik* resulted in accepting a polyglot circumstance. However, the dealings here were between colonists who were agriculturalists and princes, chiefs and peasants who were also agriculturalists. Thus the pattern of language change in North Africa or in the Dutch colonies in Indonesia, for example, where a new bureaucratic class who indeed spoke French or Dutch (just as they spoke English in India) did not have the task of eradicating the tribal languages of the territories, even where there were fierce conflicts with the tribes.

191 **Hunter-gatherer literacy** The links between literacy and tracking have been developed by Ted Chamberlin. Essays of his on the subject are soon to be published in *For the Geography of a Soul: Festschrift in Honour of Kamau Brathwaite,* edited by Timothy Reiss and Rhonda Cobham (Trenton, N.J.: Africa World Press) and *Literacy, Narrative and Culture,* edited by Jens Brockmeier, David Olsen and Min Wang (London: Curzon Press). In *A Story As Sharp As a Knife,* Robert Bringhurst also makes the link between oral culture and a literacy that is not to do with writing. He observes: "Reading, like speech, is an ancient, preliterate craft. We read the tracks and scat of animals ... We read the horns of sheep, the teeth of horses ... We read the speech of jays, ravens, hawks ... and, in infinite detail, the voices, faces, gestures, coughs and postures of other human beings. This is a serious kind of reading, and it antedates all but the earliest, most involuntary form of writing, which is the leaving of prints and traces, the making of tracks" (p. 14). Writing about the Haida, great carvers and artists of the Northwest Coast, Bringhurst could also add to this the ways in which the crests and figures of totem poles are "read."

194 **Words and truth** In the 1960s, R. D. Laing and his associates pointed to possible connections between mental illness (or aspects of behaviour that were classed as illnesses) and the contradictions between what children were told and what they knew to be the case. Laing and Esterson's *Sanity, Madness and the Family,* for example, describes a case in which parents contradict their child's memory, in the process revealing that they systematically deny the child's account of reality, implying or saying that they know the facts but the child does not. Laing's *Knots* is an exploration of the kind of mismatching and disarray that take place when words are used not to tell the truth but to manipulate. However, the Laingian pressing of these insights into the diagnosis of schizophrenia has not stood the test of scientific time. The presence or absence of schizophrenia in hunter-gatherer societies is not to be linked to the issue of words and truth. The proposition here is that disorders pertaining to control and parental manipulation are strikingly few in the hunter-gatherer societies I have known.

Interestingly, Laing looked to the work of an anthropologist, Gregory Bateson, for some of his ideas about "double bind." I am grateful to Kirk Tougas for pointing out the importance of Bateson to Laing. In fact, Bateson developed double-bind ideas when studying the peoples of New Guinea in the 1930s. In the essay "Towards a Theory of Schizophrenia," Bateson, in collaboration with others, "brought the paradigm of an insoluble, 'can't win' situation, specifically destructive of self-identity, to bear on the internal family pattern of communication of diagnosed schizophrenics" (Kirk Tougas, personal communication, May 2000).

It is also important to note that hunter-gatherer truthfulness and forthrightness are closely connected with the ways in which Inuit (and other hunter-gatherer) parents do not distinguish between a child's needs and its wants. On a very wide range of matters, the child is trusted to know what is right for it—its word is accepted rather than opposed. This may go a long way to creating and securing confident and healthy personalities. However, there are also very secure attachments between mothers and infants. Typically, a mother keeps a child close to her at all times, often in actual physical contact, until the next child is born or the baby is weaned. Since hunter-gatherer parents space children further apart than is the case in agricultural societies, mother-baby contact is sustained for a norm of about three years. This too may have importance consequences for mental health.

All that said, it is significant that in the breakdown of hunter-gatherer society as a result of displacement and colonial subordination, high incidences of social and individual pathology do occur. Suicide and violent death rates become very high, in some cases reaching five or more times the regional or national average. I documented one such example among Ojibway of two reserves in western Canada in "An Analysis of Violent Death," part of a study coordinated by Peter Usher of the impacts of mercury poisoning on the water and fisheries of the people of Grassy Narrows and Whitedog. The evidence revealed that even in the case of mercury-related illness, notorious for causing acute mental disturbance, extremely high rates of social and individual disarray were linked to colonial rather than environmental factors.)

High rates of individual pathology have always been associated with alcoholism and spree drinking. Whatever the psychological strengths secured by the hunter-gatherer way of dealing in truth and caring for children, the dissolution of their society and economy, and especially the separation from their land, resources and languages, create intense psychological difficulties.

96 **Nisga'a treaty** The Nisga'a settlement was ratified by the Canadian Parliament in April 2000. Its terms include the designation of 2,000 square kilometres of land as Nisga'a territory, in which the Nisga'a have ownership and control of all natural resources, and a share of the Nass River salmon stocks. It also establishes a new form of local government, with the Nisga'a Tribal Council taking control of many aspects of social services and related administration. In return, the Nisga'a have agreed to give up claims to aboriginal title and to forgo any further claims against the Canadian government. These two concessions have caused the deal to be greeted with hostility by some other aboriginal groups pursuing claims. Many non-aboriginal Canadians have expressed resentment towards the deal, arguing that it is an impediment to the development of both aboriginal and colonial societies, and a far less desirable option than more or less complete assimilation of aboriginal peoples into national and regional economic and social life.

An important milestone on the route towards the Nisga'a settlement was the court case argued by the Nisga'a in the Supreme Court of British Columbia in 1967. The lawyer representing this action, which led in due course to the crucial Supreme Court of Canada decision in 1971, was Thomas Berger, who subsequently led the milestone Mackenzie Valley Pipeline Inquiry. (The findings of this inquiry, published as *Northern Frontier, Northern Homeland,* are an important contribution in their own right to the understanding of hunter-gatherer society in northern North America.) In his book *Fragile Freedoms,* Berger himself wrote an account of the Nisga'a case, documenting the series of protests and actions that long preceded it (pp. 219–55).

197 **Details and categories** There are categories in Inuktitut, for example: *puijit,* the animals that have to surface to breathe; *umajuit,* living animals, often used to mean land mammals; *sijjarsiutit,* the creatures of the shoreline, used to mean wading birds; and a word for things that fly, *tingmijat,* used to mean birds in general. Under the influence of English speakers, however, many Inuit now use *qupenuaq,* the word for snow bunting, to mean bird; *iqaluq,* the word for arctic char, to mean fish; and *natiq,* the word for an adult ringed seal, to mean seal.

The first paragraphs of Claude Lévi-Strauss's *Savage Mind* draw attention to the way in which the lack of categories such as "tree" or "animal" in a society or language is "cited as evidence of the supposed ineptitude of 'primitive people' for abstract thought" (p. 1). He goes on to explore the concepts and categories at work in the "savage mind," making a set of arguments about the relationship between what he terms the "*bricolage*" of knowledge without categories and the ability that categories secure for mythic and magical forms of expression, as well as for the discovery (and expression) of connections in nature that allow people to make crucial predictions about the movements of wildlife.

Lévi-Strauss's work was immensely influential—a point I discuss briefly in part 6 of this book. But his focus was not the hunter-gatherer mind or society. His starting point is the nature of knowledge created with and for "magic"—by which term he means many, but not all, of the things that shamans in hunter-gatherer societies know and do. But Lévi-Strauss is very clear about the importance of the neolithic transformation, and he could well be speaking of hunter-gatherers rather than magic when he says: "Magical thought is not to be regarded as a beginning, a rudiment, a sketch, a part of a whole which has not yet materialised. It forms a well-articulated system, and is in this respect independent of that other system which constitutes science, except for the purely formal analogy which brings them together and makes the former a sort of metaphorical expression of the latter. It is therefore better, instead of contrasting magic and science, to compare them as two parallel modes of acquiring knowledge" (*The Savage Mind,* p. 13.)

199 **Hunter-gatherers and counting** Northwest Coast societies do use higher numbers. The Canadian anthropologist Marianne Boelscher Ignace has pointed out that the Secwepemc of the Kamloops area in Interior British Columbia, for example, have a counting system that goes to 1,000 (personal communication, August 1998). Her essay "The Secwepemc (Shuswap)" is an important contribution to any generalisations about hunter-gatherers.

203 **Speakers of Athabaskan** Missionaries are an important exception to the general rule that Europeans have failed to learn hunter-gatherer languages. Throughout the world, various Christian denominations have established missions in hunter-gatherer territories. The

Moravians, originating in Germany, were the first to spend time in Greenland and Labrador Inuit communities. Oblates, from France and Belgium, went to convert Athabaskans and Inuit in many parts of northern North America. There are Catholic and Protestant missionary "stations" and communities in Amazonia, Australia and southern Africa. There have also been Christian missionaries in remote parts of India and Sri Lanka. Many missionaries became fluent speakers of the local languages, and often established orthographies with which to create written translations of the Bible. In virtually every case where missionaries achieved language skills among hunter-gatherers, however, and even where they joined the peoples' struggles to hold onto or recover their lands, the mission endeavour has been an element in a wider process whereby hunter-gatherers are "sedentarised" and "civilised"—meaning that they adopt the economies and languages of the colonists. Missionaries often appear as important figures in the long history of campaigns for land rights; they do not often appear, especially in the early and critical periods of colonial incursions into hunter-gatherer lands, as advocates on behalf of indigenous languages.

204 **Robin Ridington** Robin Ridington's work began in the 1950s, when he first travelled in Dunne-za country adjacent to the Alaska Highway, and it continues to the present. Throughout this period he has sustained close relationships with Dunne-za families and has been active in working with their communities on many cultural and political projects. He has also brought to teaching and broadcasting remarkable attempts to convey Dunne-za and hunter-gatherer cosmology, relying on their words and music to achieve what more linear and "rational" kinds of discourse tend to obscure or ignore. His writings include a compelling essay, "The Medicine Fight," *Swan People* (a publication of Canada's National Museums) and his book *Trail to Heaven*.

213 **The judge's "embarrassment"** Leslie Hall Pinder, the Canadian lawyer and novelist, has written about Mary Johnson's evidence in the Delgamuukw case. Reflecting on the judge's language and the patterns of non-comprehension between Western ideals of objectivity and Gitxsan oral culture, she speaks of the judge's embarrassment. (See her pamphlet *The Carriers of No*. In a revised and updated version of this pamphlet published in *Index on Censorship*, she writes: "In a trial the word 'embarrassed' describes the feelings you have when something is being presented which is unacceptable ... It is the visceral effects of the disallowed, of that which does not fit our idea of information, knowledge, fact. It is the body's discomfort at being at the edge of the path" [pp. 67–69].)

Pinder's book *On Double Tracks* is a fictional account of an encounter between a hunter-gatherer society and the law. Inspired by the Apsassin case of the 1980s, in which a Dunne-za band argued that part of their territory had been taken from them illegally, the novel explores the implications of the divide between world views for all involved in such cases.

A detailed study of the Delgamuukw case, with emphasis on the judgement, can be found in Dara Culhane's *Pleasure of the Crown*, especially parts 4 to 6. Culhane's book situates the case in the history of legal decisions pertaining to aboriginal rights.

216 **Talking on film** When Utuva, Qanguk and Anaviapik talked on film about the way Inuit feared whites, they also expressed deep unhappiness about the way young Inuit seemed to prefer to speak English even among themselves. At the same time, these elders often talked of the importance of speaking English, for that was where the power lay, and hence was the means by which Inuit could best negotiate their future—as individuals, because with

an ability to speak English came the best-paying jobs, and as a society, because English-speaking leaders would be able to speak up for Inuit rights and needs. The contradiction is one that anthropologists and land claims negotiators have encountered over and over.

218 **Qanguk's story, "*Helloraluk*" and "*Polisialuk*"** The trader the Inuit killed was called Robert Janes. At the end of their trial, one man was acquitted, a second was found guilty of manslaughter and sentenced to ten years in a southern prison, and a third was convicted of aiding and abetting and sentenced to two years' "close confinement" in Pond Inlet. I give details about these and other such cases in *The People's Land.*

The "i" sound on the end of the word "police" (Inuktitut orthography has no *c*) is required in order for the word to be built on: the rules for linking sounds require a specific range of final consonants or vowels. The affix *-uluk* signifies largeness or impressiveness; the meaning is the opposite of the Scottish "wee," which has an exact Inuktitut equivalent in the form of *-kuluk* or *-apik.*

Qanguk's way of saying "I thought I was going to be beheaded" was a wonderful example of what agglutinative languages can do. The word Qanguk used was "*niaquirqtauniarnasugilaurpunga.*" It breaks down as: *niaqu(q)* ("head") + *irq* ("not/without") + *tau* (creates passive mode) + *niar* ("will soon, within the day") + *nassu(k)* ("thought") + *gi* (creates directness, as does "to" in "give something *to* someone") + *laur* (indicates past tense, more than a day ago) + *punga* ("I"). Thus, in the English that shows the shape: I / a while ago / did / think / within the day / would be / without / head.

219 **Loss of language** Inuktitut continues to be spoken in the central and eastern Arctic and throughout Greenland. This is a region that has escaped the hunter-gatherer holocaust.

In other parts of the North, from the land of the Yuit in eastern Siberia to the islands of the Aleut across the Bering Sea, throughout Arctic Alaska and into the land of the Inuvialuit of the Mackenzie Delta, native languages—Aleut, Inupiaq, Inupik, as well as Inuktitut—have not remained strong. In some communities, a small number of the oldest men and women speak their native language. In many places, the younger generations are monolingual in English. So the Inuit of central and northeast Canada may be the people who have assumed the task of maintaining that beautiful language, those words, and the ways of those words, the oral culture of the people of the Arctic.

Ensuring the survival of Inuktitut may be among the most important projects for Nunavut, the newly created territory in the Canadian North. Yet decolonisation can take unexpected and paradoxical turns, twisting the deepest needs into knots of modern "progress." This progress can be seen in other parts of the world where farmers and herders have seized their opportunities.

220 **Language and metaphor** Ted Chamberlin has written very compellingly about the surprising and important ways in which oral culture, hunter-gatherer languages, metaphor and poetry may be related to one another. He has made the overarching point that "language confuses what is present and what is not," whereas "metaphors confuse what is true and what is not." With this observation, he indicates the significance of metaphor in cultures where the demarcation of the true and the untrue is imprecise. He also makes this point: "The acceptance of the contradiction—of how a thing is and is not all at the same time—is part of an ancient understanding of how things are represented in words and images. Perhaps that's why so many central figures in the storytelling traditions of the world, from

the classical god Hermes to the crafty Coyote of native America to the west African Anansi, are tricksters. They are constantly drawing us into a world where things are and are not, all at the same time." (All quotes are taken from personal communication, June 2000.) Chamberlin's essays on the importance of metaphor and related issues of language include "Putting Performance on the Page," "Cowboy Songs, Indian Speeches and the Language of Poetry" and "Doing Things with Words."

A general point about tribal peoples' languages (and cultures) is well made and richly illustrated in Jerome Rothenberg's *Technicians of the Sacred,* his 1968 collection of what he calls "poetries." Rothenberg notes in his preface: "There are no half-formed languages, no underdeveloped or inferior languages ... People who have failed to achieve the wheel will not have failed to invent & develop a highly wrought grammar." And he says later: "What is true of language in general is equally true of poetry & of the ritual-systems of which so much poetry is a part." He makes the central point that in "primitive" societies people shape and reshape things, achieving unity and poetic coherence (and metaphor, therefore), because of "some constant or 'key' against which all disparate materials can be measured. A sound, a rhythm, a name, an image, a dream, a gesture, a picture, an action, a silence: any or all of these can function as 'keys.' Beyond that there's no need for consistency, for fixed or discrete meanings" (p. xxii).

This kind of reflection links the hunter-gatherer ability to create and share metaphor to hunter-gatherer ways of using language and to the form of language's use or performance as oral culture and in related rituals. It must be added that this link reaches beyond hunter-gatherers to include many agricultural societies that are "primitive" in the old sense of not using writing and of living apart from the hegemonies of large nation-states.

FIVE: **GODS**

224 **Harold Wright's story** Harold Wright told his stories in English. He had a good knowledge of the Nisga'a language, though he was among the first of his people to be sent to residential school. I am aware that elements in Harold Wright's narrative take a different form in other versions. I am also conscious of relying too much on memory. The original recording seems to have been lost.

229 **Moravians in Labrador** The Moravians created three mission stations in the eighteenth century. These were at Nain (in 1771), Okak (in 1776) and Hopedale (in 1782). In the nineteenth century, Moravians established four more missions at which the Inuit of the region were encouraged to trade and to base their religious lives. This history and its links to economic changes along the Labrador coast are described in Carol Brice-Bennett's essay "Land Use in the Nain and Hopedale Regions" (see especially p. 101ff).

31 **Inuktitut and Innu'aimun** In Labrador, the word used for the Inuit language is Inuttut. The infix *-ti-*, which acts as a sort of activator showing purpose and agency, is dropped. The meaning of the words Inuktitut and Inuttut is the same, i.e., "in the manner of a person." The word for the Innu language (Innu'aimun) has a completely different linguistic origin.

32 **Innu and Inuit uses of land** In the 1970s the Labrador Inuit Association contracted Carol Brice-Bennett to undertake a comprehensive land-use and occupancy study of Inuit relationships to their lands and resources throughout Labrador. This work included an essay

by William Fitzhugh, "Indian and Eskimo/Inuit Settlement History in Labrador," which sets out the background to Inuit-Innu overlaps and complementarity in the region. This and other essays important for any understanding of the history and anthropology of Labrador are to be found in *Our Footprints Are Everywhere,* edited by Carol Brice-Bennett.

243 **Reverend Tomlinson and Father Morice** The excesses of Tomlinson and Morice are recorded in stories and memories collected by Marius Barbeau and his assistant William Beynon, and available to scholars as the Barbeau-Beynon archive at the National Archives of Canada. (The anthropologist John Cove has put together an inventory of this archive.) David Mulhall's *Will to Power* is a study of Father Morice and gives many details about his highly authoritarian mission among the Carrier and the Wet'suwet'en.

Many Gitxsan and Wet'suwet'en elders spoke to me about the destruction of their regalia, house posts and totem poles. The memory of great bonfires on which masks, rattles and powerful ornaments were heaped is carried in stories. Knowledge of totem poles being cut down and taken away is also part of modern oral testimony. At the same time, awareness of how much Northwest Coast and other indigenous art is displayed and stored in museums around the world gives rise to continuing indignation: the people were deprived of these parts of their heritage because they were deemed to represent the heathen, yet these same items are put on show as great artistic and cultural achievements. There is a strong movement to get some of these objects transferred from museums back to the communities where they were made and used. These efforts are part of a widespread renewal of pride in both the symbols and the beliefs of shamanistic spirituality.

244 **Missionaries** It is important to put on the record that some of the missionaries I met when working in the North were hospitable, thoughtful and sensitive. Father Guy Mary-Rousselière was a man who devoted far more time to his pioneering archaeological and genealogical research, and to sharing this research with both Inuit and outsiders, than he did to any obvious kind of missionary work. The Bracewell and Dexter families were generous to everyone. Also, in the course of land-claims movements, missionaries have at times allied themselves strongly with their communities in lodging protests against incursions by colonists. At the same time, it is also important to distinguish the personalities of individual missionaries from the part missions play and have played in the fate of indigenous cultures.

245 **Tricksters** In many myths of central importance to hunter-gatherers, tricksters play creative roles. Raven and Coyote—in Northwest Coast, subarctic and Arctic societies—contribute to the emergence of human beings and to the world in which human beings can live. They do so by playing tricks as well as by using the power of transformation. In other stories, tricksters cause all kinds of trouble. They embody, in this way, the ambiguities of power and the uncertainty of the division between the real and the unreal or between truth and falsehood. They also often embody the humour of myths—an aspect of indigenous culture that is often neglected.

248 **Chief Jerry Jack's statement** Jerry Jack's observation about the difference between his own people's spirituality and Christianity was recorded in an interview in the film *The Washing of Tears,* a documentary made with the Mowachaht-Muchalaht band in 1992 (distributed by the National Film Board of Canada). In fact, Jerry was speaking about a ceremony that marked the raising of two house posts inside the community church at Friendly

Cove on Nootka Island, the place on the Pacific Coast where Captain James Cook first landed in 1778. Those preparing for the ceremony removed statues and other Catholic symbols and cleared the altar area of the church to make room for the house posts.

254 **Gathering and protection of plants** The work of Alan Marshall with the Nez Perce and that of Marianne Ignace and Nancy Turner with the Secwepemc (Shuswap) describes the importance of plants in a hunter-gatherer economy. Marshall has emphasised the ways in which Nez Perce gatherers manage areas of plants in their territories in the northwest United States (see "Nez Perce Social Groups"), while Ignace and Turner have carried out extensive studies of which plants are used, and in what ways, in Secwepemc territories in Interior British Columbia (see Turner's *Plants in British Columbia Indian Territory* and Ignace's "Anthropological Expert Evidence Report," prepared for the Kamloops Indian Band).

255 **Hunter-gatherer categories and the law of excluded middle** Of course, there are categories in all languages and in all social and intellectual systems. The operation of grammar seems to require the law of excluded middle, making critical distinctions between things that are and things that are not the case. Grammar can be said to create a complex rationality that, while unconscious, is nonetheless shared by all human beings. However, the point here is that the hunter-gatherer way of using language to describe the world makes allowance for realities that are full of ambiguity, and for places where rationality and irrationality, the facts and the supernatural, intertwine. This is not a point about language or grammar, but about how people know, understand and relate to their surroundings. It is possible to speak with great precision about the ambiguities of the world, and this is what hunter-gatherers may well be specialists at doing.

263 **Women's role in the economic life of hunter-gatherers** Many anthropologists have emphasised the significance of women's contributions to the economic well-being of hunter-gatherer societies. The work of Richard Lee among the !Kung and that of Megan Biesele among the Ju/'hoan are important in this regard. Similarly, the work done by Marianne Ignace, Alan Marshall and Nancy Turner on gathering in the Plateau or Sahaptin culture area shows the crucial contributions made by women to everyday supplies as well as to surpluses essential for winter. Even in economies that are dependent on meat, women's harvesting is important. In the Arctic and subarctic areas where I have worked, many women are expert at fishing and snaring. Their hunting and trapping is also often more routinised than the men's: this means the women can be relied on to bring in food on a regular basis, whereas the men undertake the less predictable kinds of hunting for larger species. At the same time, harvest levels are often determined by a community's ability to process food. The tasks of slicing, cleaning, filleting, drying and smoking meat or fish are at least as time-consuming as the actual hunting or fishing, and women tend to be relied on to do much of this vital work.

265 **Innu protests against low-level flying by NATO** A detailed and compelling account of the Innu protest movement is Marie Wadden's *Nitassinan*.

266 **Elisabeth Penashue** I am grateful to Josephine Bacon for the translation of this and other quotes from the film *Hunters and Bombers*.

267 **Pien Penashue and hunter-gatherer science** One thing Western science and shamanic forms of knowledge have in common is a reliance on stories. This characteristic is not always evident in the way biologists and others in similar academic traditions

present their findings. But "the story of evolution" is a telling phrase often used in relation to biological sciences. In an important sense, this kind of story is a myth in the same way that shamanic accounts of the early days and inner relationships of the natural world are myths. When it comes to big-bang theory and other hypotheses about the earliest origins of the universe, or to palaeontological schema to show the transition from earliest hominids to *Homo sapiens,* the mythic quality of the stories being told is very striking. The tellers of these myths—the "experts"—with their technical and highly sophisticated access to the steps in the narrative, may also evoke the kinds of specialisation found among shamans. However, these analogies, pursued to some extent in the work of Claude Lévi-Strauss and his followers, neglect the importance of scientific as opposed to shamanic *method.* It is here that the profound differences emerge and that the hunter-gatherer reliance on a combination of specificity and relationship can be seen as distinctive. A remarkable exploration of how shamanism overlaps with aspects of modern science, and a speculation that shamans have access to the fundamental data with which geneticists concern themselves, can be found in Jeremy Narby's *Cosmic Serpent.*

SIX: **MIND**

277 **People who are not people** In *Frontiers,* a book that looks at the web of intergroup conflicts that constitutes the early history of the southern African Cape, Neil Mostert gives many details about how both settlers and Xhosa herders thought about and behaved towards Khoikhoi and San peoples. The quote I use is from comments made by John Ovington, master of the *Benjamin,* on seeing the Khoikhoi of the Cape in 1693. The words of Pyrard de Laval, written in 1610, are also revealing: "Of all people they are the most bestial and sordid ... They eat ... as do dogs ... they live ... like animals" (*Frontiers,* p. 108). The fate of Tasmania's hunter-gatherers is described in Mark Cocker's *Rivers of Blood.* See p. 115ff. and p. 127 for the quotes I use.

The literature reveals many voices that concur in their suggestion that the "savages" Europeans encountered during the early years of exploration and colonisation were not quite human. Of course, many societies and nations dehumanise their enemies and competitors. Yet to say that men, women and children who have languages, beliefs, clothing, implements and art are nonetheless something other than human is both strange and troubling. When aboriginal women are also the objects of sexual desire and the bearers of colonists' children, the suggestion that they are not human appears to be as ridiculous as it is grotesque. All the more striking, therefore, to find it occur in so much commentary by settlers and explorers as they arrived in new frontiers.

There are, of course, some counterexamples, among which are the observations in G. F. Lyon's *Private Journal* about the Inuit of the Igloolik area, whom he visited often when icebound in north Hudson Bay in the 1820s. Lyon was impressed in many ways by the people, whose snowhouses he visited, and who, in turn, visited Lyon's ship. Yet even in these sympathetic and seemingly accurate reports, there are troubling descriptions of the Inuit visitors' taste for sugar and salt and their reactions to liquor.

278 **Las Casas and Sepulveda** An account of the Las Casas/Sepulveda debate is given in Lewis Hanke's *Aristotle and the American Indians.* There is also an article on the subject in Thomas Berger's *Long and Terrible Shadow.*

280 **Aztec and Maya** Mexica (or Aztec), Maya and Inca societies constituted the "high civilisations" of the Americas. The Mexica world—including its use of human sacrifice—is described by Inga Clendinnen in her book *Aztecs.* Marvin Harris's *Cannibals and Kings* documents, and speculates about the rationales for, sacrifice and cannibalism in the region. It is important to emphasise here again that Aztec society was very distant from hunter-gatherer modes of life.

281 **Hobbes** Thomas Hobbes's famous dictum about life being "nasty, brutish, and short" comes from his *Leviathan,* first published in 1651. For a compelling account of how the philosophical views of both Hobbes and John Locke related to the development of capitalism, see C. B. Macpherson, *The Political Theory of Possessive Individualism.*

281 **Vico, Montesquieu, Rousseau, Hegel, Engels and Marx** Vico, writing in the 1740s, sought to create a philosophy of history and a theory of knowledge in which the artifacts of humans (from chairs to social and economic events) could be known with certainty—unlike the creations of God (i.e., the natural world), which could not thus be known. His speculations about a "savage state" rather surprisingly include a theory about a transition from hunters to herders, representing a shift from "bestial wandering" to a reliance on pasture. Vico's work is cited in Alan Barnard's essay "Images of Hunters and Gatherers in European Thought."

The idea that human beings are born free and then enslaved by the evils of civilisation is associated with Rousseau and the romantic tradition within the Enlightenment school of philosophy. But Hugo Grotius, the Dutch theorist of law, noted in the early 1600s that "foragers" lived "without toil," and Engels, writing in the 1880s, saw in hunter-gatherer systems evidence of both "primitive communism" and matriarchy. In this regard, Engels's *Family, Private Property and the State* stands as a fascinating attempt to use anthropology as the basis for a socialist critique of nineteenth-century society. For a review of the history of this issue, see Alan Barnard, op. cit., pp. 375–83. Barnard also quotes Montesquieu's 1748 *De l'esprit des lois;* Hegel, writing in 1820; and Marx, writing in 1857 (pp. 378–79).

The question about where a theory of hunter-gatherer society (albeit not always thus named) arises in European intellectual history can perhaps be answered by examining the purposes such a theory may have served. In his paper "On the Origins of Hunter-Gatherers in Seventeenth-Century Europe," Mark Pluciennik of the University of Wales speculates about these purposes (see note for page 302, below).

282 **Ralph Standish** Ralph Standish is quoted in Neil Mostert's *Frontiers* (p. 108). Mostert also quotes a Reverend Terry, who wrote in 1616 of the Khoikhoi that they were: "Beasts in the skins of men rather than men in the skins of beasts" (pp. 107–8). In fact, Khoikhoi were pastoralists, relying on herds of fat-tailed sheep; early settlers, however, did not make a significant distinction in their judgement of degrees of human development between Khoikhoi sheep herders and San (or Bushmen) hunter-gatherers. Mostert writes very compellingly of the relationships between early settlers—black and white—and the San, describing both the degree of dependence that settlers had on San knowledge and the low level of humanity to which the San were nonetheless consigned. He quotes the South African historian George MacCall Theal as representative of his time in observing that Bushmen habits "were not much more elevated than those of animals," and that it was difficult to conceive of a human being "in a more degraded condition" (pp. 31–32).

283 **The Trudeau government's view of land claims** The Trudeau government's position on the place of Indians in Canadian society was first set out in a 1969 white paper, issued on the authority of the then minister for Indian Affairs, now Canadian prime minister, Jean Chrétien. The white paper argued that there should be a gradual ending of treaty rights of Indians, and that the dissolution of their reserves and related special status would be an appropriate and necessary form of development. This position caused outrage throughout the country's aboriginal communities, mobilising a newly radical form of First Nation politics. The Calder case entered the courts in the same year, reaching the Supreme Court (after comprehensive defeats in the two lower courts) two years later. The shift in Trudeau's policy may be said to have resulted from the combined impacts of the new Indian politics and the Supreme Court judgement.

In fact, the Calder case was not the first such consideration of aboriginal rights in Canada. Its significance lay in the Supreme Court's split decision—the sign of a changing legal atmosphere. The sequence of events, and the uses of evidence, that determined this atmosphere are set out in Dara Culhane's *Pleasure of the Crown* (see especially pp. 72–89). Culhane quotes the Canadian historian Peter Kulchyski, who has observed that the legal forum is a strange and even incoherent one for resolving issues of rights. Kulchyski comments that the history of the court cases is one of "losses and gains, of shifting terrain ... a history where the losers often win and the winners often lose" (Culhane, p. 73).

283 **Cases of importance** The modern case that most clearly defined the tests for aboriginal title and right in Canadian courts was *Baker Lake v. the Crown*. This case also led to a view that aboriginal people could only claim rights to resources they actually use. This created uncertainty about the question of resources people did not know about in their territories, for example, minerals, oil and gas. The judgement in Baker Lake suggested that if people did not know about their assets and were therefore not using them, they had no rights to them. The Delgamuukw case, which reached the courts in 1994 and was decided in 1997, has led to a redefinition of the tests, and recognises much more of indigenous people's potentially inherent rights in all aspects of their territories, including the potential for development without reference to particular use. One of the issues that Delgamuukw settled was the depth of time that evidence must show. It ruled that for title, the relevant date was 1846; for rights (i.e., use and resource, including fishing, hunting and gathering), the date should be time of first contact with Europeans.

The St. Catherine's Milling case of 1885 is an early judgement that has been brought to bear in modern land-rights cases. It is important because it speaks to which agency has the right to extinguish aboriginal title, the federal or provincial Crown. It also speaks to the consequences of extinguishment. It is not a case, however, that establishes a precedent for any protection of aboriginal rights or titles to either land or resources.

The earliest case that does address the nature of aboriginal rights, and which also bears on modern cases, was decided in Southern Rhodesia in 1919. Judgement in this case introduced the notion that you could have aboriginal people who roam and have no society, and therefore led to the requirement that native litigants show they have an organised society. It was here that the idea that human beings can be placed at different levels of an evolutionary ladder entered into modern jurisprudence, with the attendant notion that some peoples are so low on this ladder—as evidenced by their way of "just wandering around"—

that they have no rights to lands or resources. (See Dara Culhane's *Pleasure of the Crown* for these and other cases, especially chapter 6.)

84 **Organised society** The "organised society" test depends on the idea that some humans are not humans. This is social-scientific nonsense, of course, and originates in an attempt to justify expropriation on racial or cultural grounds. At the end of the twentieth century this was given the name "ethnic cleansing"—a new term for what many hunter-gatherers have long experienced.

89 **Animals and humans** Biologists spend a great deal of time testing the intelligence of many species, from rats and pigeons to chimpanzees and parrots. Many forms and levels of intelligence are described as a result of this research. But biologists who assess the achievements of animals in experimental tests, or measure their speed and effectiveness at learning new skills, are not creating models of internal linguistic abilities. It may be that some human actions that we believe arise from consciously made decisions, processed in language, are more directly controlled by our genetic hard-wiring than we like to admit—though this is an area of ongoing contention. The central point is that what happens with language, and therefore in the mind, pertains to a uniquely human form of intelligence.

91 **Apes and making noises with our mouths** The observation on apes is from Derek Bickerton's *Language and Human Behaviour,* p. 7. The quote about noises from our mouths is from the first page of Steven Pinker's *Language Instinct*.

93 **Spread of cultures and languages** A question arises in relation to the spread of humans across the world: what was the process that resulted in so many *different* languages? Linguists have identified correlations between the length of time a society is isolated from its neighbours and the evolution of a distinct language. By analysing vocabulary and grammar, it is possible to analyse links between peoples and even sequences of movement. The human geographer James L. Newman has suggested that farming, by causing people to settle in scattered but definite places, increased the degree of isolation between communities. He uses this to explain the multiplicity of agricultural languages in, for example, much of Africa (see his *Peopling of Africa,* especially p. 4).

Obviously, hunter-gatherers settled in large territories give rise to a language map with accordingly large areas in which people speak the same language. But where one hunter-gatherer society's territories meet those of another hunter-gatherer society, languages differ—sometimes at the level of dialect, sometimes at the level of language family. The boundary between Athabaskan and Algonquian speakers in subarctic North America, for example, suggests deep conservatism with regard to language, despite closely related economic systems. Similarly, a map of languages for the North Pacific Coast would show great varieties of languages in relatively small regions. The region that came to be California once contained approximately eighty different aboriginal languages. By comparison, the spread of Indo-European languages is a homogenizing process resulting from farming. It may be that when it comes to the variety of human languages, Newman has attributed to agriculturalists some of the features of hunter-gatherers.

93 ***Homo erectus* and language** The quote comes from Desmond Clark of the University of California at Berkeley. It is cited in a summary of the findings pertaining to *Homo erectus* and language in Robin McKie's *Ape Man* (pp. 82–84). McKie reviews the evidence for the links between language and *Homo Heidelbergiensis* on pp. 122–24. In fact, the evidence he

gives of humans who probably had language comes from European sites. Much related evidence argues strongly in favour of the view that these language users in Europe came from somewhere in Africa, and that their dispersal was itself a result of speech. This means that the date of *Heidelbergiensis* is a very late date to use in pinpointing the beginnings of language. The matter remains unresolved, but it would seem reasonable to say that our ancestors made the evolutionary step necessary for language at least one million years ago.

294 **Hard-wired faculty for language** The process by which human beings acquired language is bound to involve speculations about the nature of evolution, the structure of the human brain, and the functional advantages of language itself. The science that pertains to these things is constantly evolving. An account of the evidence, and the state of the relevant scientific art, is to be found in Steven Pinker's *Language Instinct*. Pinker reviews the results obtained by experts who have attempted to teach apes to use language (see pp. 335–49), revealing how unsuccessful these attempts ultimately are as language learning, then considers evolution and the probable development of a "hard-wired" faculty for language in human brains (p. 349ff.). On the question of how long it took for humans to evolve language, he notes: "We expect a fade-in, but we see a big bang" (p. 343).

295 **Acrobats of the mouth** Anthony Traill made the observation that San speakers are "the acrobats of the mouth." His essay "*!Khwa-Ka Hhouiten Hhouiten,* The Rush of the Storm" offers many insights into both the nature of San language and the causes of language loss. I am grateful to the ethnolinguist Nigel Crawhall for details about the sound elements in different languages (personal communication, 1999).

296 **Noam Chomsky** Noam Chomsky's work on language began in the 1950s and continued into the 1970s. The quotes here come from *Reflections on Language*. These passages are quoted at greater length in Steven Pinker's *Language Instinct*, pp. 22–23.

296 **Nature and nurture in language learning** This issue is discussed in Steven Pinker's *Words and Rules* (p. 198), where he cites the work of Arnold Zwicky, James Morgan, Lisa Travis, J. Tooby and Irven DeVore. For Pinker's comments on grammar learning and identical twins, see *Words and Rules*, p. 203. For Pinker's concluding remarks, see p. 210.

300 **Lévi-Strauss** Lévi-Strauss first wrote about the Nambikwara in the 1940s. He contributed an essay to volume 3 of the 1948 *Handbook of South American Indians*. The first edition of *Tristes Tropiques* was published in 1955; it was translated into English as *A World on the Wane*, first published in 1961. *The Savage Mind (La Pensée sauvage)*, published in French in 1962 and in English in 1966, is the book of his that had the most widespread impact on European intellectuals. Lévi-Strauss's work on kinship and totemism, especially *Les Structures elémentaires de la parentée* and *Totemism*, published in 1962 and 1963, respectively, examined core questions of anthropology.

300 **The myth of Asdiwal** The Lévi-Strauss version of the myth, which takes place in the Nass Valley in Nisga'a territory (though Lévi-Strauss identified it by the broader term Tsimshian, now used to refer to the language family of the region), was drawn from written accounts. His analysis of the myth, "The Story of Asdiwal," was published with an accompanying essay by Edmund Leach, in which Leach gave himself the task of explaining and celebrating structuralism to English anthropologists. Neither Lévi-Strauss nor Leach ever did field work in the region.

302 **Postmodernism** In his paper "On the Origins of Hunter-Gatherers in Seventeenth-

Century Europe," Mark Pluciennik of the University of Wales suggests that the new science of political economy divided human societies into hunters, herders and farmers as part of a need to rationalise and justify individualism, inequality and the unquestionable obligation of labour. Natural laws were sought that would show that certain kinds of individual owed service to society but could not expect to be an equal member of it. Pluciennik suggests that the purported "laziness," "ignorance" and "inferiority" of "savages" was useful to this project because it laid a "natural" foundation for hierarchy and the possibility of fundamental differences among people.

He also suggests that these categories—hunter, herder and farmer—are contingent and heuristic: they were invented to meet a particular intellectual and political need, and are not grounded in sociological or anthropological reality. A difficulty with this reductionist account of the categories lies in the extent to which they are to be found in other languages and times. At the Cambridge conference where Pluciennik put forward his view that the categories belonged to seventeenth- and eighteenth-century social theory, not to reality, James Woodburn pointed out that the same categories are to be found in the Bantu languages of southern Africa, with a provenance far from the political economists of seventeenth-century Europe. Woodburn also observed that these categories correspond to what people can most readily observe. Hence it is not surprising to find that a distinction between hunter-gatherers and the herders or farmers who are their neighbours is of great importance to both the hunter-gatherers and the agropastoralists themselves.

303 **Irish history** The history of peasant life in Ireland has yielded a number of "traditions." In reality, the domination of peasant families by landlords seeking to maximise rents created the Ireland that is most often said to be "traditional." This is the peasant life described in Conrad Arensberg's *Irish Countryman,* then in Conrad Arensberg and Solon Kimball's famous *Family and Community in Ireland.* This particular peasant society, whose features made it so vulnerable to potato blight, developed in the period between 1700 and 1850. An ancient Irish society, reaching back to Urse myths and the "golden age" of Irish ruling families, was destroyed by British invasion, beginning in the thirteenth century and ending with Cromwell's conquest of Ireland in the mid–seventeenth century. I discuss Arensberg and Kimball's sociology in *Inishkillane,* pp. 4–7.

The literature of Ireland is not to be reduced, of course, to a reflection of any single aspect of Irish (or any other) history. Yet there is a connection between the moods of much Irish writing and the successive transformations and disasters in the history of the country's peasantry. This idea is something I use throughout the text of *Inishkillane,* setting literature alongside sociology.

305 **Fear of hunter-gatherers** James Woodburn's summary of attitudes to hunter-gatherers in southern Africa is in his essay "Indigenous Discrimination." There Woodburn details a set of negative attitudes that their neighbours appear to hold towards hunter-gatherers, then goes on to identify reasons for the discrimination. These include the *appearance* of hunter-gatherers in the eyes of agriculturalists or pastoralists, with a related identification of hunter-gatherers with the wild, and the apparent absence among hunter-gatherers of institutions of authority and social control. Woodburn also cites evaluations of hunter-gatherers that are ambiguous or even positive, identifying the extent to which they are believed to have exotic powers; but he adds a fascinating observation about the extent to which in

Africa and many other places, "there is a fear that those who are weak and impotent have mysterious supernatural powers which threaten those who exercise power. Power holders tend to be ambivalent about their right to power, wealth and prestige and they fear that resentment of those who lack power, wealth and prestige may mysteriously threaten their health, well-being and prosperity." He adds that hunter-gatherers can, at times, use this fear to redress some of the political imbalance between themselves and others. Other writing by James Woodburn that bears on these issues includes the introduction he co-authored with Alan Barnard to *Hunters and Gatherers* and his essay "Egalitarian Societies."

BIBLIOGRAPHY

Following are the sources I cite in the text and endnotes of this book, as well as work that has been of importance in writing it.

Alter, Robert. *Genesis, Translation and Commentary.* New York and London: W. W. Norton, 1996.

Arensberg, Conrad. *The Irish Countryman.* New York and London: Macmillan, 1937.

Arensberg, Conrad, and Solon Kimball. *Family and Community in Ireland.* Boston: Harvard University Press, 1968.

Arroyo-Kalin, Manuel. "Non-Pristine Hunter-Gatherers or Theoretical Strait-Jackets: Foraging, 'Encapsulation' and Social Hierarchy in the Neotropics." Paper presented at Global Hunters and Gatherers: After Revisionism, Department of Archaeology, University of Cambridge, May 2000.

Balikci, Asen. *The Netsilik Eskimo.* With a foreword by Margaret Mead. New York: Natural History Press, 1970.

———. "Netsilik." In *Handbook of North American Indians.* Vol. 5, *Arctic,* edited by D. Damas, pp. 415–31. Washington, D.C.: Smithsonian Institution, 1984.

Barbeau, Marius, and William Beynon. Tsimshianic Notes and Manuscripts, 1915–1957. National Archives of Canada. Catalogued by John Cove as *A Detailed Inventory of the Barbeau Northwest Coast Files.* Canadian Centre for Folk Studies, paper 54. Ottawa: National Museum of Canada, 1985.

Barnard, Alan. "Images of Hunters and Gatherers in European Thought." In *The Cambridge Encyclopedia of Hunters and Gatherers*, Richard B. Lee and Richard Daly, 375–83.

Barnett, Anthony, and Roger Scruton, eds. *Town and Country.* London: Jonathan Cape, 1998.

Bateson, G., D. D. Jackson, J. Haley, and J. Weakland. "Towards a Theory of Schizophrenia." *Behavioural Science* 1, no. 251 (1956).

Bauman, Richard. *Let Your Words Be Few: Symbolism of Speaking and Silence among Seventeenth-Century Quakers.* Cambridge and New York: Cambridge University Press, 1983.

Berger, Thomas. *Northern Frontier, Northern Homeland: The Report of the Mackenzie Valley Pipeline Inquiry.* Ottawa: Ministryof Supply and Services, 1977. Rev. ed. Vancouver: Douglas & McIntyre, 1988.

———. *Fragile Freedoms: Human Rights and Dissent in Canada.* Toronto and Vancouver: Clarke, Irwin, 1981.

———. *A Long and Terrible Shadow: White Values, Native Rights in the Americas, 1492–1992.* Vancouver: Douglas & McIntyre, 1991.

Bickerton, Derek. *Language and Human Behaviour.* London: University College, 1996.

Biesele, Megan. *Women Like Meat: The Folklore and Foraging Ideology of the Kalahari Ju/'hoan.* Bloomington, Ind.: Indiana University Press, 1993.

Bodenstein, Christel, Hans Bodenstein, and Linda Rode, comps. *Stories South of the Sun.* Cape Town: Tagelberg, 1993.

Bowman, Glenn. "Radical Empiricism: Anthropological Fieldwork after Psychoanalysis and the *Année Sociologique.*" *Anthropological Journal on European Cultures* 6, no. 2 (1998): 79–107.

Brice-Bennett, Carol. "Land Use in the Nain and Hopedale Regions." In *Our Footprints Are Everywhere: Inuit Land Use and Occupancy in Labrador,* edited by Carol Brice-Bennett, 97–203. Nain, Nfld. and Lab.: Labrador Inuit Association, 1977.

———, ed. *Our Footprints Are Everywhere: Inuit Land Use and Occupancy in Labrador.* Nain, Nfld. and Lab.: Labrador Inuit Association, 1977.

Briggs, Jean. *Utkuhikhalingmiut Eskimo Emotional Expression.* Ottawa: Northern Science Research Group, 1968.

———. *Never in Anger.* Cambridge, Mass.: Harvard University Press, 1970.

Bringhurst, Robert. *A Story As Sharp As a Knife: The Classical Haida Mythtellers and Their World.* Vancouver and Toronto: Douglas & McIntyre, 1999.

Brody, Hugh. *Inishkillane: Change and Decline in the West of Ireland.* London: Allen Lane/Penguin, 1973.

———. *The People's Land: Eskimos and Whites in the Eastern Arctic.* London: Penguin Books, 1975. Rev. with new introduction. Vancouver: Douglas & McIntyre: 1991.

———. "Permanence and Change among the Inuit and Settlers of Labrador." In *Our Footprints Are Everywhere: Inuit Land Use and Occupancy in Labrador,* edited by Carol Brice-Bennet 311–47. Nain, Nfld. and Lab.: Labrador Inuit Association, 1977.

———. "An Analysis of Violent Death." In *The Economic and Social Impact of Mercury Pollution on the Whitedog and Grassy Narrows Indian Reserves, Ontario,* by Peter Usher et al., 305–63. Report to the Anti-Mercury Ojibway Group, Kenora, Ontario, 1979.

———. *Maps And Dreams: Indians and the British Columbia Frontier.* Vancouver: Douglas & McIntyre, 1981.

———. "Nomads and Settlers." In *Town and Country,* edited by Anthony Barnett and Roger Scruton, 3–18. London: Jonathan Cape, 1998.

———. Introduction to *Seasons of the Arctic.* Photographs by Paul Nicklen. Vancouver: Greystone Books, 2000.

Buliard, Roger. *Inuk.* London: Macmillan, 1953.

Chamberlin, J. Edward. "Doing Things with Words." *Voices: A Journal of Oral Studies* 1 (January 1998).

———. "Putting Performance on the Page." In *Talking on the Page: Editing Aboriginal Oral Texts,* edited by Laura J. Murray and Keren D. Rice, 69–90. Toronto: University of Toronto Press, 1999.

———. "Cowboy Songs, Indian Speeches and the Language of Poetry." In *Postcolonising the Commonwealth: Studies in Literature and Culture,* edited by Rowland Smith, 5–27. Waterloo, Ont.: Wilfred Laurier University Press, 2000.

Bibliography

Chomsky, Noam. *Reflections on Language.* New York: Pantheon, 1975.

Clendinnen, Inga. *Aztecs: An Interpretation.* Cambridge: Cambridge University Press, 1991.

Cocker, Mark. *Rivers of Blood, Rivers of Gold: Europe's Conflict with Tribal Peoples.* London: Jonathan Cape, 1998.

Connell, K. H. *The Population of Ireland, 1750–1845.* Oxford: Clarendon Press, 1950.

Correll, Thomas C. "Language and Location in Traditional Inuit Societies." In *Inuit Land Use and Occupancy Project.* Edited by Milton Freeman. Vol. 2, 171–86. Ottawa: Ministry of Supply and Services, 1976.

Culhane, Dara. *The Pleasure of the Crown: Anthropology, Law and First Nations.* Burnaby, B.C.: Talonbooks, 1989.

Daly, Richard. Expert Opinion Evidence. *Delgamuukw et al. v. the Queen.* Court Transcripts, Vols. 184–89. Supreme Court of British Columbia, 0843, Smithers Registry, 1985–91.

Damas, David, ed. *Handbook of North American Indians.* Vol. 5, *Arctic.* Washington, D.C.: Smithsonian Institution, 1984.

Delgamuukw et al. v. the Queen. Court Transcripts of *Delgamuukw v. the Queen in the Right of the Province of British Columbia and the Attorney General of Canada.* Supreme Court of British Columbia, 0843, Smithers Registry, 1985–91.

Diamond, Jared. *Guns, Germs and Steel.* London: Vintage, 1998.

Engels, Friedrich. *The Family, Private Property and the State.* 1884. Reprint, London: Lawrence & Wishart, 1972.

Fanon, Frantz. *The Wretched of the Earth.* London: Penguin, 1967.

———. *Black Skin, White Masks.* London: MacGibbon & Kee, 1968.

Fitzhugh, William W. "Indian and Eskimo/Inuit Settlement History in Labrador: An Archaeological View." In *Our Footprints Are Everywhere: Inuit Land Use and Occupancy in Labrador,* edited by Carol Brice-Bennett, 1–40. Nain, Nfld. and Lab.: Labrador Inuit Association, 1977.

Fitzhugh, William W., and Chisato O. Dubreuil, eds. *Ainu: Spirit of a Northern People.* Washington, D.C.: Arctic Studies Center, National Museum of Natural History, Smithsonian Institution in association with University of Washington Press, 1999.

Freeman, Milton, ed. *Inuit Land Use and Occupancy Project.* 3 vols. Ottawa: Ministry of Supply and Services, 1976.

Gagné, R. C. *Tentative Standard Orthography for Canadian Eskimos.* Rev. ed. Ottawa: Department of Northern Affairs and National Resources, 1962.

Gilberg, Rolf. "Polar Eskimo." In *Handbook of North American Indians.* Vol. 5, *Arctic,* edited by David Damas, 577–95. Washington, D.C.: Smithsonian Institution, 1984.

Graburn, Nelson. *Eskimos without Igloos: Social and Economic Development in Sugluk.* Boston: Little, Brown, 1969.

Hanke, Lewis. *Aristotle and the American Indians: A Study in Race Prejudice in the Modern World.* Bloomington, Ind.: Indiana University Press, 1970.

Harris, Marvin. *Cannibals and Kings: The Origins of Cultures.* New York: Random House, 1977.

Ignace, Marianne Boelscher. "The Secwepemc (Shuswap)." In *Handbook of North American Indians.* Vol 12, *The Plateau,* edited by Deward Walker. Washington, D.C.: Smithsonian Institution, 1998.

———. "Anthropological Expert Evidence Report," prepared for the Kamloops Indian Band, British Columbia, 1995.

Ingold, Tim, David Riches, and James Woodburn, eds. *Hunters and Gatherers.* Vol. 1, *History, Evolution and Social Change.* London: Berg, 1988.

———. *Hunters and Gatherers.* Vol. 2, *Power, Property and Ideology.* London: Berg, 1988.

Jenness, Diamond. *The People of the Twilight.* New York: Macmillan, 1928.

———. *Dawn in Arctic Alaska.* Minneapolis: University of Minnesota, 1957.

———. *Eskimo Administration.* 4 vols. Montreal: Arctic Institute of North America, 1962–67.

Kaner, Simon. "Hunter-Gatherer 'Civilisation'? A View from the Japanese Archipelago." Paper presented at Global Hunters and Gatherers: After Revisionism, Department of Archaeology, University of Cambridge, May 2000.

Kulchyski, Peter, ed. *Unjust Relations: Aboriginal Rights in Canadian Courts.* Toronto: Oxford University Press, 1994.

Laing, R. D. *Knots.* London: Routledge, 1971.

Laing, R. D., and A. Esterson. *Sanity, Madness and the Family.* London: Tavistock Publications, 1964.

Leach, Edmund. *The Structural Study of Myth and Totemism.* London: Tavistock Publications, 1967.

Lee, Richard B. *The !Kung San: Men, Women and Work in a Foraging Society.* Cambridge: Cambridge University Press, 1979.

Lee, Richard B., and Richard Daly. *The Cambridge Encyclopedia of Hunters and Gatherers.* Cambridge: Cambridge University Press, 1999.

Lee, Richard B., and Irven DeVore. *Man the Hunter.* Chicago: Aldine, 1968.

Lévi-Strauss, Claude. "The Nambikwara." In *Handbook of South American Indians,* Vol. 3, edited by J. Steward, 465–86. Washington, D.C: Smithsonian Institution, 1948.

———. *Tristes Tropiques.* 1955. Translated by John Russell as *A World on the Wane.* London: Hutchison, 1961.

———. *Structural Anthropology.* 1963. London: Allen Lane/Penguin, 1968.

———. *The Raw and the Cooked.* 1964. London: Harper and Row/Jonathan Cape, 1969.

———. "The Story of Asdiwal." In *The Structural Study of Myth and Totemism,* edited by Edmund Leach, 1–40. London: Tavistock Publications, 1967.

Lindqvist, Sven. *Exterminate All the Brutes.* London: Granta Books, 1997.

Lyon, G. F. *The Private Journal.* London: J. Murray, 1824.

Macintosh, William. *Te Ika Maui: New Zealand and Its Inhabitants.* London: n.p., 1870.

McKie, Robin. *Ape Man: The Story of Human Evolution.* London: BBC Worldwide, 2000.

Macpherson, C. B. *The Political Theory of Possessive Individualism: Hobbes to Locke.* Oxford: Oxford University Press, 1964.

Mannoni, O. *Prospero and Caliban: The Psychology of Colonization.* London: Methuen, 1956.

Marcus, Alan. "Out in the Cold: Canada's Experimental Inuit Relocation to Grise Fjord and Resolute Bay." *Polar Record* 27, no. 163 (1991): 285–96.

Marshall, Alan. "Nez Perce Social Groups: An Ecological Intepretation." Ph.D. diss., Washington State University, 1977.

Martin, Laura. "Eskimo Words for Snow." *American Anthropologist* 88, no. 2 (1986): 418–23.

Bibliography

Mary-Rousselière, Guy. *Qitdlarssuaq: The Story of a Polar Migration.* Winnipeg, Man.: Wuerz Publishing, 1991.

Mauss, Marcel. *The Gift: The Form and Reason for Exchange in Archaic Societies.* 1950. Originally published as *Essai sur le don* by Presses Universitaires de France. Translated by W. D. Halls, with a foreword by Mary Douglas. London and New York: Routledge, 1990.

May, Margaret, ed. *The Orchard Book of Magical Tales.* London: Orchard Books, 1993.

Meillasoux, Claude. "On the Mode of Production of the Hunting Band." In *French Perspectives in African Studies: A Collection of Translated Essays,* edited by Pierre Alexandre. London: Oxford University Press for the International African Institute, 1973.

Miller, Alice. *For Your Own Good: The Roots of Violence in Child-Rearing.* London: Virago, 1987.

Milloy, John. *A National Crime: The Canadian Government and the Residential School System, 1879–1986.* Winnipeg, Man.: University of Manitoba Press, 1999.

Mills, Antonia. *Eagle Down Is Our Law: Witsuwit'en Law, Feasts and Land Claims.* Vancouver: University of British Columbia Press, 1994.

Mostert, Neil. *Frontiers: The Epic of South Africa's Creation and the Tragedy of the Xhosa People.* London: Jonathan Cape, 1992.

Mowat, Farley. *The People of the Deer.* Boston: Little, Brown, 1951.

———. *The Desperate People.* Boston: Little, Brown, 1959.

Mulhall, David. *Will to Power: The Missionary Career of Father Morice.* Vancouver: University of British Columbia Press, 1986.

Munk, Rabbi Michael L. *The Wisdom in the Hebrew Alphabet: The Sacred Letters As a Guide to Jewish Deed and Thought.* New York: Mesorah Publications, 1983.

Narby, Jeremy. *The Cosmic Serpent: DNA and the Origins of Knowledge.* New York: Tarcher/Putnam, 1998.

Newman, James L. *The Peopling of Africa.* New Haven and London: Yale University Press, 1995.

Padel, Felix. *The Sacrifice of Human Being: British Rule and the Konds of Orissa.* Oxford and Delhi: Oxford University Press, 1995.

Pei, Mario. *The Story of Language.* New York: J. B. Lippincott, 1949.

Pinder, Leslie Hall. *The Carriers of No: After the Land Claims Trial.* Vancouver: Lazara Press, 1991. Revised and updated in "Tribes Battle for Land and Language." *Index on Censorship,* no. 4 (1999): 65–75.

———. *On Double Tracks.* London: Bloomsbury, 1990.

Pinker, Steven. *The Language Instinct: The New Science of Language and Mind.* London and New York: Penguin Books, 1994.

———. *Words and Rules: The Ingredients of Language.* London: Weidenfeld & Nicolson, 1999.

Pluciennik, Mark. "On the Origins of Hunter-Gatherers in Seventeenth-Century Europe." Paper presented at Global Hunters and Gatherers: After Revisionism, Department of Archaeology, Unversity of Cambridge, May 2000.

Pullum, Geoffrey K. *The Great Eskimo Vocabulary Hoax.* Chicago: Chicago University Press, 1991.

Rasmussen, Knud, and Kaj Birket-Smith. *Report on the Fifth Thule Expedition 1921–1924.* Vols. 4, 6, 7, 8, 9. Copenhagen: Gyldendalske Boghandel, Nordisk Forlag, 1930.

Renfrew, Colin. *Archaeology and Language.* London: Jonathan Cape, 1987.

Richard, Richard B. *The Dobe Ju/'hoansi.* 2d ed. New York and Toronto: Harcourt Brace College Publishers, 1984.

Riches, David. "The Netsilik Eskimo: A Special Case of Selective Female Infanticide." *Ethnology* 13, no. 4: 351–61.

Ridington, Robin. "The Medicine Fight: An Instrument of Political Process among the Beaver Indians." *American Anthropologist* 70, no. 6 (1962): 1152–60.

———. *Swan People: A Study of the Dunneza Prophet Dance.* Mercury series. Ottawa: National Museums of Canada, 1978.

———. *Trail to Heaven: Knowledge and Narrative in a Northern Native Community.* Vancouver: Douglas & McIntyre, 1988.

Rothenberg, Jerome, ed. *Technicians of the Sacred: A Range of Poetries from Africa, America, Asia and Oceania.* New York: Doubleday, 1968.

Rowley, Graham W. *Cold Comfort: My Love Affair with the Arctic.* Montreal and Kingston: McGill-Queen's University Press, 1996.

Schneider, Lucien, o.m.i. *Dictionnaire Esquimau-Français du parler de l'Ungava.* 2 vols. Nouvelle édition augmentée. Québec: Presses de l'Université Laval, 1970.

S.-Senghor, Leopold, ed. *Anthologie de la nouvelle poésie negre at malgache.* Paris: Presses Universitaires de France, 1958.

Shahn, Ben. *Love and Joy about Letters.* London: Cory, Adams & Mackay, 1964.

Smith, Derek G. *The Mackenzie Delta: Domestic Economy of the Native People.* Ottawa: Department of Indian Affairs and Northern Development, Northern Coordination and Research Centre, 1967.

———. "Mackenzie Delta Eskimo." In *Handbook of North American Indians.* Vol. 5, *Arctic,* edited by David Damas, 347–59. Washington: Smithsonian Institution, 1984.

Sperber, Dan. "Apparently Irrational Beliefs." In *On Anthropological Knowledge: Three Essays.* 6–63. New York: Cambridge University Press, 1985.

Steenhoven, G. Van Den. *Leadership and Law among the Eskimos of the Keewatin District, Northwest Territories.* Ph.D thesis, Rijswik, Uitgeverj Excelsior, 1962.

Stefansson, Vilhjalmur. *My Life with the Eskimo.* New York: Macmillan, 1941.

———. *Not by Bread Alone.* New York: Macmillan, 1946.

Tester, Frank, and Peter Kulchyski. *Tammarnit: Inuit Relocation in the Eastern Arctic, 1939–63.* Vancouver: University of British Columbia Press, 1994.

Traill, Anthony. "*!Khwa-Ka Hhouiten Hhouiten,* The Rush of the Storm: The Linguistic Death of the /Xan." In *Miscast: Negotiating the Presence of the Bushmen,* edited by Pippa Skatnas, 161–85. Cape Town: University of Cape Town Press, 1996.

Turnbull, Colin. *The Mountain People.* London: Jonathan Cape, 1973.

Turner, Nancy. *Food Plants of British Columbia Indians.* Vol. 2, *Interior Peoples.* Natural History Handbook no. 36. Victoria, B.C.: British Columbia Provincial Museum, 1975.

———. *Plants in British Columbia Indian Territory.* Natural History Handbook no. 38. Victoria, B.C.: British Columbia Provincial Museum, 1979.

Usher, Peter. *The Bankslanders: Economy and Ecology of a Frontier Trapping Community.* 3 vols. Ottawa: Department of Indian Affairs and Northern Development, 1971–73.

Usher, Peter, P. Anderson, H. Brody, J. Keck, and J. Torrie. *The Economic and Social Impact of Mercury Pollution on the Whitedog and Grassy Narrows Indian Reserves, Ontario.* Report to the Anti-Mercury Ojibway Group, Kenora, Ontario, 1979.

Vallee, Frank. *Kabloona and Eskimo in the Central Keewatin.* Ottawa: Canadian Research Centre for Anthropology, 1971.

Van der Post, Laurens. *The Heart of the Hunter.* London: Hogarth Press, 1961.

Wadden, Marie. *Nitassinan: The Innu Struggle to Reclaim Their Homeland.* Vancouver: Douglas & McIntyre, 1988.

Walker, Deward, ed. *Handbook of North American Indians.* Vol. 12, *The Plateau.* Washington, D.C.: Smithsonian Institution, 1998

Whorff, Benjamin Lee. "The Relation of Habitual Thought and Behaviour to Language." In *Language, Thought and Reality: Selected Writings of Benjamin Lee Whorff,* edited by John B. Carroll, 134–270. Cambridge, Mass.: Massachusetts Institute of Technology Press, 1964.

Widlock, Thomas. *Living on Mangetti: "Bushman" Autonomy and Namibian Independence.* Oxford: Oxford University Press, 1999.

Willcox Smith, Jessie (illustrator). *A Child's Book of Stories: Best-Known and Best-Loved Tales from around the World.* New York: Dilithium Press, 1986.

Wittgenstein, Ludwig. *The Blue and the Brown Books.* Oxford: Basil Blackwell, 1962.

———. *Philosophical Investigations.* Oxford: Basil Blackwell, 1963.

———. *Lectures and Conversations on Aesthetics, Psychology and Religious Belief.* Oxford: Basil Blackwell, 1965.

Woodburn, James, dir. *The Hadza.* Documentary film. London: Department of Anthropology, 1966.

———. "An Introduction to Hadza Ecology." In *Man the Hunter,* edited by Richard B. Lee and Irven DeVore, 49–55. Chicago: Aldine, 1968.

———. "Minimal Politics: The Political Organization of the Hadza of Tanzania." In *Politics in Leadership: A Comparative Perspective,* edited by P. Cohen and W. Shack, 244–66. Oxford: Clarendon Press, 1979.

———. "Hunter-Gatherers Today and Reconstruction of the Past." In *Soviet and Western Anthropology,* edited by E. Gellner, 95–117. London: Gerald Duckworth, 1980.

———. "Egalitarian Societies." *Man* n.s., 17 (1981): 31–51.

———. "African Hunter-Gatherer Social Organization: Is It Best Understood As a Product of Encapsulation?" In *Hunters and Gatherers: History, Evolution and Social Change.* Vol. 1, edited by Tim Ingold, David Riches and James Woodburn, 31–64. London: Berg, 1988.

———. "Indigenous Discrimination: The Ideological Basis for Local Discrimination against Hunter-Gatherer Minorities in Sub-Saharan Africa." *Ethnic Studies and Racial Studies* 20, no. 2 (April 1997): 345–51.

Young, Arminius. *One Hundred Years of Mission Work in the Wilds of Labrador.* London: Stockwell, 1931.

INDEX

Aborigines, 148, 183, 185, 256, 257, 278, 280, 312

Africa, 121–22, 148, 187, 202, 258, 351, 352; southern, 155, 156, 186, 188, 189, 277–78, 305, 309, 334, 337, 348, 353 (*see also* South Africa). *See also* North Africa

Agriculture, 157–58, 185, 257, 329, 330, 337–38; conditions of, 76, 85, 87, 101, 150, 157, 158, 269; and Genesis, 73, 74, 75, 76, 77, 82, 89, 90, 101; and languages, 152, 153, 154, 202, 203; origins of, 149, 152–53, 156; and shaping of the world, 82, 101, 149, 307, 313; spread of, 149, 150, 153; and warfare, 89, 154, 155, 157. *See also* Farmers; Farm(s), family; Farming

Ainu, 334

Aiyansh, 167, 168, 223, 224

Alaska Highway, 131, 164, 331

Alcohol, 152, 251–54, 341; in Genesis, 77, 78, 79, 327; and indigenous peoples of Canada, 138–39, 234, 235, 249, 250, 285–87, 341

Aleph-beis, 70, 325

Aleut, 317, 344

Algonquian, 52, 106, 129, 199, 231, 256, 287, 319, 351. *See also* Cree

Alter, Robert, 77, 326–27

Amazonia, 121, 185, 262, 312, 334

Anaviapik, Simon, 89, 100, 101, 125–26, 140, 142, 249, 250–51, 321; and Arctic journeys, 45–46, 53–64; as author's teacher, 37–44, 60, 64, 93, 251; children of, 38, 44, 142, 192; and documentary film, 91, 92, 93, 94, 96, 216, 217, 218, 219, 331; in London, 94–100, 331; on southerners, 126–27, 217, 273

Anaviapik, Ulajuk, 38, 44, 249, 273

Andaman Islands, 324

Andrew, Alex, 234, 236, 237, 238, 241, 276

Andrew, Mary Adele, 276–77

Animals, 12, 158, 229, 254, 255, 256, 268–69, 288–90, 351. *See also* Livestock; *names of animals*

Anthropologists, 4–5, 48–49, 315–17, 320, 322, 338; and hunter-gatherers, 122–24, 139, 144, 145–46, 334–35, 337, 347. *See also* Structuralism

Apache, 185

Archaeology and Language (Renfrew), 151

Archaeologists, 113, 114, 115, 116, 152–53, 213, 293, 337

Arctic, 3, 4, 16, 92, 127–28, 312, 316, 324, 333; animals, 12; birds, 24, 37, 322; conditions, 45, 50–51, 128; flora, 37; High, 26, 37, 230, 321; spring in the, 24–25, 26, 27, 30, 37, 50–51, 140–41. *See also* Journeys, author's; Tundra

Arctic Bay, 47, 54, 317

Arctic Quebec, 316, 317, 321

Arensberg, Conrad, 353
Aristotle, 278
Arnatsiak (Anaviapik's cousin), 38
Arroyo-Kalin, Manuel, 334
Athabaskan languages, 52, 106, 197, 198, 199, 203–4, 208, 274, 351
Athabaskan peoples, 105, 110, 129, 131–32, 136, 140, 148, 197, 240, 319, 336; and missionaries, 244, 343; territories of, 131, 331, 332. *See also* Dunne-za; Wet'suwet'en
Atiq system, 11–12, 13, 31, 140, 142
Australia, 120, 144, 155, 156, 186, 188, 189. *See also* Aborigines
Aymana, 336
Aztec, 121, 185, 280, 282, 333, 334, 349

Baffin Island, 27, 34, 35, 317
Balikci, Asen, 316, 333
Barbeau, Marius, 346
Barnard, Alan, 349, 354
Barren Lands, 22, 23, 324. *See also* Tundra
Basarwa, 337
Bateson, Gregory, 340, 341
Bauman, Richard, 322
Bears, 107, 256; polar, 42, 56, 61, 93, 96
Beaver Indians, 105, 106. *See also* Dunne-za
Berger, Thomas, 342, 348
Bering Strait theory, 113–14
Beynon, William, 346
Bible, 75, 81, 228, 301, 327, 343. *See also* Genesis
Bickerton, Derek, 291, 351
Biesele, Megan, 347
Bilingualism, 199, 201–2
Binary pairs. *See* Dichotomies
Biologists, 94, 265, 291, 351
Birds, 67, 94, 158, 197, 198, 210; of the Arctic, 21, 22, 23, 24, 34, 37, 49, 265, 322; eggs, 34, 67, 93; of Labrador, 229, 237–38
Birket-Smith, Kaj, 315
Blacks, 336
Bodenstein, Christel and Hans, 329
Boorman, John, 262
Boreal forest, 129, 148
Bowman, Glenn, 320
Brice-Bennett, Carol, 316, 345, 346
Briggs, Jean, 318, 322, 335
Bringhurst, Robert, 326, 340
British Empire, 186, 187, 340
British North America Act, 182
Buliard, Father Roger, 318
Bushmen, 122, 144, 148, 197, 256, 280. *See also* Kalahari Desert; Khoisan; San
Bylot Island, 3, 27, 62, 93

Camps, native, 237–38, 240, 274, 275–76, 277. *See also* Journeys, author's: within the Arctic
Canaan(ites), 78, 79, 115, 327
Canada, 189; Department of Indian and Northern Affairs, 25, 90, 316, 322
Capitalism, 163, 257, 258, 281, 330
Caribou, 45, 47, 100, 232, 238, 239
Cartier, Jacques, 229–30
Cattle, 73, 107, 152, 158, 256. *See also* Livestock
Central America, 258
Chamberlin, Ted, 112, 340, 344–45
Childe, V. Gordon, 151–52
Childhood, 67–69, 318. *See also* Child raising; Children
Child raising, 194–95, 201, 310, 339–40; among hunter-gatherers, 5, 121, 195, 297, 298, 341; among Inuit, 11–15, 29, 30–31, 318, 341
Children, 82, 85, 194, 201–2, 329, 340; and hunter-gatherers, 119, 192, 333; Inuit, 11–15, 29, 30–31, 38, 96, 218. *See also* Residential schools
China, 257, 329
Chippewyan, 240
Chomsky, Noam, 296, 299, 352
Christianity, 228, 230–31, 233, 242, 243, 247, 255, 278. *See also* Missionaries
Clark, Desmond, 351
Clendinnen, Inga, 333, 349
Clothing, caribou skin, 3, 45, 47, 58, 95, 100, 125

Cocker, Mark, 348
Colonialism, 43–44, 46, 86, 126, 143, 187, 251; and hunter-gatherers, 185–85, 189–90, 203, 277–80, 282, 307; and language change, 186–90, 340. *See also* Frontier(s): colonial
Columbus, Christopher, 281
Conflict(s), 185, 229. *See also* Violence; Warfare
Connell, K.H., 329
Correll, Thomas C., 318
Counter-cultural movements, 88
Crawhall, Nigel, 352
Creation stories, 114, 115, 204–5, 223–27, 326, 328; Judaeo-Christian, 70–71, 72, 246–47, 248, 313, 327 (*see also* Genesis)
Cree, 113, 130, 134, 231, 240, 287, 312, 331, 333
Creole, 336
Culhane, Dara, 343, 350, 351
Cultural differences, 6–7, 37, 48–49, 52, 53, 121, 127, 188
Cultural relativism, 320

Daly, Richard, 333, 335
David (Lisa's friend), 249–50, 251
Day schools, Indian, 181, 182
Decolonisation, 91, 317, 344
Deer, 107, 110, 137, 190, 259
Delgamuukw v. the Queen, 206–14, 215, 265, 333, 334, 343, 350
Development, 6, 122, 144–45, 183–84, 185, 262, 263, 280, 336. *See also* Frontier development
DeVore, Irven, 334
Diamond, Jared, 158, 337
Dichotomies, 246–47, 255, 269, 299–307
Diet: of hunter-gatherers, 123, 263, 333; of Inuit, 32, 34, 37, 53, 57, 58, 59, 315, 325. *See also* Fishing; Food gathering; Hunting; Starvation
Differentiation, 52
Disappearing World series, 91, 330–31. *See also The People's Land* (film)
Diseases, 157, 158, 181, 182, 229, 242, 248, 337–38
Dogs, 34, 53, 55, 56, 57, 59, 61, 152, 158, 324
Dogteams, 27, 59, 319. *See also* Journeys, author's: within the Arctic
Dreams, 107, 108, 118, 132–33, 164, 220, 238, 247, 259, 261–62, 336
Drunkenness. *See* Alcohol
Dubreuil, Chisato O., 334
Dunne-za, 105, 119, 130–31, 194, 331; and alcohol, 138–39, 253; and animals/birds, 107, 132, 197, 198, 276; anthropology of, 204, 343; as author's teachers, 108, 111, 135, 203; creation story, 204–5; and cultural pride, 106; and dreams/dreamers, 107, 108, 132–34, 135–36, 336; dry-meat camp, 274, 275–76; and English, 198, 204, 205; and humour, 106, 111, 138, 139; and hunting, 132, 274; knowledge, 105, 106, 132; and land, 105–6, 132, 204, 333, 343; language, 197, 198, 203–4, 274 (*see also* Athabaskan languages); losses of, 105–6, 333, 343; origins of, 110; and outsiders, 135–36, 138–39, 204; territories, 105–6, 109, 111, 131, 331, 333; and treaties, 134–35

Edmonton, 4, 285–87
Education, 30–31, 181, 183–84, 185, 189, 204, 218, 219, 233, 243, 277. *See also* Day schools, Indian; Language learning; "Poisonous pedagogy"; Residential schools
Egalitarian individualism, 13, 14, 31, 46, 105, 118, 121, 264, 298, 299, 308, 333, 336
Elders, 135, 136, 140, 173, 181, 182, 204, 250; and history workshop, 112, 113, 114, 133; hunter-gatherer, 144, 145, 147–48; Inuit, 38, 41, 44, 60, 118, 142, 216
Elisapik (of Rankin Inlet), 21, 25
The Emerald Forest (film), 262

England, 67, 68, 95–96, 281
English language, 180, 196, 197, 253, 295, 321; grammar, 186, 187, 319; and the Inuit, 29, 30, 38, 44, 45–46; and Inuktitut, 199; and Jews, 68, 202; and residential schools, 176, 180; words, 45–46, 51, 321
English people, 252–53
Equanimity, 60, 61, 63, 195
Eskimo, 315, 317. *See also* Inuit
Ethnocentrism, 157
Ethnocide, 189. *See also* Genocide
Ethnonyms, 317, 331
Europe, 153, 305–6
Europeans, 87, 113, 229–30, 278–80, 325, 342; ideas of the North, 127–28; views of hunter-gatherers, 281, 353
Evolution, 278, 352; human, 116, 119, 262, 293, 294, 348, 351; and hunter-gatherers, 6–7, 121, 122, 126, 159, 185, 279
Excluded middle, law of, 255, 299, 347
Executives, 89, 191

Family, 21, 38, 76–77, 82–84, 85–86, 158, 322, 330, 338. *See also* Farm(s), family
Fanon, Frantz, 143, 336
Farmers, 157, 158–59, 353; and alcohol, 251–52, 253; characteristics of, 83, 150, 202, 255, 269, 307, 330; and Christianity, 228; conservatism of, 83, 330; and contradiction, 76, 87, 90, 329–30; and dichotomies, 255, 269; and domination, 7, 185, 187, 189, 257, 269, 306, 307, 308; and Genesis, 74, 76, 77, 79, 89, 150, 185; and home, 90, 329–30; and hunter-gatherers, 6, 7, 52, 100–101, 120, 128, 150, 154, 155–56, 157, 190, 202, 215, 257–59, 269, 307, 309, 334, 337; and hunting, 306; indigenous, 228, 258, 305; and inheritance, 85, 265; Irish, 48, 86, 303; and labour, 150, 329; and land, 88, 120; and language(s), 150–54, 187, 199, 200; and migration, 85, 86–87, 119, 153, 304; and mind, 255; as nomads, 7, 76, 87, 89, 90, 160; and population, 85–86, 187, 303, 329; and reproduction, 101, 341; in the Subarctic, 129; and wilderness, 100–101, 120, 190; world view of, 89. *See also* Agriculture; Family; Farm(s), family; Farming; Herders; Pastoralists
Farm(s), family, 83, 84–85, 86, 87, 90, 185, 189, 329, 330; and hardship, 87; and migration, 87, 90, 329–30
Farming, 6, 88–89, 90, 143, 153, 257. *See also* Agriculture
Field, Abalie, 111, 136, 137, 139, 162, 204, 205
Field, Jimmy, 106–11, 112, 114, 131, 136, 137, 139, 160–64, 338
Field, Rosie, 111, 139
Field work, 145, 315, 322, 338, 352
Film crews, 173, 179, 217, 236, 237, 330–31, 339, 347. *See also names of films*
Films, wildlife, 288–90
Fish, 255, 259; of Arctic, 35, 36, 37, 93; of Labrador, 229; of Northwest Coast, 169–71, 172–73, 192, 196, 223–24
Fishing, 24, 32–37, 93, 169–70, 263, 347
Fitzhugh, William, 334, 346
Fleetwood (film), 91
Food cache(s), 55, 61–62, 108, 325
Food gathering, 254, 259, 263–64, 335, 338, 347
Food processing, 263, 274, 347
Fort Churchill, 15
Fort St. John, 106, 131, 136, 138, 162, 164, 252, 274
Freeman, Milton, 316
French language, 336
Frobisher Bay, 47
Frontier development, 162, 190
Frontier(s), 7, 8, 15, 27, 88, 100, 155–56, 161, 163, 203, 252; agricultural, 86–87, 90, 100, 128, 160, 252, 258, 337; colonial, 87, 90, 143–44, 187–89, 257, 279–80, 282

Fulmars, 49, 322
Fur trade, 110, 204, 232, 233, 331

Gagné, Raymond, 317, 320
Garden of Eden, 73–74, 75, 80, 101, 322
Geese, 94, 237, 265
Genesis, 71–82, 114–15, 326, 327; and agriculture, 73, 74, 75, 76, 77, 82, 89, 101, 327; and animals, 73, 74, 75, 76, 77, 289; and artistic legacy, 72, 81, 328; chapters 1 to 11, 71–81, 101, 325, 327–28; curses of, 75, 76, 77, 79, 80, 185, 193, 304, 327; and dichotomies, 246–47; and the human condition, 89, 328; and morality, 74, 246, 247; as myth, 71–72, 100, 228; as poetry, 72, 101, 247, 327–28; social system implied by, 77, 79, 82; translations of, 81, 326–27; and universality, 71, 72, 73, 81, 89, 101, 115
Genocide, 144, 189, 200
Germany, 339
Ghosts, 43, 44, 224, 227
Gilberg, Rolf, 333
Gitxsan, 147, 172, 206–14, 215, 243, 256, 312, 333, 334, 346
Gosnell, George, 172–79, 180, 223
Gosnell, James, 172
Gosnell, Joe, 172
Government (Canada): and animals, 96, 238; and Indians, 182, 350; in Labrador, 235, 238; and land claims, 282–84, 311–12, 350 (*see also Delgamuukw v. the Queen*); and northern administration, 26, 29, 96, 125, 197, 315; and residential schools, 182. *See also* Canada: Department of Indian and Northern Affairs
Grains, 152, 157
Grant, Peter, 210
Gravesites, 34, 109
Greece, 149, 153, 154, 278
Greenland, 229, 320, 333
Grief, 60–61, 140, 142, 173, 180, 186, 297
Grigsby, Michael, 91, 93, 215, 217, 330
Grosz, Stephen, 329
Grouse, 197, 210, 237–38
Guns, Germs and Steel (Diamond), 158

Hadza, 183, 335
Haida, 172, 340
Halfway River, 136–37
Halfway River Reserve, 105, 131, 132, 138, 274, 331
Harper, Jim, 331
Harris, Marvin, 349
Harris, Olivia, 335
Harry (of Edmonton), 286–87
Health, 182, 235, 267, 316, 325. *See also* Diseases; Psychological disorders
Health care, 29, 141–42, 275
Hebrew, 68–71, 201, 325, 327
Hegel, Georg Wilhelm Friedrich, 280
Henke, Lewis, 348
Herders, 73, 74, 76, 77, 120, 129, 337, 353. *See also* Pastoralists
Hindi, 186, 187
History, 6, 8, 110, 112, 114, 144, 145, 159; oral, 112, 123, 136, 147, 229
Hobbes, Thomas, 281, 349
Holocaust, 68, 70, 214, 215, 344. *See also* Genocide
Homo erectus, 116, 293, 351
Homo Heidelbergiensis, 351, 352
Homo sapiens, 116, 119, 294
Horses, 106, 107, 109, 136–37, 256, 276, 304
Housing, 11, 28, 38, 276–77. *See also* Snowhouses; Sod houses
Hudson Bay, 127, 128
Hudson's Bay Company, 28; store, 27, 32
Human condition, 8, 89, 90, 101, 284, 307–8, 313, 326, 328
Hunter-gatherer(s), 4–8, 111, 244–46, 261, 331–32; and alcohol, 251, 252, 253–54, 285–87; and animals, 254, 255, 256, 268–69; anthropology of, 122–24, 334–35; and boundaries, 244–45, 254, 256, 259, 261, 264, 267; characteristics of, 117, 118, 121,

Index

123, 139–40, 146–47, 266–67, 280–81, 299, 308, 309, 333, 336; and child raising, 5, 121, 192, 195, 297, 298, 333, 341; and Christianity, 228, 230, 248; and control, 258, 260, 261, 269, 311; counting systems, 14, 179, 199, 342; and cultural differences, 116–17, 332; and decision making, 118, 259–60; and development, 6, 143, 148; and dichotomies, 302; diet of, 123, 263, 333; and discrimination, 121–22, 123, 126, 334; and diseases, 158–59, 242; and farmers/farming, 6–7, 52, 150, 154, 202, 257–58, 269, 305–6, 307, 309, 337; and feelings, 143, 195; and gender, 5, 261, 262–67, 335, 347; and Genesis, 89–90; genius of, 5, 7, 122, 148, 309; and hierarchy, 118, 192–93, 268, 311, 338, 339; and home as Eden, 90, 117, 119; and humour, 254; and hunting rituals, 117, 132; and journeys, 259, 260;
knowledge, 7, 140, 148, 189, 193, 198, 255, 259–61, 268, 309; and prosperity, 35, 117; and shamanism, 220, 244–45, 246, 254, 258, 267, 268–69, 347; of territories, 35, 115, 119, 123, 133, 320;
and land, 120, 246, 256, 320; and land rights, 122, 123, 124, 160, 189, 195, 206, 311–12; and land use, 5, 89, 117, 123, 188, 190, 232, 254–55, 256–57; and language(s), 155, 156, 190, 195, 198–99, 200, 201, 202, 205, 214, 215, 220, 244, 261, 314, 319, 342, 343, 347, 351; literacy, 191, 340; losses of, 144, 154, 212, 214, 313–14; at margins, 6, 119–20, 148, 195, 332; and migration, 113–14, 119; and mind(s), 298; murder of, 280 (*see also* Field, Jimmy; Genocide); and oral culture, 191–92, 339; origins of, 113–14, 115, 116; and outsiders, 120, 121–22, 124, 126, 128, 143, 147, 161, 188–89, 269–70, 282, 308–9, 336;
outsiders' views of, 181, 182, 184–85, 305, 333–34, 353–54; as inhuman, 277–82, 285, 348, 351; as myth, 5, 302; as natural, 281; as primitive, 121, 128, 182, 184, 185, 248, 262–63, 280, 307, 333, 335 (*see also* Stereotypes: of hunter-gatherers);
and population, 118, 119, 148, 149, 294, 333; present-day, 5, 195, 214, 309–13; and psychological disorders, 194–95, 340–41; and realism vs. romanticism, 143, 144–45, 146, 147, 336; and reproduction, 89, 118–19, 333, 341; and residential schools, 189; and resources, 229, 309, 332; seasonal round, 148, 232, 259; self-conception of, 126, 143, 254, 335, 337; and self-destruction, 143, 195; as settled, 7, 90, 160; and shaping of the world, 115, 313; and sharing, 118, 119, 192, 193, 232, 332–33; and social breakdown, 195, 254, 341; and society, 160, 192, 283, 284, 332, 350, 351; spirituality, 118, 228, 244–45, 247–48, 252, 258 (*see also* Shamanism; Shamans); of the Subarctic, 129, 230; and survival, 146, 214, 259, 267, 269, 313; territories, 148, 240, 332–33; and truth, 192, 193, 194, 195, 207, 244, 246, 340–41; and violence, 147, 324–25, 336; ways, 147–48, 309, 313–14
Hunters. *See* Hunter-gatherers
Hunters and Bombers (film), 234–40, 266, 276, 347
Hunter, Thomas, 111, 139, 274–76, 331
Hunting, 329; author's, 21, 22–23; and dreaming, 259, 260 (*see also* Dreams; Trails); Dunne-za, 105, 106, 107–8, 137, 197, 274; Innu, 238–39; Inuit, 14, 40, 45, 47, 48, 100, 347 (*see also* Seal hunting); supplies, 32–33, 47, 53, 55, 109, 259. *See also* Diet
Hunting territories, 109, 111, 131, 188, 206, 332

Igloolik, 54–55, 56, 323–24, 348
Igloos. *See* Snowhouses
Ignace, Marianne Boelscher, 342, 347
Igunaaq, 62, 325
Ik, 146, 147, 320
Ilira, 42, 43–44, 46, 60, 217–18
Inca, 121, 282, 334
India, 186, 187, 188, 196, 201, 258, 329, 340
Infanticide, 119, 333
Ingold, Tim, 335
Innu, 119, 231–32, 234–42, 268; and alcohol, 266; and Christianity, 232, 233–34, 239–40, 241, 242, 277; and English, 267; and gender, 264; on government, 235, 238, 266; and hunting, 235, 237–39, 240, 277; and the Inuit, 231; knowledge, 238–39, 240, 266, 267; and land use, 232, 277; losses of, 277; and NATO test flights, 234, 235, 238, 265–66, 267–68, 347; and outsiders, 233, 239, 265; seasonal round, 232–33, 235–36, 238; and settlements, 234, 238, 276–77; spirituality, 236, 238–40, 241, 242, 264; territories, 231, 232, 233, 240, 346; ways, 235, 242. *See also Hunters and Bombers* (film); Innu'aimun
Innu'aimun, 231, 238, 268, 345
Intuition, 220, 268, 269, 309, 311
Inugu (of Pond Inlet), 32, 33, 34, 35, 36, 37, 44–45, 140, 141–42
Inuit, 119, 148, 317, 320; and aggression, 249–50, 273; and alcohol, 249, 250, 251, 253, 254; and animals, 12, 256; anthropology of, 315–17, 335, 348, *atiq* system, 11–12, 13, 31, 140, 142; characteristics of, 13, 14, 29, 31, 40, 44, 46, 55, 64, 298–99; child raising, 11–15, 29, 30–31, 318, 341; and Christianity, 96, 230, 244, 343; and decision making, 59, 62, 63; family life, 21, 38, 322; and feelings, 45–46, 60–61, 63, 250, 321, 322, 324; and gender, 264; and Genesis, 89, 101; and home, 11, 35, 128; and humour, 42, 44, 58, 59, 60, 63–64, 141, 251, 318; hunting rituals, 14; and knowledge, 35, 48, 92–93; and land, 14–15, 35, 94, 120, 128, 218, 317–18; land claims, 313; and language(s), 12, 38, 40, 196, 218, 219, 249, 318, 342, 343–44; and laughter, 20, 39, 41, 60, 217, 273; mechanical skills of, 45, 127, 319; and mind, 299; and missionaries, 244, 343; and mystery, 4, 13; navigation, 35, 47, 53, 54, 96; and neighbours, aboriginal, 147, 319; outsiders' views of, 128, 320, 348; personality traits, 31, 40, 44, 60, 61, 63, 249–50, 319; relocation of, 317; seasons, 128, 336; self-conception of, 335; and settlements, 21, 28–29, 218; social life, 29; and southerners, 28, 29, 42–44, 92–93, 94, 98, 126–27, 128, 217–19; and southern law, 218–19, 324, 344; stories, 12–13, 15, 218, 324; territories, 39–40, 318, 332; travel, measurement of, 322; and truth, 192, 194, 319; and violence, 147, 324, 336; ways, 16, 28–29, 40–41, 42, 44, 64, 96, 125, 316, 320; and weather, 51. *See also* Inuktitut; Journeys, author's; *The People's Land* (film)
Inuit, Labrador, 229, 230–31, 232, 345, 346
Inuit Tapirisat of Canada, 316
Inuja (Anaviapik's niece), 38
Inuk (Inuja's husband), 38
Inuktitut, 253, 295, 317, 319; and ambiguity, 244, 246, 297–98; and author, 16–21, 22, 24, 25, 34, 37–44, 60, 64, 249, 251, 319, 320; and categories, 12, 14, 198, 342; dictionaries, 322–23; and gender, 18, 261; grammar, 17–19, 220, 261, 318–19, 320, 344; and hierarchy, lack of, 14, 46; and intertranslatability, 53, 323; and the Inuit, 29, 38, 321, 344; language course, 16–21, 25; and languages, 21, 199, 219, 231, 345; and metaphor, 220, 251; numbers, 14, 199, 231;

orthography, 39–40, 317, 320, 321, 344; place names, 35, 317, 319, 323–24; and politeness, 41, 46, 321; precision of, 18–19, 40, 197–98, 318; reading of, 40; and subtleties, 318, 319; as way of being, 39, 64, 218; and "wilderness," 22, 35, 55
words: for animals, 12, 198, 342; for birds, 24, 37, 342; for danger, 42; and Europeans, 325; for fear, 42, 43–44, 321; for fish, 35, 197–98, 342; for God and Jesus, 231; for hello and good-bye, 321; for hunting, 40, 41; for kinship, 249; lacking, 14, 46, 216; for landmarks, 34, 35, 324; for mind, 297, 298, 299 (*see also Isuma*); for money, 14; for ownership, 199; for snow, 48, 49–50, 51–52, 53, 57, 99, 198, 199, 322–23; that sound similar, 41–42; of welcome, 11, 273, 321; for whites, 218 (*see also* Qallunaat)
Inukuluk, Paulussie, 3, 92, 318
Inummarittitut, 39
Inupiaq, 317, 344
Inupik, 344
Inuvialuit, 312, 324, 333, 344
Iqaluit River, 35, 36
Ireland, 86, 303–4, 329, 353
Irkloo, Elijah, 28
Iroquois, 185
Isuma, 46, 47, 92, 273, 297, 298, 299, 322

Jack, Chief Jerry, 248, 346
Janes, Robert, 344
Jenness, Diamond, 315, 335
Jews, 67–68, 201, 202. *See also* Genesis; Hebrew
Johnson, Mary, 208–12, 265, 343
Jomon, 334
Jones, Sir William, 150–51, 337
Journeys, 121, 288; spirit, 134, 245, 248, 251, 254, 261 (*see also* Shamanism; Shamans)
Journeys, author's: to the Arctic, 25–26, 67; within the Arctic, 3–4, 21–24, 32–37, 47–48, 53–64, 140–41, 142; to hunter-gatherers, 124, 304; to the Northwest Coast, 167–68; onto Innu lands, 236–37; to the Subarctic, 129–31
Judaism, 71, 81, 247. *See also* Genesis; Hebrew; Jews
Ju/'hoan, 347

Kabloona, 320. *See also* Qallunaat
Kalahari Desert, 120, 148, 256, 257, 324, 332, 337. *See also* Bushmen; San
Kaner, Simon, 334
Karlik (of Rankin Inlet), 21, 24, 25
Khoikhoi, 348, 349
Khoisan, 277–78, 295. *See also* Bushmen; San
Kimball, Solon, 353
Kispiox, 208
Kulchyski, Peter, 317, 350
Kumangapik, Inuja, 265
!Kung, 324, 347. *See also* Bushmen; San

Laboucane, Arlene, 162
Labour, 28, 29, 82, 83, 84–85, 121, 143, 150, 304
Labrador, 229, 230, 316, 345
Laing, R. D., 340–41
Land, 76–77, 82, 189
Land claims, 282–84, 285, 312–13, 350–51
Land rights, 14, 122, 123, 124, 134–35, 160, 195, 206–14, 215, 311–12, 343; and slaves, 278, 279
Land use, 5, 14, 89, 117, 123, 188, 190, 232, 254–55, 256–57
Land use and occupancy studies, 111, 130–31, 132, 331, 345–46
Language(s), 8, 292–97; Adamic, 321–22; and animals, 290–91; and archaeology, 337; and categories, 342, 347; and colonialism, 186–90, 340; and culture(s), 53, 117, 196–203, 292–93, 323; and diversity, 351; and environment, 323, 326; and Genesis, 71, 80; and grammar, 19, 52, 69, 294, 295,

Language(s): and grammar (*continued*) 297, 299, 323, 345, 347; history of, 150–51; and humans, 73, 290–91, 294–95, 296–97, 308, 351, 352; of hunter-gatherers, 52, 116–17, 188; Indo-European, 52, 151–52, 153, 154, 156, 196, 198, 199, 203, 295, 337, 351; and intertranslatability, 49, 52, 53, 198–99, 208; loss of, 154, 156, 179, 180, 189–90, 200–201, 220, 254, 314, 336, 344; and meaning, 49, 323, 326; and metaphor, 220, 344; and mind, 323; non–Indo-European, 155; of North America, 117, 119, 351; origins of, 293–94, 351–52; and reality, 48–53, 323, 325–26; and society, 284–85, 295; and truth, 191, 192, 193–94, 195, 340–41. *See also* Words; *specific languages*
Language learning, 12, 16–21, 68–71, 183; by children, 291, 294, 296–97; and nature vs. nurture, 296–97, 352; by primates, 291, 352
Lapps, 155
Las Casas, Bartolomé de, 279, 348
Leach, Edmund, 300, 301, 352
Lee, Richard, 123, 334, 335, 347
Levi, Primo, 215
Lévi-Strauss, Claude, 300, 302, 337, 342, 348, 352
Liquor. *See* Alcohol
Lisa (Anaviapik's relative), 249, 250
Literature, 82–83, 84, 127–28, 292, 303–4, 315–17, 329, 330, 335, 353
Livestock, 76, 152, 158, 338. *See also* Cattle; Horses
Lubragge, Michael T., 338
Lyon, G.F., 318, 348
Lytton, 174

Macdonald, John A., 182
McEachern, Chief Justice Allan, 206, 209–11, 212–14, 334, 343
Macintosh, William, 329
McKay, Alvin, 168–69
Mackenzie Delta, 197, 312, 316, 333
Mackenzie Highway, 130
McKie, Robin, 351
Macpherson, C.B., 349
Madagascar, 144
Maku, 337
Mallon, Mick, 16, 17, 18, 19, 20, 40, 317
Man the Hunter (Lee and DeVore), 123
Man the Hunter conference, 123, 334, 335
Manifest destiny, 159, 279, 338
Mannoni, Octave, 143, 336
Manuel, George, 130
Maori, 328–29
Maps, 53, 111, 114, 132, 237, 324, 331, 332
Maps and Dreams (Brody), 132, 135, 331
Marcus, Alan, 317
Marriage, 232, 264–65
Marshall (Jimmy Field's nephew), 111, 136
Marshall, Alan, 347
Martin, Laura, 322, 323
Mary-Rousselière, Guy, 316, 346
Mauss, Marcel, 332
May, Margaret, 329
Maya, 185, 334, 349
Meillasoux, Claude, 320
Men, 5, 77, 263, 264
Migration, 85, 86–87, 88–89, 90, 113–14, 119, 153, 304, 329–30
Mi'kmaq, 229
Miller, Alice, 184, 339
Milloy, John, 182, 183, 339
Mills, Antonia, 333
Mind(s), 163, 190, 243, 273, 282, 287, 290–91, 300, 301; agricultural, 306, 311; animal, 289–90; and culture(s), 293; hunter-gatherer, 306, 310; and language, 294, 323; and reality, 326; rival forms of, 311
Missionaries, 203–4, 229, 230, 232, 233, 243, 342–43, 345, 346; and hunter-gatherers, 228, 229, 230, 242–44, 247–48, 257, 346; and the Inuit, 28, 44, 93, 147, 230, 244, 316, 333; and shamanism, 231, 239–40, 248; and truth, 243, 247, 248

Index

Miut groups, 14, 317–18
Moose, 105, 107, 108, 274
Morality, 74, 184, 245, 246, 247, 290, 307
Mortality, 183. *See also* Genocide
Moser, Brian, 91, 330
Mostert, Neil, 348, 349
The Mountain People (Turnbull), 146, 147, 320
Mowat, Farley, 315
Muckpah (of Pond Inlet), 47, 53, 56, 57, 58, 59, 60, 61, 62
Mulhall, David, 346
Müller, Max, 291
Munk, Rabbi Michael L., 325
Muttuk, 61–62, 141, 325
Myths, 112, 196, 210, 223, 245, 261, 262, 289, 292, 299, 301, 302, 326, 346, 348. *See also* Creation stories; Genesis: as myth; Literature; Stories

Na-cho Nyak Dun, 331
Nambikwara, 300, 352
Narby, Jeremy, 348
Nass River/Valley, 167, 169, 170, 223, 283, 300, 352
A National Crime (Milloy), 182, 339
NATO test flights, 233, 234, 235, 265–66, 267–68, 347
Nature, 288
Neolithic system, 101, 149, 156, 342. *See also* Agriculture; Farming
Netsilik, 316, 333
Netsilingmiut, 316
New Guinea, 341
Newman, James L., 351
New Zealand, 115, 328, 329
Nez Perce, 324, 347
Nicklen, Paul, 336
Nisga'a, 168, 179, 224, 230, 341; and documentary film, 168–69, 170, 171, 223–27; and fish, 169–71, 172, 223–24, 256; land claims, 169, 195–96, 283, 312–13; and language(s), 173, 176, 177–78, 179, 180, 196, 224, 345; myths, 170–71, 196, 223–27, 300, 352
Nisga'a Land Settlement Agreement, 172, 196, 341–42
Nomadism: and capitalism, 163, 330; of farmers, 7, 76, 87, 89, 90; and the human condition, 89, 101; and hunter-gatherers, 159, 160; of urban dwellers, 88, 89, 97–98
North Africa, 340
North America, 116, 155, 156, 185, 186, 188, 229, 230, 309; exploration and settlement, 86–87, 127–28, 181, 229–30, 279, 280; original inhabitants, 113–14
Northern Science Research Group, 25, 90, 316, 322
Northwest Coast peoples, 172, 324, 332, 334, 338–39, 342, 346; and chiefs, 206–13; and Christianity, 230; and gender, 263, 264; knowledge of, 205; and land rights, 191, 205, 206, 207; and languages, 351; and oratory, 191. *See also* Gitxsan; Haida; Nisga'a; Tsimshian
Northwest Passage, 127
Northwest Territories, 313, 316
Norwegian, 295
Nunavut, 313, 344

Ojibway, 231, 341
Oral culture(s), 191, 196, 205, 207, 213, 216, 340, 345; and aboriginal origins, 113, 339; Dunne-za, 105; and Genesis, 81; and history workshop, 112, 114; Inuit, 12–13, 15, 40; and reincarnation, 140; and spirit journeys, 136. *See also* History: oral
"The Original Affluent Society" (Sahlins), 123, 335
Original sin, 75, 193
O'Sullivan, John L., 159
Otters, 237
Overstall, Richard, 331

Padel, Felix, 333
Pagon, Andrew, 334

Parenting. *See* Child raising
Pastichi, Sebastien, 234–35
Pastoralists, 48, 155, 329, 330, 337, 349. *See also* Herders
"Patriarchal belt," 330. *See also* Social systems: patriarchal/patrilineal
Peace River, 110, 163
Pei, Mario, 337
Penashue, Elisabeth, 235, 266, 277, 347
Penashue, Pien, 236, 237–40, 241–42, 267, 268, 269
Pennines, 67
The People's Land (film), 91–92, 93–94, 96, 215–19, 330–31, 343
Philip II (king of Spain), 278, 279
Philosophy, 281, 325–26, 349
Pinder, Leslie Hall, 338, 343
Pinker, Steven, 297, 319, 351, 352
Plains peoples, 324
Plants, 254–55, 256, 263–64, 347
Pluciennik, Mark, 349, 353
Poetry, 78, 205, 328–29, 336, 345. *See also* Genesis: as poetry
"Poisonous pedagogy," 184, 310, 339–40
Police. *See* Royal Canadian Mounted Police
Politics, 191, 193, 265, 266
Pond Inlet, 25, 26–31, 44, 140, 141, 142, 273, 317, 324; and documentary film, 91–92, 93–94, 331
Population, 85–86, 116, 118, 148, 149, 153, 293, 303, 329, 330, 333, 351. *See also* Migration
Porcupines, 110, 137–38, 139, 239
Postmodernism, 302, 323, 352–53
Potatoes, 303, 304
Potlatch, 168, 169, 191, 193, 332, 338, 339
Poverty, 28, 143, 183, 248, 262, 333
Progress. *See* Development
Property. *See* Land
Psychological disorders, 194–95, 340–41
Psychology, 339
Ptarmigan, 21, 22, 23, 197
Pullum, Geoffrey K., 322
Pygmies, 121, 148, 320, 334, 337

Qallunaat, 31, 37, 44, 46, 47, 92, 93, 94, 126–27, 128, 217, 273, 319–20
Qallunaatitut, 44
Qanguk (of Pond Inlet), 216–17, 218–19, 343, 344
Quakers, 46, 311, 321–22

Racism, 126, 160, 184, 252, 254, 282
Rankin Inlet, 15, 17, 20, 324
Rasmussen, Knud, 315, 335
Reincarnation, 140, 142
Religion, 145, 185, 190, 197, 242, 247, 257, 336. *See also* Christianity; Judaism; Missionaries; Residential schools
Renfrew, Colin, 151, 152, 153–54, 337
Reproduction, 77, 78, 79, 85, 87, 89–90, 101, 115, 118–19, 333, 341
Reserves, 331. *See also* Halfway River Reserve
Residential schools, 92, 173–79, 180, 182–83, 184, 186, 189, 200, 243, 339, 345
Resolute Bay, 25–26
Revisionism, 332
Riches, David, 333, 335
Ridington, Robin, 204, 205, 336, 343
Rituals, 14, 118, 256, 258, 264, 299, 301, 345
Robinson, Rod, 168–69, 170, 223
Rode, Linda, 329
Rothenberg, Jerome, 345
Rowley, Graham, 317
Royal Canadian Mounted Police, 28, 39, 42, 44, 93, 218–19, 321
Royal Commission on Aboriginal Peoples, 112, 113, 114, 339
Rules of evidence, 206, 207, 209
Russia, 201, 330
Rwanda, 334, 337

Saami, 155, 183
Sahlins, Marshall, 123, 335
Saint George's Residential School, 173, 174–79

Index

S.-Senglor, Leopold, 336
Saladin D'Anglures, Bernard, 316
San, 185, 199, 257, 295, 332, 337, 348, 349, 352. *See also* Bushmen; Kalahari Desert; Khoisan
Sanskrit, 150–51
Sapir, Edward, 323
The Savage Mind (Lévi-Strauss), 300, 352
Scapulamancy, 239, 240, 241
Schneider, Lucien, 322
Schooner, Dinah, 331
Schultz-Lorentzen, C.W., 322
Science, 268, 347–48
Scullion, Collie, 325
Scullion, John, 26
Scurvy, 127
Seal hunting, 3–4, 24, 33–34, 41, 48, 55–56, 58, 59, 93, 140–41, 256
Seal meat, 62
Sealskin, 95, 125
Secwepemc (Shuswap), 342, 347
Sepulveda, Juan Gines de, 278, 348
Settlement(s): agricultural, 84, 185, 190, 269; Arctic, 21, 26, 28–29, 31, 317; Innu, 234, 276–77. *See also* North America: exploration and settlement
Settlement manager, 26, 27
Settlers, 77, 88, 156, 159, 188, 190. *See also* Colonialism; Farmers
Shahn, Ben, 69, 70, 325
Shamanism, 230, 239–40, 245–46, 258, 260, 261, 268–69, 289, 316, 324, 347–48. *See also* Dreams
Shamans, 44, 118, 133, 230, 243, 245, 247–48, 251, 258–59, 264, 269, 288, 289, 342
Sheffield, 67, 68
Shelley, Mary, 127
Sheshashiu, 234
Skeena River, 167
Skidoos, 45, 140, 141, 239, 319
Skid row (Edmonton), 4, 285–87
Slavery, 202, 278, 279
Sledges, 22, 319, 325
Smith, Derek, 316, 333
Snowhouses, 48
Snowmobiles. *See* Skidoos
Social systems, 79, 82, 257, 332, 339; patriarchal/patrilineal, 77, 201, 202, 330
Society, 191, 192, 193–94, 195, 257, 284
Sod houses, 96, 125
Songs, 133, 135, 205, 210–13, 220, 238, 241, 287, 330
South Africa, 188, 312, 329, 337
South America, 258, 278–79, 309, 335–36, 337
Southeast Asia, 187, 340
Southerners. *See* Qallunaat; Whites
Spain, 278–79, 282
Sperber, Dan, 320
Standish, Ralph, 282, 349
Starvation, 324
Steenhoven, G. Van Den, 324
Stefansson, Vilhjalmur, 315, 335
Stereotypes, 7, 48, 52, 60, 128, 252, 324; of hunter-gatherers, 6, 7, 121, 122, 123, 124, 261–63
Stone Age Economics (Sahlins), 123
Stoney Indian Band, 113
Stories, 112, 313, 329; of hunter-gatherers, 133, 140, 191, 192, 245, 246, 314, 347–48; Inuit, 12–13, 15, 216, 324; of Northwest Coast peoples, 170–71, 205–6, 207, 210, 211. *See also* Creation stories; History: oral; Literature; Myths; Oral culture(s)
Structural Anthropology (Lévi-Strauss), 300
Structuralism, 299–302, 352
Subarctic, 105, 128–30, 197, 198, 230
Swahili, 186, 187

Talmud, 68, 71
Tasmania, 348
Tautungnik (of Rankin Inlet), 20
Tester, Frank, 317
Time, 33, 110, 123, 139–40, 161, 162
Time Immemorial (film), 168–69, 170, 171, 172, 173–79, 196, 223–27, 339
Time travel, 124, 133, 134, 135–36, 164
Tocharian, 151, 152, 337

Torah, 69, 71, 325
Tracking, 107, 109, 191, 340
Trade goods, 28, 55, 232, 233
Traders, 28, 39, 44, 93, 110, 147, 218, 344
Traill, Anthony, 352
Trails, 109, 110, 132, 134, 135, 136, 259, 336
Transformation, 224, 226, 227, 245, 253, 254, 256, 289, 346
Transportation, 25, 26, 92, 94, 236. *See also* Dogteams; Skidoos; Sledges
Trappers and trapping, 47, 99, 197, 347. *See also* Fur trade
Treaties, 134–35, 196
Treaty Number 8, 130, 134–35, 331
Tricksters, 227, 245, 345, 346
Tristes Tropiques (Lévi-Strauss), 300, 352
Trudeau, Pierre Elliott, 283, 350
Truth, 191, 193–94, 208, 269; and hunter-gatherers, 192, 193, 194, 195, 207, 244, 246, 340–41; and the Inuit, 192, 194, 319; and missionaries, 243, 247, 248
Tseax River, 167, 171–72, 224
Tsimshian, 208, 212
Tuberculosis, 21, 92, 158, 274, 275
Tundra, 21–24
Tungus, 245
Tunisian man, 98–99
Turnbull, Colin, 146, 320
Turner, Nancy, 347

Ungava, 321
Union of B.C. Indian Chiefs, 130, 132, 162, 331
United States, 144, 181, 189
Urban dwellers, 84, 88, 89, 98, 309–10, 330
Urn, 335–36
Usher, Peter, 316, 341
Utuva (of Pond Inlet), 216, 217, 218, 219, 265, 343

Vallee, Frank, 320
Vaniallaatto, 334
Vikings, 229
Violence, 147, 234, 324–25, 336. *See also* Conflict(s); Warfare

Wadden, Marie, 347
Warfare, 89, 154, 181
The Washing of Tears (film), 346
Weinstein, Martin, 331
Welland, Tony, 316
Wet'suwet'en, 206–14, 215, 243, 312, 331, 333, 334, 346
Whalebone, 22, 96, 319
Whales, 61–62, 93, 127, 140, 141, 256, 325
Whites, 17, 20, 27–28, 39, 42, 43, 110, 126, 161, 163–64. *See also* Qallunaat
Whorff, Benjamin Lee, 323
Widlock, Thomas, 337
Wilderness, 35, 67, 100–101, 120, 163, 190, 214
Willie (Inugu's son), 33, 34, 35, 36, 37
Witchcraft, 258
Wittgenstein, Ludwig, 49, 53, 323, 326
Women, 5, 77, 263, 264, 265–66, 322, 335, 347, 348
Woodburn, James, 123, 139, 305, 320, 334, 335, 353, 354
Words, 190; and Genesis, 71, 73, 74, 75, 76; and truth, 191, 192, 193–94, 195, 340–41. *See also* Language(s)
Wright, Harold, 223–27, 230, 345
Wright, Jessica, 334

Xhosa, 337, 348

Yiddish, 202
Young, Arminius, 318
Yuit, 317, 344
Yupik, 317